Staphylococcus aureus

Staphylococcus aureus

Edited by

Alexandra Fetsch

Academic Press is an imprint of Elsevier
125 London Wall, London EC2Y 5AS, United Kingdom
525 B Street, Suite 1800, San Diego, CA 92101-4495, United States
50 Hampshire Street, 5th Floor, Cambridge, MA 02139, United States
The Boulevard, Langford Lane, Kidlington, Oxford OX5 1GB, United Kingdom

Library of Congress Cataloging-in-Publication Data
A catalog record for this book is available from the Library of Congress

British Library Cataloguing-in-Publication Data
A catalogue record for this book is available from the British Library

ISBN: 978-0-12-809671-0

For information on all Academic Press publications visit our website at
https://www.elsevier.com/books-and-journals

Publisher: Andre G. Wolff
Acquisition Editor: Patricia Osborn
Editorial Project Manager: Jaclyn Truesdell
Production Project Manager: Poulouse Joseph
Designer: Christian Bilbow

Typeset by TNQ Books and Journals

Contents

List of Contributors

Claire B. Andreasen
College of Veterinary Medicine, Iowa State University, Ames, IA, United States

Karsten Becker
Institute of Medical Microbiology, University Hospital Münster, Münster, Germany

Paula Bourke
School of Food Science and Environmental Health, Dublin Institute of Technology, Dublin, Ireland

Sara Ceballos
Department of Food and Agriculture, University of La Rioja, Logroño, Spain

Marco Ebert
University of Applied Sciences, Neubrandenburg, Germany

Rendong Fang
Southwest University, Chongqing, P. R. China

Andrea T. Feßler
Freie Universität Berlin, Berlin, Germany

Alexandra Fetsch
German Federal Institute for Risk Assessment (BfR), Berlin, Germany

Paula Gómez
Department of Food and Agriculture, University of La Rioja, Logroño, Spain

Delia Grace
International Livestock Research Institute (ILRI), Nairobi, Kenya

Jacques-Antoine Hennekinne
ANSES, French Agency for Food, Environmental and Occupational Health & Safety, Maisons-Alfort Cedex, France

Dong-Liang Hu
Kitasato University School of Veterinary Medicine, Towada, Aomori, Japan; College of Veterinary Medicine, Jilin University, Changchun, Jilin, P. R. China

Kristina Kadlec
Friedrich-Loeffler-Institut (FLI), Neustadt-Mariensee, Germany

Franziska Layer
Robert Koch Institute, Wernigerode, Germany

Jun Li
Friedrich-Loeffler-Institut (FLI), Neustadt-Mariensee, Germany; College of Veterinary Medicine, China Agricultural University, Beijing, P. R. China

Catherine M. Logue
College of Veterinary Medicine, Iowa State University, Ames, IA, United States

Agata Los
School of Food Science and Environmental Health, Dublin Institute of Technology, Dublin, Ireland

Masashi Okamura
Kitasato University School of Veterinary Medicine, Towada, Aomori, Japan

Hisaya K. Ono
Kitasato University School of Veterinary Medicine, Towada, Aomori, Japan

Maria de Lourdes Ribeiro de Souza da Cunha
Botucatu Institute of Biosciences, UNESP – University Estadual Paulista, Botucatu, Brazil

Pedro Rodríguez-López
Instituto de Investigaciones Marinas (IIM-CSIC), Vigo, Spain

Stefan Schwarz
Freie Universität Berlin, Berlin, Germany

John F. Sheehan
U.S. Food and Drug Administration (FDA), Silver Spring, MD, United States

Birgit Strommenger
Robert Koch Institute, Wernigerode, Germany

Sandra M. Tallent
U.S. Food and Drug Administration (FDA), Silver Spring, MD, United States

Carmen Torres
Department of Food and Agriculture, University of La Rioja, Logroño, Spain

Daniel Vázquez-Sánchez
"Luiz de Queiroz" College of Agriculture (ESALQ), University of São Paulo (USP), Sao Paulo, Brazil

Lizhe Wang
Zhuhai Foodawa Food Technology Ltd., Zhuhai, P. R. China

Yang Wang
College of Veterinary Medicine, China Agricultural University, Beijing, P. R. China

Guido Werner
Robert Koch Institute, Wernigerode, Germany

Myriam Zarazaga
Department of Food and Agriculture, University of La Rioja, Logroño, Spain

Dana Ziuzina
School of Food Science and Environmental Health, Dublin Institute of Technology, Dublin, Ireland

Preface

"Safer food saves lives," the World Health Organization states in its report on the global burden of foodborne diseases. It is with this in mind that the authors of this book have attempted to summarize the progress in *Staphylococcus (S.) aureus* research. Foodborne diseases are an important cause of morbidity and mortality and a significant impediment to socioeconomic development; without doubt *S. aureus* is among the leading causes of foodborne diseases around the globe. To reduce the impact of *S. aureus* foodborne illness, a full understanding of the pathogen is required along with detailed knowledge of its mode of transmission, detection, biology, and impact. In this book the current perspectives on the characteristics of *S. aureus* are provided, i.e., the pathogenesis of *S. aureus*, its virulence, and antimicrobial resistance properties as well as the biofilm formation ability of *S. aureus*. The book further provides information on methods used for identification and characterization of *S. aureus* including *S. aureus* enterotoxins. Special emphasis is given to food safety aspects of *S. aureus*, on the extent of foodborne outbreaks of disease, the molecular epidemiology of *S. aureus*, and methicillin-resistant *S. aureus* clonal lineages in the livestock industry and in the animal–human interface as well as on prevention and control options of *S. aureus* on the whole farm to fork food chain.

As the editor of this book, I am honored to have worked with such an outstanding group of scientists, and I would sincerely like to thank each of them for helping me to bring this baby of mine to life. *S. aureus* is and will always be my research passion. I would like to invite you now to join me on my adventure journey aiming to understand and combat this fascinating bug.

The Editor
Dr. Alexandra Fetsch, MSc
Berlin, Germany

PART

INTRODUCTION

I

CHAPTER

STAPHYLOCOCCUS AUREUS—A FOODBORNE PATHOGEN: EPIDEMIOLOGY, DETECTION, CHARACTERIZATION, PREVENTION, AND CONTROL: AN OVERVIEW

1

Delia Grace[1], **Alexandra Fetsch**[2]

[1]*International Livestock Research Institute (ILRI), Nairobi, Kenya;* [2]*German Federal Institute for Risk Assessment (BfR), Berlin, Germany*

1. INTRODUCTION

The genus *Staphylococcus* currently comprises more than 50 species. These small, hardy bacteria are normal inhabitants of the skin and mucous membrane in many animal species including humans; they are also ubiquitous in the environment. However, *Staphylococcus aureus* is also an important pathogen of humans and animals. It is a common cause of skin infections and foodborne disease (FBD) in people, as well as sepsis in hospitals and nurseries. It is also an important cause of mastitis in dairy animals and of bone and joint lesions in poultry (bumblefoot) as well as an occasional cause of skin infections in livestock. Companion animals, such as dogs, cats and horses, may play a role in *S. aureus* transmission; they are also vulnerable to *S. aureus* infections (Bierowiec et al., 2016).

This chapter aims to provide a brief introduction of the versatile bacterial organism *S. aureus*, with special focus on its role as foodborne pathogen both, from the perspective of the industrialized and the developing word. Moreover, this chapter briefly outlines the content of the whole book.

2. *STAPHYLOCOCCUS AUREUS*—A BRIEF OVERVIEW

Staphylococci were first isolated from human pus in 1880 by the Scottish surgeon Alexander Ogston: the name derives from *staphyle* (bunch of grapes) and *kokkos* (berry) because the bacteria resembled bunches of grapes when viewed microscopically (Licitra, 2013). In 1886, two strains of *Staphylococcus* were isolated in pure culture by Anton J. Rosenbach, a German surgeon (Rosenbach, 1884). One of these was *S. aureus*, so-named because of the color of the pigmented colonies (aureus means gold-colored in Latin). In the 1920s, it was recognized that presence of coagulase (an enzyme that clots plasma) was associated with pathogenicity, and a coagulase test developed in the late 1930s was an important advance in diagnosis. During the Second World War, penicillin was introduced to clinical use: *S. aureus* was highly susceptible. However, by the late 1940s, penicillin use was common in hospitals and resistant

Staphylococcus aureus. http://dx.doi.org/10.1016/B978-0-12-809671-0.00001-2

strains started to outnumber susceptible strains. Increased resistance stimulated development of semi-synthetic penicillins, such as methicillin, but by the 1990s hospitals were reporting high levels of methicillin-resistant *S. aureus* (MRSA) as well as emerging cases of community-associated MRSA, now an important problem in some parts of the world (Chambers and DeLeo, 2009; David and Daum, 2010). Vancomycin is a common treatment for MRSA but resistance to this has also developed. In the early 21st century a new strain of MRSA was identified in pigs and swine workers, and subsequently other livestock-associated MRSA (LA-MRSA) strains have been found in livestock and raw meat. These LA-MRSA are sporadic causes of human infections in several countries but there is also concern that LA-MRSA could readapt to humans resulting in widespread endemic and epidemic human disease (Cuny et al., 2015).

The largest ecological reservoir of human strains of *S. aureus* is the human nose; however, the skin, hair, and mucous membranes may also be colonized. These resident bacteria do not normally cause disease but nasal carriage is strongly associated with infection; although, only a tiny minority of carriers will ever fall ill from *S. aureus* infection (Brown et al., 2013).

Molecular analysis and pathogenesis suggest that the ancestral host of *S. aureus* is humans (Ng et al., 2009) and phylogenetic data estimate a root age of 25,000 to 142,000 years ago (Weinert et al., 2012). As such, *S. aureus* can be considered an "heirloom disease", i.e., one that has been passed down for millennia from person-to-person (Grace and McDermott, 2011). The health of humans and animals is closely interdependent and many human diseases are shared with animals and vice versa. Molecular epidemiology suggests that *S. aureus* has jumped from humans to livestock several times in the past and has more rarely switched species from livestock back to people (Shepheard et al., 2013). The first jump from humans to cattle took place around 5500 years ago, coinciding with the expansion of cattle domestication throughout the Old World. The first jump to poultry is estimated at around 275 years ago (Weinert et al., 2012). It is currently thought there were at least 13 jumps from humans and animals and two jumps from animal populations back to humans (Shepheard et al., 2013).

As for many diseases, *S. aureus* infections tend to appear in cyclical outbreaks when new epidemic strains appear, spread widely, and then fade again (Shinefield and Ruff, 2009). The last century saw four major waves of drug resistant *S. aureus* in human medicine (Chambers and DeLeo, 2009). Similarly, the clone of *S. aureus* responsible for an ongoing pandemic of lameness in broiler poultry is likely the descendants of a single human-to-poultry host jump that occurred about 40 years ago probably promoted by industrialization of the poultry industry (Lowder et al., 2009).

This book focuses on the role of *S. aureus* as a foodborne pathogen, but the organism causes a wide range of important diseases in people and animals. In humans, *S. aureus* can infect damaged skin resulting in superficial or deep infections of skin and soft tissue. Less commonly but with more severe consequences, *S. aureus* causes bacteremia; this is often hospital associated, following breaching of the skin barrier with surgical devices or implants. Bacteremia can be complicated septicemia and localization in organs, joints, bones, and elsewhere. Staphylococcal pneumonia may be a primary infection or result from hematogenous spread: it is not common in community-acquired infections but it is in hospital-acquired infections. Neonates, especially those born prematurely, are also susceptible to staphylococcal disease, resulting in scalded skin syndrome and other manifestations. *S. aureus* is also a cause of toxic shock syndrome, first described in children and later associated with use of super-absorbent tampons in menstruating women.

S. aureus is not only a human (facultative) pathogen but a commensal bacteria and pathogen of several animal species, too. However, the strains carried by animals usually are host-specific and not normally present in humans. In countries with a developed dairy sector, *S. aureus* is one of the most important causes of mastitis, which in turn is the most common and costly disease on dairy farms (Keefe, 2012); goats and sheep are also affected. *S. aureus* is also responsible for a range of skin diseases in livestock including impetigo and udder dermatitis in dairy cattle, dermatitis in sheep, and secondary skin infection in goats (Foster, 2012). *S aureus* is an important cause of omphalitis, joint infections, and bumblefoot in poultry (Andreasen, 2013) and can cause virulent epidemics of skin abscesses, mastitis, and septicemia in farmed rabbits (Harcourt-Brown, 2001). Food of animal origin is also a potential source of human FBD, although most cases are because of human strains.

3. *STAPHYLOCOCCUS AUREUS* AS FOODBORNE PATHOGEN—THE INDUSTRIALIZED WORLD PERSPECTIVE

Worldwide, *S. aureus* is one of the most common FBDs. It was first reported when investigation a food poisoning resulting from eating cheese in Michigan (United States) revealed the presence of micrococci (Hennekinne et al., 2012). Since then, *S. aureus* has been implicated in numerous outbreaks and cases of *S. aureus* food poisoning have been reported from many countries and in association with a wide variety of foods.

FBD has been a major concern for millennia and empirical understanding of food safety may have been the root of ancient laws regarding clean and unclean foods. Currently, consumer surveys show high levels of concern over food safety in rich and poor countries (Grace and McDermott, 2015). The true burden of FBD is difficult to assess because of widespread underreporting and challenges in attributing diseases with multiple transmission routes to food consumption. Despite this, there have been considerable advances in understanding the burden of FBD, with robust estimates from several developed countries suggest a substantial proportion of the population are affected each year (12%–33%). Many of these multipathogen disease burden assessments have indicated an important role for *S. aureus*. For example, it is responsible for an estimated 292,000 cases a year in the Netherlands (Mangen et al., 2015), 241,000 illnesses in the United States (Scallan et al., 2011), 25,000 cases in Canada (Thomas et al., 2013), and 1300 cases in Australia (Kirk et al., 2014). This very wide incidence range may reflect dietary and ecological differences but also reflects challenges in assessing the incidence of FBD. Although *S. aureus* is typically among the top 10 causes of bacterial FBD, some FBD burden surveys did not include this pathogen in their analysis including estimates for the United Kingdom and the first global assessment of the health burden of FBD (Havelaar et al., 2015).

In addition to global health burden comparable to malaria, HIV/AIDs, or tuberculosis (Havelaar et al., 2015), FBD has important economic impacts. These include the cost of treatment, lost productivity from illness, the cost to industry, the cost of public health, and the cost of food safety governance (McLinden et al., 2014). Economic studies use different methodologies, but the cost of FBD is high: it is estimated to cost the United States up to $152 billion, Australia over $1 billion, and Sweden $171 million a year (Hoffmann and Anekwe, 2013; Hall et al., 2005; Toljander et al., 2012). While fewer studies have looked at the cost of *S. aureus*, its position among the most important FBD suggests its economic impact is similarly high: one assessment of 14 important food-associated pathogens in the Netherlands found *S. aureus* intoxications accounted for the highest share of costs attributed to food (€ 47.1 million/year) (Mangen et al., 2015).

4. *STAPHYLOCOCCUS AUREUS* AS FOODBORNE PATHOGEN—THE PERSPECTIVE FROM THE DEVELOPING WORLD

In contrast to developed countries, *S. aureus* in poor countries ranks low on the public health agenda, and the limited literature focuses mainly on systemic infections and drug resistance (Nickerson et al., 2009). It is known, however, that FBD disproportionately affects poor countries, which bear around 98% of the total health burden (Havelaar et al., 2015) and so the importance of *S. aureus* is likely greatly underestimated.

Most studies on *S. aureus* in food report that it is present in at least some samples, and prevalence levels of 20%–50% are not uncommon, especially in milk (Akindolire et al., 2015; Ngasala et al., 2015; Tigabu et al., 2015). *S. aureus* has also been isolated from the fingernails of food handlers in Nigeria and Botswana (Ifeadike et al., 2012; Loeto et al., 2007) and from nasal passages of food handlers in several countries (Dagnew et al., 2012; Wei and Chiou, 2002). Outbreaks of staphylococcal FBD have been reported from Asia (Nema et al., 2007; Wei and Chiou, 2002) and in Kenya where *S. aureus* was the hazard most often responsible for confirmed outbreaks of FBD (38% of the total) (Ombui et al., 2001). The only quantitative assessment of staphylococcal foodborne illness from developing countries estimated that *S. aureus* was responsible for two cases of FBD per 100 people; however, without traditional fermentation the risk would be 16 times higher (Makita et al., 2012).

Given its important role as a cause of FBD in developing countries, the livelihood and economic burdens of *S. aureus* are likely similarly high. One study from Nigeria estimated that costs of all FBD were around $2 million a year (ILRI, 2011). Moreover, the presence of hazards in food is an increasingly important barrier to market access and trade (Unnevehr and Ronchi, 2014).

5. *STAPHYLOCOCCUS AUREUS*—A FOODBORNE PATHOGEN: EPIDEMIOLOGY, DETECTION, CHARACTERIZATION, PREVENTION, AND CONTROL—WHAT THE BOOK IS ABOUT

In this book we cover various aspects of *S. aureus* as follows:

Part I gives an overview of *S. aureus* with special focus on its role as foodborne pathogen both from the perspective of the industrialized and the developing word (Chapter 1).

Part II focuses on the characteristics of *S. aureus* itself. This includes a comprehensive overview of the pathogenesis of *S. aureus* (Chapter 2), also providing information on major aspects of the genetic basis of the pathogen's virulence and mechanisms for colonizing and entering host cells. In Chapter 3 the superfamily of *S. aureus* enterotoxins is described, i.e., their gene locations, molecular structure, superantigenic activity, and their biological characteristics and emetic activity. The latest information on resistance genes and resistance-mediating mutations detected in *S. aureus* is presented in Chapter 4. In Chapter 5 the major processes known to date regarding biofilm development in *S. aureus* are presented; this includes a detailed description of the processes involved in biofilm formation and what the advantages of multicellular structures are. Moreover, the impact and consequences of biofilm producing *S. aureus* in the food industry are discussed. Finally, Part II provides an overview of the broad variety of different methods used for identification, characterization, and tracking the spread of

S. aureus (Chapter 6); this includes information on rapid commercial test systems as well as on methods to detect staphylococcal enterotoxins in food.

Part III focuses on food safety aspects of *S. aureus* from farm to fork. Chapter 7 emphasizes why *S. aureus* is among the leading causes of FBDs worldwide. It also describes examples of food poisoning events because of staphylococci, mainly *S. aureus* from the history to the present and provides information on the characteristics and behavior of *S. aureus* in the food environment as well as data from reporting systems of staphylococcal food poisoning outbreaks. Chapter 8 provides an overview of *S. aureus* and MRSA in the livestock industry, describing its impact on human health from a veterinary public health perspective; this also includes information on *S. aureus* as a cause of animal disease. Moreover, current knowledge about the *S. aureus* population colonizing humans, including those in close contact to animals and food, is summarized (Chapter 9); this subchapter also reviews data on the molecular characterization of *S. aureus* isolates related to staphylococcal FBD and the elucidation of staphylococcal foodborne outbreaks. Finally, in Chapter 10 different *S. aureus* lineages that have expanded in the animal–human interface are described in detail; particular attention will be payed to the emergence, characteristics, and molecular epidemiology of the livestock-associated lineage CC398, amongst others.

Finally, in Part IV prevention and control options for *S. aureus* along the farm to fork food chain are described. This includes hygiene principles such as good hygiene practice and hazard analysis critical control point (HACCP) principles, temperature-related food hygiene principles, as well as personal hygiene and training to avoid contamination and/or cross-contamination in commercial kitchens and private households during food processing (Chapter 11). Moreover, special attention is given to thermal and nonthermal control strategies, as well as the combination of both, for *S. aureus* inactivation (Chapter 12); particular emerging nonthermal inactivation strategies show high potential for *S. aureus* control to ensure food safety. Finally, a comprehensive overview on mitigation strategies, international food standards, guidelines, and codes of practices put in place around the globe and tackling *S. aureus* with the overall goal to ensure consumers' safety is provided (Chapter 13).

Taken together, this book provides a comprehensive overview and latest research on *S. aureus* as one of the major foodborne pathogens worldwide. Contributions are from leading national and international experts including those from world renowned institutions.

6. SUMMARY AND CONCLUSION

Without doubt, *S. aureus* is among the most famous and also one of the most important bacterial organisms worldwide, both, from the human and animal health perspective, but *S. aureus* also need to be considered as veterinary public health issue.

Control of staphylococcal FBD is based on hygiene measures to avoid contamination of food. Because foodborne outbreaks have largely been attributed to faulty food handling, improving the knowledge and skills of food handlers and ensuring that incentives are in place to translate knowledge and skills to behavior change are important. The widespread application of approaches such as risk assessment and hazard analysis and critical control points and good hygienic practice can help prevent contamination (Kadariya et al., 2014). However, these are not so applicable to the informal markets of developing countries where most fresh and cooked foods are sold: in these contexts, participatory and stakeholder-led approaches have had more success (Grace, 2015).

Although the prevalence and burden of staphylococcal FBD remains high, recent decades have seen several advances in detection, prevention, and control of *S. aureus*. The explosion of sequencing data for *S. aureus* is transforming our understanding of population diversity, disease spread, and emergence (Fitzgerald and Holden, 2016). Molecular and immunological-based methods are increasingly used in diagnosis of *S. aureus* (Hennekinne et al., 2012). Advances in disease burden assessment are showing the enormous burden of FBD, including *S. aureus*, in developing countries and are stimulating donor investment in disease management. The issue of antimicrobial resistance is receiving widespread attention with important initiatives to improve reporting and develop new strategies for prevention and control (Robinson et al., 2016).

REFERENCES

Akindolire, M.A., Babalola, O.O., Ateba, C.N., 2015. Detection of antibiotic resistant *Staphylococcus aureus* from milk: a public health implication. In: Uyttendaele, M., Franz, E., Schlüter, O. (Eds.), International Journal of Environmental Research and Public Health, vol. 12 (9), pp. 10254–10275. http://dx.doi.org/10.3390/ijerph120910254.

Andreasen, C.B., 2013. Staphylococcsis. In: Swayne, D. (Ed.), Diseases of Poultry, thirteenth ed. Wiley Blackwell.

Bierowiec, K., Płoneczka-Janeczko, K., Rypuła, K., 2016. Is the colonisation of *Staphylococcus aureus* in pets associated with their close contact with owners? PLoS One. 11 (5), e0156052. https://dx.doi.org/10.1371/journal.pone.0156052.

Brown, A.F., Leech, J.M., Rogers, T.R., McLoughlin, R.M., 2013. *Staphylococcus aureus* colonization: modulation of host immune response and impact on human vaccine design. Front. Immunol. 4, 507. http://dx.doi.org/10.3389/fimmu.2013.00507.

Chambers, H.F., DeLeo, F.R., 2009. Waves of resistance: *Staphylococcus aureus* in the antibiotic era. Nat. Rev. Microbiol. 7 (9), 629–641. http://dx.doi.org/10.1038/nrmicro2200.

Cuny, C., Wieler, L.H., Witte, W., 2015. Livestock-associated MRSA: the impact on humans. In: Woodward, M.J. (Ed.), Antibiotics, vol. 4 (4), pp. 521–543. http://dx.doi.org/10.3390/antibiotics4040521.

Dagnew, M., Tiruneh, M., Moges, F., Tekeste, Z., 2012. Survey of nasal carriage of *Staphylococcus aureus* and intestinal parasites among food handlers working at Gondar University, Northwest Ethiopia. BMC Public Health 12, 837. http://dx.doi.org/10.1186/1471-2458-12-837.

David, M.Z., Daum, R.S., 2010. Community-associated methicillin-resistant *Staphylococcus aureus*: epidemiology and clinical consequences of an emerging epidemic. Clin. Microbiol. Rev. 23 (3), 616–687. http://dx.doi.org/10.1128/CMR.00081-09.

Fitzgerald, J.R., Holden, M.T.G., 2016. Genomics of natural populations of *Staphylococcus aureus*. Annu. Rev. Microbiol. 70, 459–478. http://dx.doi.org/10.1146/annurev-micro-102215-095547.

Foster, A.P., 2012. Staphylococcal skin disease in livestock. Vet. Dermatol. 23, 342–351, e63. http://dx.doi.org/10.1111/j.1365-3164.2012.01093.x.

Grace, D., 2015. Food safety in low and middle income countries. Int. J. Environ. Res. Public Health 12 (9), 10490–10507.

Grace, D., McDermott, J., 2011. Livestock epidemics and disasters. In: Kelman, I., et al. (Ed.), Handbook of Hazards and Disaster Risk Reduction. Routledge.

Grace, D., McDermott, J., 2015. Food safety: reducing and managing food scares. In: International Food Policy Research Institute. 2014–2015 Global Food Policy Report. International Food Policy Research Institute, Washington, DC, pp. 41–50.

Hall, G., Kirk, M., Becker, N., Gregory, J., Unicomb, L., Millard, G., 2005. Estimating foodborne gastroenteritis, Australia. Emerg. Infect. Dis. 11, 1257–1264. https://dx.doi.org/10.3201/eid1108.041367.

Harcourt-Brown, F., 2001. Textbook of Rabbit Medicine. Butterworth-Heinemann, Oxford, UK.

Havelaar, A.H., Kirk, M.D., Torgerson, P.R., Gibb, H.J., Hald, T., Lake, R.J., et al., 2015. World Health Organization global estimates and regional comparisons of the burden of foodborne disease in 2010. PLoS Med. 12 (12), e1001923. https://dx.doi.org/10.1371/journal.pmed.1001923.

Hennekinne, J.-A., De Buyser, M.L., Dragacci, S., 2012. *Staphylococcus aureus* and its food poisoning toxins: characterization and outbreak investigation. FEMS Microbiol. Rev. 36 (4), 815–836. http://dx.doi.org/10.1111/j.1574-6976.2011.00311.x.

Hoffmann, S., Anekwe, T.D., 2013. Making Sense of Recent Cost-of-Foodborne-Illness Estimates, Economic Information Bulletin No. (EIB-118). United States Department of Agriculture, Economic Research Service.

Ifeadike, C.O., Ironkwe, O.C., Adogu, P.O.U., et al., 2012. Prevalence and pattern of bacteria and intestinal parasites among food handlers in the Federal Capital Territory of Nigeria. Niger. Med. J. 53 (3), 166–171. http://dx.doi.org/10.4103/0300-1652.104389.

ILRI, 2011. Assessment of Risks to Human Health Associated with Meat from Different Value Chains in Nigeria: Using the Example of the Beef Value Chain. Nigeria Integrated Animal and Human Health Management Project Draft Report. ILRI, Nairobi, Kenya.

Kadariya, J., Smith, T.C., Thapaliya, D., 2014. *Staphylococcus aureus* and staphylococcal food-borne disease: an ongoing challenge in public health. BioMed Res. Int. 2014, 827965. http://dx.doi.org/10.1155/2014/827965.

Keefe, G., 2012. Update on control of *Staphylococcus aureus* and *Streptococcus agalactiae* for management of mastitis. Vet. Clin. N. Am. Food Anim. Pract. 28, 203–216. http://dx.doi.org/10.1016/j.cvfa.2012.03.010.

Kirk, M., Ford, L., Glass, K., Hall, G., 2014. Foodborne illness, Australia, Circa 2000 and Circa 2010. Emerg. Infect. Dis. 20 (11), 1857–1864. https://dx.doi.org/10.3201/eid2011.131315.

Licitra, G., 2013. Etymologia: *Staphylococcus*. Emerg. Infect. Dis. 19 (9), 1553. http://dx.doi.org/10.3201/eid.1909.ET1909.

Loeto, D., Matsheka, M.I., Gashe, B.A., 2007. Enterotoxigenic and antibiotic resistance determination of *Staphylococcus aureus* strains isolated from food handlers in Gaborone, Botswana. J. Food Prot. 70 (12), 2764–2768.

Lowder, B.V., Guinane, C.M., Ben Zakour, N.L., Weinert, L.A., Conway-Morris, A., Cartwright, R.A., et al., 2009. Recent human-to-poultry host jump, adaptation, and pandemic spread of *Staphylococcus aureus*. PNAS. 106 (46), 19545–19550. http://dx.doi.org/10.1073/pnas.0909285106.

Makita, K., Desissa, F., Teklu, A., Zewde, G., Grace, D., 2012. Risk assessment of staphylococcal poisoning due to consumption of informally-marketed milk and home-made yoghurt in Debre Zeit, Ethiopia. Int. J. Food Microbiol. 153 (1–2), 135–141. http://dx.doi.org/10.1016/j.ijfoodmicro.2011.10.028.

Mangen, M.-J.J., Bouwknegt, M., Friesema, I.H.M., Haagsma, J.A., Kortbeek, L.M., Tariq, L., Wilson, M., van Pelt, W., Havelaar, A.H., 2015. Cost-of-illness and disease burden of food-related pathogens in the Netherlands, 2011. Int. J. Food Microbiol. 196, 84–93. http://dx.doi.org/10.1016/j.ijfoodmicro.2014.11.022.

McLinden, T., Sargeant, J.M., Thomas, M.K., Papadopoulos, A., Fazil, A., 2014. Component costs of foodborne illness: a scoping review. BMC Public Health 14, 509. http://dx.doi.org/10.1186/1471-2458-14-509.

Nema, V., Agrawal, R., Kamboj, D.V., Goel, A.K., Singh, L., 2007. Isolation and characterization of heat resistant enterotoxigenic *Staphylococcus aureus* from a food poisoning outbreak in Indian subcontinent. Int. J. Food Microbiol. 117 (1), 29–35.

Ng, J.S., Holt, D.C., Lilliebridge, R.A., Stephens, A.J., Huygens, F., Tong, S.Y.C., Currie, B.J., Giffard, P.M., 2009. A phylogenetically distinct *Staphylococcus aureus* lineage prevalent among Indigenous communities in Northern Australia. J. Clin. Microbiol. 47, 2295–2300. http://dx.doi.org/10.1128/JCM.00122-09.

Ngasala, J.U., Nonga, H.E., Mtambo, M.M., June 2015. Assessment of raw milk quality and stakeholders' awareness on milk-borne health risks in Arusha City and Meru District, Tanzania. Trop. Anim. Health Prod. 47 (5), 927–932. http://dx.doi.org/10.1007/s11250-015-0810-y.

Nickerson, E.K., West, T.E., Day, N.P., Peacock, S.J., 2009. *Staphylococcus aureus* disease and drug resistance in resource-limited countries in south and east Asia. Lancet Infect. Dis. 9 (2), 130–135. http://dx.doi.org/10.1016/S1473-3099(09)70022-2.

Ombui, J.N., Kagiko, M.M., Arimi, S.M., 2001. Foodborne diseases in Kenya. East Afr. Med. J. 78 (1), 40–44.

Rosenbach, A.J., 1884. Mikro-Qrganismen bei den Wund-Infektions-Krankheiten des Menschen. J.F. Bergmann, Wiesbaden, p. 18.

Robinson, T.P., Bu, D.P., Carrique-Mas, J., Fèvre, E.M., Gilbert, M., Grace, D., Hay, S.I., Jiwakanon, J., Kakkar, M., Kariuki, S., Laxminarayan, R., Lubroth, J., Magnusson, U., Thi Ngoc, P., van Boeckel, T.P., Woolhouse, M.E.J., 2016. Antibiotic resistance is the quintessential one health issue. Trans. R. Soc. Trop. Med. Hyg. 110 (7), 377–380.

Scallan, E., Hoekstra, R.M., Angulo, F.J., et al., 2011. Foodborne illness acquired in the United States—major pathogens. Emerg. Infect. Dis. 17 (1), 7–15.

Shepheard, M.A., Fleming, V.M., Connor, T.R., Corander, J., Feil, E.J., Fraser, C., et al., 2013. Historical Zoonoses and other changes in host tropism of *Staphylococcus aureus*, identified by phylogenetic analysis of a population dataset. PLoS One 8 (5), e62369. http://dx.doi.org/10.1371/journal.pone.0062369.

Shinefield, H.R., Ruff, N.L., 2009. Staphylococcal infections: a historical perspective. Infect. Dis. Clin. N. Am. 23, 1–15. http://dx.doi.org/10.1016/j.idc.2008.10.007.

Thomas, M.K., Murray, R., Flockhart, L., Pintar, K., Pollari, F., Fazil, A., Nesbitt, A., Marshall, B., 2013. Estimates of the burden of foodborne illness in Canada for 30 specified pathogens and unspecified agents, Circa 2006. Foodborne Pathog. Dis. 10, 639–648.

Tigabu, E., Asrat, D., Kassa, T., Sinmegn, T., Molla, B., Gebreyes, W., 2015. Assessment of risk factors in milk contamination with *Staphylococcus aureus* in urban and peri-urban small-holder dairy farming in Central Ethiopia. Zoonoses Public Health 62 (8), 637–643. http://dx.doi.org/10.1111/zph.12199.

Toljander, J., Dovärn, A., Andersson, Y., Ivarsson, S., Lindqvist, R., 2012. Public health burden due to infections by verocytotoxin-producing Escherichia coli (VTEC) and *Campylobacter* spp. as estimated by cost of illness and different approaches to model disability-adjusted life years. Scand. J. Public Health. 40 (3), 294–302. http://dx.doi.org/10.1177/1403494811435495.

Unnevehr, L., Ronchi, L., 2014. Food Safety and Developing Markets: Research Findings and Research Gaps. International Food Policy Research Institute, Washington, DC. Retrieved from: https://books.google.com/books?id=xbOBAAAQBAJ&pgis=1.

Weinert, L.A., Welch, J.J., Suchard, M.A., Lemey, P., Rambaut, A., Fitzgerald, J.R., 2012. Molecular dating of human-to-bovid host jumps by *Staphylococcus aureus* reveals an association with the spread of domestication. Biol. Lett. 8 (5), 829–832. http://dx.doi.org/10.1098/rsbl.2012.0290.

Wei, H.L., Chiou, C.S., 2002. Molecular subtyping of *Staphylococcus aureus* from an outbreak associated with a food handler. Epidemiol. Infect. 128 (1), 15–20.

PART

CHARACTERISTICS AND DETECTION OF *STAPHYLOCOCCUS AUREUS* II

CHAPTER

PATHOGENESIS OF *STAPHYLOCOCCUS AUREUS* 2

Karsten Becker
Institute of Medical Microbiology, University Hospital Münster, Münster, Germany

1. INTRODUCTION

The opportunistic pathogen *Staphylococcus aureus* is a common colonizer of the human skin, but, once overcoming the skin barrier, it may cause a variety of pyogenic and systemic infections, acute and chronic infections, and toxin-mediated syndromes in both health care and community settings. In addition to its classical conception as an extracellularly acting microorganism, *S. aureus* has been also recognized as intracellular pathogen. This property potentially contributes to bacterial persistence, protection from antibiotics, and evasion of immune defenses. Its genetic plasticity as basis for a striking adaptation potential enables this versatile species to express strain-dependently an enormous, often redundant arsenal of virulence factors, including adhesins, enzymes, toxins, and immune evasion proteins. Staphylococcal virulence factors are characterized by often overlapping roles in different pathogenic processes leading to adherence on and aggression to host structures, followed by internalization, intracellular persistence, and immune evasion. This chapter is particularly focused on major aspects of the genetic basis of the pathogen's virulence and its mechanisms to colonize and enter host cells.

2. IMPACT ON HEALTH

Within the large genus *Staphylococcus*, the species *S. aureus* represents obviously the most versatile species in terms of host spectrum, equipment with virulence factors, and pathogenic capacity (Becker et al., 2014; Tong et al., 2015). Although every *S. aureus* isolate is able to cause even life-threatening diseases, there is much evidence that strain-specific equipment with virulence factors facilitates certain clones to be more virulent than others (Melles et al., 2004; Peacock et al., 2002). Thus, it is not surprising that this opportunistic pathogen causes an enormous health-related burden characterized by high morbidity and mortality for human and animal in- and outpatients as well as enormous socioeconomic costs worldwide (Lowy, 1998; Rasigade and Vandenesch, 2014; Antonanzas et al., 2015; Humphreys et al., 2016). For *S. aureus* as an animal-adapted pathogen in farm and companion animals and its role as zoonotic pathogen, please refer to respective literature (Pantosti, 2012; Becker et al., 2015; Morgan, 2008; Peton and Le Loir, 2014) or Chapter 10 of this book.

The pathogenic capacity of this microorganism is additionally aggravated by the acquisition of often multiple resistances to antibiotics and other agents with antimicrobial activity (see Chapter 5 for more details). In particular, the various manifestations of methicillin-resistant *S. aureus* (MRSA), i.e.,

Staphylococcus aureus. http://dx.doi.org/10.1016/B978-0-12-809671-0.00002-4

health care–associated, community-associated (CA-MRSA), and livestock-associated (LA-MRSA) clonal lineages, are responsible for drastically lowered therapeutic options, serious courses of infection, and dramatically increased costs for prevention measures (Gould et al., 2010; Köck et al., 2010; Cuny et al., 2013; Chambers and Deleo, 2009; Diederen and Kluytmans, 2006). In this chapter, an overview limited to main features and mechanisms that contribute to the pathogenesis of *S. aureus*–caused diseases will be given.

2.1 COLONIZATION AND ITS CLINICAL IMPACT

S. aureus colonizes the human host and has been also isolated from different animal species, in particular from mammalians, including farm and companion animals. Habitually, the pathogen has been isolated from healthy individuals not only from different localizations of the nasopharyngeal area but also from other parts of the skin, the vaginal tract, and the intestine, in particular from the perineum (Williams, 1963).

The nasal cavity is the primary ecological reservoir, i.e., the main habitat of *S. aureus* (Kloos and Musselwhite, 1975; Kluytmans et al., 1997). Although the anterior nares are commonly considered in literature, in fact, the posterior regions of the vestibule revealed most frequent colonization based on in-depth studies on the nasal culturome (Kaspar et al., 2016). The results of earlier longitudinal studies led to the assumption that three types of *S. aureus* nasal carriers exist, namely noncarriers, intermittent carriers, and persistent carriers (Kluytmans and Wertheim, 2005). Meanwhile, van Belkum et al. (2009) reclassified *S. aureus* carriage into persistent carriers and others only. Permanent nasal colonization with *S. aureus* is exhibited by up to 30%–40% of the human population, and the other part may show a variable status of carriage (Köck et al., 2016; Wertheim et al., 2005; van Belkum et al., 2009). Newer investigations have shown that colonization by *S. aureus* is not strictly clonal and polyclonality of the resident type within single body niches occurs (Muenks et al., 2016; Paterson et al., 2015). Spreading from the nasal habitat, the surfaces of other skin and mucous membrane areas such as the axillae, groin and throat, and the gastrointestinal and vaginal tract might be colonized as secondary niches.

Nasal *S. aureus* carriage plays a key role in the pathogenesis of infection by representing the source and the independent risk factor for subsequent infections (von Eiff et al., 2001a; Wertheim et al., 2004). Consequently, studies that were able to show that nasal *S. aureus* decolonization is effective in reducing the incidence of infections caused by this pathogen have been initiated (Kalmeijer et al., 2002; Perl et al., 2002; Bode et al., 2010; Yu et al., 1986). According to metaanalyses, the best evidence of decolonization therapy is hitherto available for dialysis and adult intensive care patients and those undergoing orthopedic or cardiothoracic surgery (Köck et al., 2014; Nair et al., 2016).

In addition to colonizing humans, *S. aureus* may colonize the skin and upper respiratory tract of other mammals and has been also isolated from birds, including industrially raised poultry, and snakes (Cuny et al., 2010). Although animal lineages are closely related to human lineages, some relatively host-specific clonal lineages, e.g., for ruminants, occur (Sung et al., 2008). The former *S. aureus* sequence type ST2022, recently separated as *Staphylococcus schweitzeri*, has been preferentially found as colonizer of monkeys (Schaumburg et al., 2015). A sheep-adapted microaerophilic *S. aureus* subspecies, described already in 1985, has been shown to be clonal in nature (de la Fuente et al., 2011).

2.2 CLINICAL SYNDROMES

In principle, two different kinds of *S. aureus*–caused diseases must be distinguished: first, pyogenic and/or systemic infections that can affect virtually all organs and organ systems, respectively, and, second, toxin-mediated diseases.

S. aureus is the causative agent of a multitude of acute and chronic infections, often with severe consequences. Skin and soft tissue infections are the most frequent clinical entities, ranging from infections of the epidermis (e.g., impetigo), infections of the superficial and deep dermis (e.g., folliculitis, furuncles, and carbuncles), and infections of subcutaneous tissues (e.g., limited cellulitis [i.e., phlegmon]) and severe cellulitis and necrotizing fasciitis) (Sunderkötter and Becker, 2015). Encapsulated cutaneous abscesses are found in the dermis and subcutis. Other important infections due to *S. aureus* include wound infections, mastitis, meningitis, community-acquired and hospital-acquired pneumonias, bone and joint infections (e.g., arthritis, bursitis, and osteomyelitis), and primary pyomyositis (i.e., tropical myositis) (Lowy, 1998). Staphylococcal sepsis syndrome is the most feared consequence of *S. aureus* community-onset and nosocomial bacteremia, which might be additionally complicated by metastatic infections (e.g., hematogenous osteomyelitis and septic arthritis) and endocarditis of native and prosthetic valves (Holland et al., 2014). Particularly in acute *S. aureus* endocarditis, septic emboli lead to multiple microabscesses.

The *S. aureus*–caused toxin-mediated diseases comprise the following clinically defined entities: (1) the menstrual and nonmenstrual staphylococcal toxic shock syndrome caused by the toxic shock syndrome toxin-1 (TSST-1) and some staphylococcal enterotoxins (SEs), mainly by SEB and SEC, (2) the SE-caused staphylococcal food poisoning (SFP), and (3) the staphylococcal scalded skin syndrome caused by exfoliative toxins (ETs) (Spaulding et al., 2013; Bukowski et al., 2010). For further details on SE-caused SFP, Chapter 4 of this book can be referred.

3. GENETIC BASIS FOR VIRULENCE

S. aureus is characterized by a highly clonal structure (Lindsay, 2010). Although an enormous diversity of *S. aureus* clones have been described (Becker et al., 2017), a limited number of pandemic or regional clonal complexes dominate those isolates recovered from clinical specimens from colonized and infected patients (Schaumburg et al., 2012; Melles et al., 2004; Monecke et al., 2011).

The *S. aureus* genome ranges in size from ca. 2.8 to 2.9 Mb (Lindsay and Holden, 2004). The core genome of *S. aureus*, which is conserved among the different clonal lineages, comprises up to three quarters of its genetic resources. In addition to housekeeping genes, it also contains genes that are important for the pathogens' virulence and are produced by the vast majority of all strains as illustrated by genes encoding, e.g., aureolysin (*aur*), the clumping factors A and B (*clfA, clfB*), the coagulase (*coa*), the extracellular adherence protein (*eap*), the extracellular matrix protein–binding protein (*emp*), the fibronectin-binding protein A (*fnbPA*), α-hemolysin (*hla*), lipase (*lip*), phenol-soluble modulins (*psms* genes, *hld*), protein A (*spa*), and von Willebrand factor–binding protein (*vWbp*). However, some of these genes and/or variant genes are also carried by mobile genetic elements (MGEs) such as plasmids, bacteriophages, *S. aureus* pathogenicity islands (SaPIs), chromosomal cassettes, and transposons.

Besides this core genome, *S. aureus* owns a relatively broad accessory (auxiliary) genome defining the intraspecies diversity. This accessory genome is mainly occupied by MGEs that contribute largely to the vast phenotypic and genetic plasticity of *S. aureus* (Lindsay and Holden, 2004; Foster, 2004). These mobile or once mobile genetic elements may carry a multitude of different genes involved in virulence and in resistance toward antimicrobial agents, biocides, heavy metals, and metalloids. Thus, horizontal gene transfer appears to be the key to the pathogen's adaptation to different hosts and environments and to its epidemiological success, and the resulting enormous genomic plasticity impacts significantly on the virulence of *S. aureus*. This fact plus the existence of very complex regulatory systems (see below) are reflected by an extreme heterogeneity of the pathogen's exoproteome (Ziebandt et al., 2010).

3.1 *STAPHYLOCOCCUS AUREUS* PHAGES

Allover, more than 250 staphylococcal phages are reported, and all belong to the order *Caudovirales* (tailed phages) (Xia and Wolz, 2014). Most *S. aureus* strains possess at least one temperate bacteriophage (prophage) of the *Siphoviridae* family (genome size of 39–43 kb) but not several of the same integrase gene family (Ackermann and Dubow, 1987; Deghorain and Van Melderen, 2012; Pantůček et al., 2004). A few staphylococcal phages belong to the *Podoviridae* (genome size of 16–18 kb) and *Myoviridae* (genome size of 127–140 kb) families (Kwan et al., 2005; Lobocka et al., 2012). Classically, *S. aureus* has been assumed as being not naturally competent; however, expression of alternative sigma factor H (SigH) and SigH-controlled competence genes—found in a minor fraction of an *S. aureus* cell population under certain circumstances—lead to competence for transformation by plasmid or chromosomal DNA (Morikawa et al., 2012). Because of recombination and horizontal gene transfer, the genomes of staphylococcal phages and prophages are very diverse and extensively mosaic (Kwan et al., 2005). Equipping *S. aureus* strains with additional genes, lysogeny drives short-term evolution (Xia and Wolz, 2014).

Clinically important toxins encoded by bacteriophages are the SFP-causing staphylococcal enterotoxin A (SEA), the SFP-causing exfoliative toxin A, and the Panton–Valentine leukocidin (PVL), a bicomponent synergohymenotropic toxin encoded by the *lukSF-PV* genes. In particular, *lukSF-PV*-carrying phages are of special clinical interest because *S. aureus* isolates "armed" with this virulence factor are strongly linked with skin and soft tissue infections including necrotizing entities (necrotizing fasciitis and myositis). Although PVL has been known for a long time (Panton and Valentine, 1932), there was renewed interest in this toxin because of the emergence and ongoing global spread of the CA-MRSA clonal lineages characterized by highly frequent PVL possession (Schaumburg et al., 2012; Maguire et al., 1996; Groom et al., 2001; Herold et al., 1998). Typical CA-MRSA clonal lineages are USA300 (sequence type (ST) 8; t008), USA400 (ST1; t175/t558), and the "European CA-MRSA" ST-80 (t044) (Monecke et al., 2007)—see Chapter 7 of this book for further details. The necrotizing pneumonia is a further PVL-associated disease that has emerged with the advent of the CA-MRSA lineages (Lina et al., 1999; Dufour et al., 2002). However, the possession and expression of PVL is not restricted to MRSA strains.

Other bacteriophage-associated virulence factors are SEE, SEG, SEK and SEP, lipase, staphylokinase, and β-hemolysin. Transducing bacteriophages may mobilize the transfer of SaPIs as shown for SaPI1 and SaPI2 (Lindsay et al., 1998; Ruzin et al., 2001).

3.2 *STAPHYLOCOCCUS AUREUS* PATHOGENICITY ISLANDS

Contributing to bacterial evolution as a repository of virulence genes, most *S. aureus* isolates carry one or more highly mobile SaPIs, which are distinct MGEs inserted at specific sites (att_S) of the chromosomes of bacterial pathogens (Hacker and Kaper, 2000; Novick and Subedi, 2007). SaPIs show a strong association with temperate phages (Ubeda et al., 2008).

Besides core genes, which are necessary for bacteriophage and SaPI interaction, many genes are carried by SaPIs that are involved in pathogenesis and resistance (Malachowa and DeLeo, 2010). For example, most of the known SEs/staphylococcal enterotoxin-like toxins (SEls) are located on SaPIs, which may possess one or more superantigen-encoding genes including *tst* and various SEs/SEls (e.g., *sea, seb*, different *sec* subtypes, *sek, sel, sep, seq*). In addition, genes coding for ETs and further virulence factors (e.g., *aad, bap, ear, mdr, vwb*) as well as resistance genes (e.g., *fosB*) have been detected as being part of SaPIs.

So far, most SaPIs have been identified from *S. aureus* strains of human origin (SaPI1 to SaPI5, SaPIn1/m1, SaPI1mw2), whereas others have been found in camel (SaPIcam1 and SaPIcam2) and bovine isolates (SaPIbov1 to SaPIbov5), the latter in association with mastitis cases (Zubair et al., 2015; Viana et al., 2010; Fitzgerald et al., 2001; Ubeda et al., 2003; Novick et al., 2010). SaPIS0385 (SaPIpig) has been discovered analyzing an LA-MRSA ST398 isolate from a case of human endocarditis (Schijffelen et al., 2010). This SaPI was found to be composed at the 5′-end homologous to SaPI5 (from an USA300 strain) and SaPIbov expanded by a unique region at its 3′-end, which contained genes encoding the protein vWbp and the staphylococcal complement inhibitor (Schijffelen et al., 2010). In addition, SaPIs have been detected from isolates of food-poisoning origin (SaPishikawa11, SaPIno10, SaPI68111, and SaPIhirosaki4) (Suzuki et al., 2014; Sato'o et al., 2013; Li et al., 2011). Recently, further SaPIs have been identified in equine (SaPIeq1) and ovine isolates (SaPIov1 and SaPIov2) (Viana et al., 2010).

4. REGULATORY SYSTEMS

4.1 GLOBAL REGULATORS

S. aureus owns very complex regulatory systems to perceive the environmental conditions and respond adequately on their changes. Those global regulatory elements may affect the expression of multiple factors at the same time. In addition, strain-dependent differences in regulation are likely (Valle et al., 2003).

Main components that regulate virulence are the accessory gene regulator (*agr*) system, which functions as bacterial density registering quorum-sensing control, and the staphylococcal accessory regulator SarA protein family comprising several DNA-binding SarA homologues (SarA, SarR, SarV, SarX, SarZ, MgrA, and the repressor of toxins Rot) (Cheung et al., 2004; McNamara et al., 2000). The transcriptional regulatory protein Rot works as general regulation antagonist of the *agr* system. Including *agr*, there are at least 16 *S. aureus* two-component regulatory systems including, e.g., the autolysis-related locus sensor *arlRS*, the staphylococcal respiratory response (*srrAB*) system, the *sae* (*S. aureus* exoproteins) system, and the vancomycin resistance–associated sensor/regulator (*vraSR*) system (Novick et al., 1993; Giraudo et al., 1997; Fournier and Hooper, 2000; Yarwood et al., 2001; Cheung et al., 1992; Walker et al., 2013). The sigma factors direct transcription in response to external stress with σ^A as *S. aureus*' housekeeping sigma factor and alternative factors σ^B and σ^H (Kullik et al., 1998; Morikawa et al., 2003).

For further information, see respective reviews Saïd-Salim et al. (2003), Cheung et al. (2008), Le and Otto (2015), and Pané-Farré et al. (2006).

4.1.1 Regulatory RNAs

In the past decade, the so-called small regulatory RNAs have been identified as relevant in *S. aureus* pathogenicity. These special RNAs represent a class of molecules that are not translated into proteins but exert numerous functions in response to host signals and environmental changes encoded by the RNAs itself or in complexes with proteins (Brosius, 2005). RNAIII, encoded by the *agr* locus, was the first of these small, regulatory and usually nonprotein-coding RNAs (npcRNAs, also referred to as sRNAs and srRNAs), which were reported for staphylococcal species (Novick et al., 1993). This molecule is one of the largest npcRNAs. RNAIII serves as the intracellular effector of the *S. aureus* quorum-sensing system and controls the expression of a number of virulence genes. Of note, in the small colony variant (SCV) phenotype of *S. aureus* (see in the following), RNAIII is absent and, in late growth phases, the expression of several SaPI-encoded sRNAs, the so-called small pathogenicity island rNAs (Spr), is turned off. Both may add to the reduced virulence of *S. aureus* SCVs compared to the normal wild-type phenotype (Chabelskaya et al., 2010; Proctor et al., 2006). By constructing specialized cDNA libraries from different growth phases of an isogenic clinical *S. aureus* strain pair displaying both the normal and the SCV phenotype, phenotype-specifically expressed npcRNAs were detected (Abu-Qatouseh et al., 2010).

S. aureus npcRNAs discovered so far are very heterogeneous in terms of their size (short till >500 nucleotides), structure, and function (Tomasini et al., 2014). They include *cis*-encoded antisense npcRNAs, *trans*-encoded npcRNAs, and *cis*-encoded antisense npcRNAs acting in *trans* (Guillet et al., 2013). Many staphylococcal npcRNAs and their associated proteins are part of complex regulatory pathways. Their—in part strain-dependent—expression seems to contribute to the diversity of phenotypes by influencing the isolate's pathogenic capacity, stress response, and metabolism (Pichon and Felden, 2005; Abu-Qatouseh et al., 2010; Bohn et al., 2007; Chabelskaya et al., 2010).

Regarding virulence gene regulation, the significant involvement of npcRNAs has been recently detected. The cytolysin called PSM-mec is an amphipathic peptide toxin belonging to the PSM family, which is—in contrast to other core genome-encoded PSMs—part of specific types (II including several subtypes, III and VIII) of the SCC*mec* (staphylococcal cassette chromosome *mec*) MGE (Queck et al., 2009). PSM-mec suppresses colony spreading and the expression of PSMα, a cytolytic toxin. Besides PSM-mec, the *psm-mec* locus encodes an npcRNA. This *psm-mec* RNA is bifunctional and suppresses *agrA* translation by binding to the *agrA* mRNA and attenuates MRSA virulence by *agrA* translation inhibition (Qin et al., 2016). The absence of PSM-mec in CA-MRSA strains has been associated with their elevated virulence (Chatterjee et al., 2011).

Sprs are npcRNAs of mostly unknown function, which are expressed from genomic pathogenicity islands. An exception is the recently discovered function of SprD. This npcRNA interacts with the *sbi* mRNA by an antisense mechanism and negatively regulates the expression of an immune evasion molecule, Sbi (Chabelskaya et al., 2010). The *S. aureus* IgG–binding protein (Sbi) acts as an immunoglobulin-binding protein and complement inhibitor (Haupt et al., 2008). It has been demonstrated in a murine intravenous sepsis model that SprD contributes to disease (Chabelskaya et al., 2010).

5. ADHESION

Adhesion on biotic and abiotic surfaces, respectively, is the first essential step in the process of biofilm generation. Thus, virulence factors that allow the survival of a given microorganism on the surfaces of the skin and/or mucous membranes are crucial to establish a commensal state. In the case of infection, adhesion to eukaryotic cells, extracellular matrix (ECM) components, and serum proteins is crucial for the pathogen's interplay with its host. In addition to their role as common target for bacterial adhesins, ECM glycoproteins play an important part not only in the regulation of adhesion but also in growth and survival processes, tissue repair, and cell migration (Hymes and Klaenhammer, 2016; Theocharis et al., 2016).

In current medicine, which is characterized by a tremendous use of temporary or permanently inserted foreign bodies, foreign body–related infections (FBRIs), including catheter-related infections as most frequent FBRIs, lead to enormously increased morbidity, mortality, and economic burden. Biofilms on the surface of colonized foreign bodies are the biological correlative for the clinical picture of FBRIs (von Eiff et al., 2005).

S. aureus is well equipped for all these host- and environment-associated challenges during colonization and infection processes. Strain-dependently, the pathogen expresses a variable, yet always redundant number of different adhesins with varying specificity to host cells and ECM components. The sheer diversity of staphylococcal proteinaceous and nonproteinaceous adhesins warrants classification. With overlaps, they may be classified regarding their function, chemical nature, and association with the bacterial cell wall (Table 2.1) (Heilmann, 2011; Foster et al., 2014). Linkage of *S. aureus* proteinaceous adhesins to the cell wall peptidoglycan can either be of covalent nature or may occur as ionic or hydrophobic interactions.

5.1 CELL WALL–ANCHORED PROTEINS

Surface proteins that are covalently attached to peptidoglycan have been designated as cell wall–anchored (CWA) proteins. They essentially mediate the complex multifactorial adhesion process and comprise members of the microbial surface component recognizing adhesive matrix molecule (MSCRAMM) family, the near iron transporter (NEAT) motif proteins, the tandemly repeated three-helical bundles with protein A as only CWA member, and the G5–E repeat family with *S. aureus* surface protein G (SasG) and its partial MRSA homologue Pls (plasmin-sensitive surface protein) (Foster et al., 2014). NEAT motif proteins include the iron-regulated surface proteins (IsdA, IsdB, and IsdH) (Clarke et al., 2007).

Among the CWA proteins, the MSCRAMMs are the most prevalent factors. Their covalent linkage is catalyzed by species-specific members of the sortase family. The sortases recognize a conserved carboxy-terminal LPXTG anchor motif and link it to the peptide bridge of the cell membrane-attached peptidoglycan (Navarre and Schneewind, 1994; Ruzin et al., 2002).

For clumping of *S. aureus* cells in the presence of plasma, a process that is independent of coagulation (see Section 7.3), a clumping factor is required (Much, 1908; Lee et al., 2002). Besides protein A and the proteins FnBPA and FnBPB, the factors ClfA and ClfB are among the best studied MSCRAMMs (O'Neill et al., 2008; Heilmann et al., 2004; Hartleib et al., 2000; Wertheim et al., 2008). These and other factors (Table 2.1) recognize macromolecular host matrix ligands, such as collagen, elastin, fibronectin, fibrinogen, laminins, proteoglycans/glycosaminoglycans, and vitronectin, with high affinity and

Table 2.1 Classification of Staphylococcal Adhesins Including Examples

Adhesin Groups and Subgroups[a]			Examples

Proteinaceous adhesins	Covalently cell wall–anchored proteins	Microbial surface components recognizing adhesive matrix molecules (MSCRAMM) family	• Serine-aspartate dipeptide (SD) repeat-containing (Sdr) protein subfamily
			  ◦ Clumping factor A (ClfA) and B (ClfB)
			  ◦ Sdrc, SdrD, and SdrE
			  ◦ Bone sialoprotein–binding protein (Bbp; isoform of SdrE)
			  ◦ Pls ("plasmin sensitive"; partial SasG homologue in MRSA)
			• Collagen adhesin (Cna)
			• Fibronectin-binding proteins A (FnBPA) and B (FnBPB)
			• Iron-regulated surface determinants (Isda, Isdb, Isdc, IsdH)
			• Serine-rich adhesin for platelets (SraP, also called SasA)
			• *S. aureus* surface protein (SasG)
			• Biofilm-associated protein (Bap)

	Noncovalently surface-associated proteins	Autolysin/adhesins	• *S. aureus* autolysin (AtlA [*S. epidermidis* AtlE homologue]) • *Staphylococcus aureus* autolysin (Aaa)
		Secretable expanded repertoire adhesive molecules (SERAM) family	• Extracellular adherence protein (Eap; also called MHC class II analogous protein (Map) or P70) • Extracellular matrix and plasma-binding protein (Emp)
		Membrane-spanning proteins	• Extracellular matrix–binding protein homologue (Ebh) • Elastin-binding protein on the surface of *S. aureus* (EbpS)
Nonproteinaceous adhesins			• Polysaccharide intercellular adhesin (PIA) • Wall teichoic (WTA) and lipoteichoic acids (LTA)

[a]*According to Heilmann (2011).*

specificity (Patti et al., 1994). Of note, single MSCRAMMs may bind several ECM ligands, and functional redundancy is typical for this molecule family (Otto, 2010).

Anchorless adhesin proteins, such as the Eap and Emp, are released into the extracellular milieu. Those factors are referred to as secretable expanded repertoire adhesive molecules (SERAMs; see Table 2.1) (Chavakis et al., 2005; Hussain et al., 2001a,b). Furthermore, noncovalently surface-associated proteins comprise *S. aureus* autolysins (AtlA, Aaa) and membrane-spanning proteins such as the extracellular matrix–binding protein homologue (Ebh) and the elastin-binding protein on the surface of *S. aureus* (EbpS) (Table 2.1) (Heilmann, 2011).

5.2 EXTRACELLULAR SUGAR-BASED POLYMERS

Electrostatic interactions are crucially involved in intercellular aggregation processes based on sugar-based polymers such as PIA (polysaccharide intercellular adhesin; also termed PNAG, polymeric *N*-acetyl-glucosamine) and teichoic acids. PIA/PNAG, whose expression is organized in the *icaADBC* operon, has been initially described for *Staphylococcus epidermidis*, linked to the pathogenesis of FBRIs and later on described as being required for immune evasion and virulence (Mack et al., 1996; Heilmann et al., 1996; Vuong et al., 2004). In addition, most clinical *S. aureus* isolates contain the *ica* operon (Cramton et al., 1999). Although in vitro expression of the *S. aureus ica* operon is tightly controlled and more stringently regulated in *S. epidermidis*, there is evidence that it is SarA-regulated and upregulated in vivo (McKenney et al., 1999; Cramton et al., 2001; Beenken et al., 2003).

5.2.1 Teichoic Acids

As most other gram-positive bacteria, *S. aureus* produces highly charged cell wall polymers. These are represented by teichoic acids, which are either covalently connected to peptidoglycan, the so-called wall teichoic acids (WTAs), or anchored via a glycolipid to the cytoplasmic membrane, i.e., lipoteichoic acids (Xia et al., 2010). The cell wall–bound WTA is one of the most important factors contributing to *S. aureus* colonization of abiotic surfaces and nasal colonization (Weidenmaier et al., 2004). In addition to (1) mediating interactions with receptors and biomaterials, further known and proposed functions of *S. aureus* teichoic acids include (2) protection against cell damage by enabling resistance to cationic antimicrobial peptides (CAMPs), cationic antibiotics (glycopeptides), antimicrobial fatty acids, lysozyme, and other factors, (3) controlling protein machineries in the pathogens' cell envelope, and (4) serving as phage receptor (Xia et al., 2010; Peschel, 2002).

6. BACTERIAL INTERFERENCE

Staphylococci secrete a diverse array of bioactive polypeptides involved in both inter- and intraspecies interactions as response to bacterial cell density.

PSMs belong to a family of short amphipathic, alpha-helical peptides (Cheung et al., 2014; Tsompanidou et al., 2013b). Receptor-independently, many of them lead to a pronounced cytolysis of leukocytes, erythrocytes, osteoblasts, and other human cell types because of membrane perturbation (Kretschmer et al., 2010). In particular, *S. aureus* PSMα3 is one of the most potent cytolytic PSMs (Wang et al., 2007). The detergent-like PSM activity is enhanced in the presence of toxins, such as PVL or β-hemolysin. In addition, PSMs contribute to *S. aureus* biofilm stabilization in vitro by forming fibrillar amyloid structures and stimulate inflammatory responses (Schwartz et al., 2012; Kretschmer et al., 2010). Because PSMs exhibit antimicrobial activity against selected other bacterial species and may have a key function in the growing and spreading of staphylococcal cells on epithelial surfaces, PSMs seem to have a primary role in the biotic community by interfering with competing members of the microbiota and switching from sessile to planktonic/motile lifestyle (Tsompanidou et al., 2013a; Joo et al., 2011). However, in highly virulent *S. aureus* strains extremely cytolytic PSMs may contribute to their aggressiveness (Cheung et al., 2014).

Lantibiotics ("lanthionine-containing antibiotic peptides") are accorded a role in bacterial interference on skin and mucous membranes surfaces (Sahl and Bierbaum, 1998; Peschel and Sahl, 2006). They belong to the class of cationic antimicrobial peptides with activity against gram-positive bacterial species. For several staphylococcal species, a strain-dependent production of those bacteriocins, such as epidermin, epidermicin NI01, epicidin 280, epilancin K7, gallidermin, hominicin, nukacin ISK-1, and Pep5, has been described (Götz et al., 2014). For *S. aureus*, the two-peptide lantibiotic staphylococcins BacR1 and C55 as well as the aureocins A53 and A70 are known (Crupper et al., 1997; Navaratna et al., 1998; Netz et al., 2002). Although it became obvious as a result of genome sequencing that *S. aureus* strains may also carry the complete biosynthetic gene cluster required to synthesize an epidermin-like lantibiotic (Diep et al., 2006), it has long been assumed that *S. aureus* is not capable of production of this kind of lantibiotics. However, recently, an AI-type lantibiotic (BacCH91) with an amino acid sequence that is similar to that of epidermin and gallidermin has been detected in a culture from a poultry-associated *S. aureus* strain recovered from a chicken with atopic dermatitis (Wladyka et al., 2013).

7. AGGRESSIVE POTENTIAL: TOXINS

S. aureus produces strain-dependently a large variety of secreted toxins that are able to directly harm the cells of its host. They comprise (1) membrane-damaging toxins, (2) toxins that interfere with receptor function but are not membrane damaging, and (3) secreted enzymes, which are able to degrade host molecules or affect defense mechanisms of the host (Otto, 2014).

7.1 PORE-FORMING PROTEIN TOXINS (MEMBRANE-DAMAGING TOXINS)

Members of this functional group act as membrane-damaging toxins and comprise cytolytic exotoxins acting by pore formation. Respective *S. aureus* cytolytic toxins are able to lyse host cell specifically red (hemolysins) and/or white (leukotoxins or leukocidins) blood cells. They form two groups: (1) the more target-specific toxins needing initial receptor interaction for subsequent cytolytic action and (2) those with receptor interaction.

The prototype for the class of small β-barrel pore-forming cytotoxins, i.e., the 34 kDa *S. aureus* alpha-toxin (α-hemolysin, Hla), is able to intoxicate many different human cell types such as epithelial and endothelial cells including hematopoietic-lineage cells (macrophages, monocytes, neutrophils, and T cells). It is lytic to red blood cells and leukocytes excluding neutrophils. At low doses, it binds specifically to the cell surface, resulting in DNA fragmentation and apoptosis via generation of small heptameric pores, which leads to the release of monovalent ions and caspase activation, whereas higher doses of Hla nonspecifically adsorbs to the lipid bilayer with the result of the formation of larger pores followed by vast necrosis (Nygaard et al., 2012; Bantel et al., 2001). However, even in the presence of active caspases, Hla induces cell death by way of a necrotic pathway (Essmann et al., 2003). Downregulation of Hla production enables intracellular persistence of *S. aureus* (see below). From a clinical perspective, this toxin is a main cause of damage in the context of both necrotizing processes of the skin and lethal infection (Berube and Bubeck Wardenburg, 2013; Bhakdi and Tranum-Jensen, 1991; Otto, 2014).

Bicomponent pore-forming toxins consist of two separate monomeric subunits that bind via specific receptors to host leukocyte membranes followed by forming β-barrel pores spanning the phospholipid bilayer, which causes an efflux of vital intracellular molecules and metabolites (Alonzo and Torres, 2014). For *S. aureus*, a total of seven different bicomponent pore-forming toxins have been described, at least three of them are encoded in the core genome and produced by almost all strains: gamma hemolysin (HlgAB), leucocidin R (HlgCB), and LukAB (also known as LukHG). Other leukocidins are part of phages (LukSF, better known as PVL found on ϕSa1 (see above) and LukMF' found on ϕSa1) or detected on PIs (LukED [vSAβ]).

Of interest, the long-known *S. aureus* delta-hemolysin (Hld), whose gene is embedded within the regulatory RNAIII molecule, has been described as member of the PSMs family; at least some of them are known to exhibit nonspecific cytolytic activities, interfere with cell membranes without receptor interaction, and trigger inflammatory responses (Wang et al., 2007). The expression of PSMα peptides causing neutrophil lysis after phagocytosis may contribute to an explanation of the disease severity due to strongly aggressive, high toxic *S. aureus* strains (Surewaard et al., 2013) (see also Section 6).

Beyond disruption of cellular adherens junctions at epithelial barriers, the leucocidins play several additional roles in pathogenesis. They are involved in the modulation of the immune response of the host, alteration of intracellular signaling events, and killing of eukaryotic immune and nonimmune cells (Yoong and Torres, 2013).

7.2 SUPERANTIGENS AND OTHER NONMEMBRANE-DAMAGING TOXINS THAT INTERFERE WITH RECEPTOR FUNCTION

According to Otto (2014), superantigens (SAgs) may be also classified as toxins that interfere with receptor function but are not membrane damaging. In contrast to "regular" antigens, SAgs bypass the specific antigen processing, which results in an uncontrolled, exceptionally potent stimulation of the immune system by triggering the activation and proliferation of host T cells. SAgs belong to a large protein family [also referred to as pyrogenic toxin superantigen (PTSAg) family], and their possession is part of the "standard equipment" of *S. aureus*. Virtually all *S. aureus* strains own at least one or more of these exotoxins (Becker et al., 2003, 2004). In staphylococci, more than 20 toxins have been described using the letters "A" till "X" comprising, e.g., the "classical" SEs, designated SEA through SEE, and the TSST-1, initially designated as SEF. In the past few years, several toxins and toxin variants that display high sequence similarity to typical SEs, such as SEG, SHE, or SEI, have been described, whereas other SEs lack their emetic properties or were not yet tested and thus were designated as "staphylococcal enterotoxin-like" (SEl) toxins according to the INCSS standard nomenclature (Lina et al., 2004; Jarraud et al., 2001; Blaiotta et al., 2006). For details, please refer to Chapter 4 of this book.

CHIPS (chemotaxis inhibitory protein of *S. aureus*) as well as FLIPr (FPR-like 1 inhibitory protein) and its homologue FLIPr-like are encoded within pathogenicity islands and contribute to immune evasion by interfering with leukocyte receptors. They block the recognition of bacterial-formylated peptides by the FPR receptor, and CHIPS blocks the activation of leukocytes via C5a, which is a terminal effector of the complement system (de Haas et al., 2004; Prat et al., 2006).

7.3 ENZYMATIC ACTING TOXINS

S. aureus owns the capacity to produce a lot of secreted enzymes that are able to degrade host molecules and/or to interfere with them.

The most famous enzyme of *S. aureus* is the staphylocoagulase (Coa) that triggers the conversion of fibrinogen to fibrin with the result of the formation of phagocytosis-impeding fibrin clots on the pathogens' surface (Cheng et al., 2010). The protein vWbp is another coagulase secreted by *S.*

aureus. Both coagulases support the formation of a fibrin shield around the bacteria. Of note, agglutination or clumping, which is mediated by cell-surface proteins that bind to fibrinogen, is different from aggregation, i.e., the formation of multicellular clusters in the absence of host factors (e.g., due to PIA) (Crosby et al., 2016). In situations, where it is advantageous for the pathogen to escape clumps, *S. aureus* has an arsenal of mechanisms for prevention of clumping processes and for leaving clumps and fibrin aggregates. With the expression, e.g., of the staphylokinase (Sak), *S. aureus* is able to activate human plasminogen into plasmin, a fibrinolytic protease, and, thus, to degrade present clumps that results into dissemination of separated cocci (Bokarewa et al., 2006). Sak also contributes to overcoming the host's skin barrier (Kwiecinski et al., 2013). The "toolbox" of *S. aureus* includes further several more or less specific proteases such as *aur*, glutamyl endopeptidase, and staphopain A and B, which interfere with complement factors to avoid complement-mediated bacterial killing (Jusko et al., 2014).

The enzymatic arsenal of *S. aureus* with impact on the pathogen's virulence is complemented by sphingomyelinases (e.g., *S. aureus* beta-toxin), lipases, and nucleases. As an example, the nucleases may facilitate escape from neutrophil extracellular traps made by stimulated neutrophils to capture and kill microorganisms (Berends et al., 2010).

8. EVASION: BLOCKING AND ESCAPING THE HOST'S IMMUNE RESPONSE

Evasion of various components of host innate immunity is one key to *S. aureus*' success as intruding pathogen. This capacity comprises direct resisting to phagocytosis and the capacity to escape the immune response by internalization and invasion, respectively, of endothelial and epithelial host cells.

8.1 RESISTING PHAGOCYTIC ACTIVITY

Activated phagocytes produce radical nitric oxide as broad-spectrum answer to pathogenic microorganisms (Mogensen, 2009). Based on a coordinated nitrosative stress response, *S. aureus* owns the capacity to resist the antimicrobial effects of radical nitric oxide (Richardson et al., 2006, 2008).

A majority of *S. aureus* strains is surrounded by an antiphagocytic polysaccharide capsule for which initially 13 serotypes have been described, but only four types have been chemically characterized (Karakawa et al., 1985; Gilbert, 1931; Karakawa and Vann, 1982). In addition to the very rare, heavy encapsulated capsule types 1 and 2 appearing as mucoid colonies on solid media, *S. aureus* strains exhibiting capsule types 5 and 8 are most prevalent, accounting for 70%–80% of strains recovered from human specimens (Arbeit et al., 1984). Nontypeable strains, i.e., strains that are not recognized with antibodies to capsule types 1, 2, 5, and 8, lost their capability to yield a capsule although still possessing the encoding genes for type 5 or 8 capsule production (Cocchiaro et al., 2006). These strains of the so-called 336 serotype are characterized by variation of a polyribitol phosphate *N*-acetylglucosamine (O'Brien et al., 2000). They account for about 10% of all isolates; however, strain-dependently, they do also occur with increased percentage (von Eiff et al., 2007). The capsules act as protective antigens. They block complement-mediated killing by the infected host and contribute to the pathogen's virulence (Thakker et al., 1998; Robbins et al., 2004).

8.2 INTERNALIZATION/INVASION AND INTRACELLULAR PERSISTENCE

Significantly contributing to infection process, it is a major property of *S. aureus* to invade professional and nonprofessional phagocytes, including endothelial and epithelial cells (Sinha and Fraunholz, 2010; Werbick et al., 2007). Internalization and invasion, respectively, are determined by both the respective pathogenicity arsenal of the *S. aureus* cells and the host defense mechanisms. Internalization is initiated by adhering to host cell surfaces via expression of various adhesins (see above). Subsequently, postinvasion processes that depend on strain-specific possession and expression of virulence factors take place and the host cell type entered. In particular, because of α-toxin activity and factors regulated by *agr* and *sigB*, *S. aureus* is able to induce host cell activation and death, which has been also linked to the escape of the pathogen from the phagolysosomes (Löffler et al., 2014). However, by downregulation of those factors responsible for inflammatory and cytotoxic effects, *S. aureus* is able to survive and dwell intracellularly for longer periods of time. Thus, *S. aureus* and also other staphylococcal species own the ability of adaptation to an intracellular lifestyle.

Switching into the SCV phenotype is the best studied strategy of *S. aureus* to escape host immune response and/or antimicrobial treatments without activity on intracellular pathogens (von Eiff et al., 2001b; Tuchscherr et al., 2011; Vesga et al., 1996). This phenotype exhibits reduced growth rates as the key feature, leading to microcolonies on solid media. Applying usual incubation times, these colonies are about 10-fold smaller in colony diameter than those of the normal phenotype (Proctor et al., 2006). The unusual colonial appearance of SCVs comprising pinpoint colonies with decreased pigment formation and reduced hemolytic activity is complemented by atypical pattern of enzymatic and other diagnostically relevant biochemical reactions, reduced or missing production of toxins, and decreased susceptibility toward selected antibiotics (aminoglycosides, trimethoprim–sulfamethoxazole, and fluoroquinolones) and antimicrobial peptides (von Eiff et al., 2006a; Garcia et al., 2013; Gläser et al., 2014).

These radical phenotypic changes are based on extensive alterations in the pathogen's chemical composition and metabolism, most frequently in pathways associated with the electron transport (hemin and menadione auxotrophic SCVs) and, in the case of thymidine-auxotrophic SCVs, with the thymidylate biosynthesis (Proctor et al., 2014; Kriegeskorte et al., 2011, 2014b,a; Seggewiß et al., 2006; Kohler et al., 2003; Chatterjee et al., 2008; Becker et al., 2006).

Especially, their ability to persist intracellularly explains the association of staphylococcal SCVs with persistent and relapsing infections (Proctor et al., 2006; Kahl et al., 2016; von Eiff et al., 2006b). Studies and case reports have noted some infections, e.g., chronic osteomyelitis, chronic arthritis, deep abscesses, and chronic ulcers (Proctor et al., 1995; von Eiff et al., 1997; Rolauffs et al., 2002; Cervantes-García et al., 2015; Borderon and Horodniceanu, 1976), to be associated with SCVs. In addition, associations with FBRIs, such as recurrent pacemaker-related bloodstream infection, infection of cardiac prosthesis, and hip prosthetic joint infection, have been described (von Eiff and Becker, 2007; Seifert et al., 2003; Quie, 1969; Spanu et al., 2005; Sendi et al., 2006). Methicillin-resistant SCVs of *S. aureus* were reported within the course of an outbreak in an intensive care unit and in patients with infected diabetic foot ulcers (Cervantes-García et al., 2015; Salgado et al., 2001). A vancomycin-intermediate *S. aureus* SCV emerged in a patient with septic arthritis during long-term treatment with daptomycin (Lin et al., 2016).

Besides detection in specimens from chronic infections, thymidine-dependent SCVs have been particularly recovered from patients suffering from cystic fibrosis (CF). Here, SCVs are highly prevalent in the airways, and the formation of the *S. aureus* SCV phenotype was found to be associated with trimethoprim–sulfamethoxazole therapy and an advanced stage of pulmonary disease (Kahl et al., 1998; Besier et al., 2007; Kriegeskorte et al., 2015). In a pediatric CF cohort, the occurrence of thymidine-auxotrophic SCVs was significantly and independently associated with a worse respiratory outcome (Wolter et al., 2013).

Other mechanisms, in part also described for other persistent organisms such as persister or viable but nonculturable cells, may also be involved in SCV formation (Proctor et al., 2014). Altered RNA processing as well as changes in the levels of specific regulatory small nonprotein–coding RNAs (npcRNAs, also known as sRNAs) may be also involved in SCV formation because some of npcRNAs are targets of the RNA degrasome (Abu-Qatouseh et al., 2010; Geissmann et al., 2009; Felden et al., 2011). Applying whole genome sequencing for an induced stable *S. aureus* SCV, the most conspicuous genetic changes pertained to the global regulator MgrA and phosphoserine phosphatase RsbU, which is part of the regulatory pathway of the sigma factor SigB (Bui and Kidd, 2015). Furthermore, defects in the stringent stress response due to mutations in *relA* encoding a synthetase (RelA) (Gao et al., 2010) may result in phenotype switch generating SCVs. RelA and SpoT, a hydrolase, control the alarmones ppGpp and pppGpp [(p)ppGpp], the intracellular accumulation of which mediates the stringent response as response to stress and nutritional starvation (Godfrey et al., 2002).

9. CONCLUSION AND PERSPECTIVE

S. aureus is richly equipped with virulence factors enabling this versatile pathogen to colonize the skin and mucous membrane surfaces of humans and animals and, subsequently breaking this barrier, to cause mild to life-threatening infections and toxin-mediated syndromes, affecting a wide range of tissues and organs. Moreover, phenotypic switch extends its capabilities to act not only as extracellular but also as intracellular pathogen. In the light of continued or even growing public health problem due to antibiotic-resistant strains such as MRSA, research on pathogenesis may lead to the development of alternative "antipathogenic" drugs to break the vicious circle of selection pressure and resistance.

REFERENCES

Abu-Qatouseh, L.F., Chinni, S.V., Seggewiß, J., Proctor, R.A., Brosius, J., Rozhdestvensky, T.S., Peters, G., von Eiff, C., Becker, K., 2010. Identification of differentially expressed small non-protein-coding RNAs in *Staphylococcus aureus* displaying both the normal and the small-colony variant phenotype. J. Mol. Med. (Berl.) 88, 565–575.

Ackermann, H.W., Dubow, M.S., 1987. Viruses of procaryotes. Natural Groups of Bacteriophages, vol. 2. CRC Press, Boca Raton.

Alonzo 3RD, F., Torres, V.J., 2014. The bicomponent pore-forming leucocidins of *Staphylococcus aureus*. Microbiol. Mol. Biol. Rev. 78, 199–230.

Antonanzas, F., Lozano, C., Torres, C., 2015. Economic features of antibiotic resistance: the case of methicillin-resistant *Staphylococcus aureus*. Pharmacoeconomics 33, 285–325.

Arbeit, R.D., Karakawa, W.W., Vann, W.F., Robbins, J.B., 1984. Predominance of two newly described capsular polysaccharide types among clinical isolates of *Staphylococcus aureus*. Diagn. Microbiol. Infect. Dis. 2, 85–91.

Bantel, H., Sinha, B., Domschke, W., Peters, G., Schulze-Osthoff, K., Jänicke, R.U., 2001. α-toxin is a mediator of *Staphylococcus aureus*-induced cell death and activates caspases via the intrinsic death pathway independently of death receptor signaling. J. Cell Biol. 155, 637–648.

Becker, K., Ballhausen, B., Kahl, B.C., Köck, R., 2015. The clinical impact of livestock-associated methicillin-resistant *Staphylococcus aureus* of the clonal complex 398 for humans. Vet. Microbiol. 33–38.

Becker, K., Friedrich, A.W., Lubritz, G., Weilert, M., Peters, G., von Eiff, C., 2003. Prevalence of genes encoding pyrogenic toxin superantigens and exfoliative toxins among strains of *Staphylococcus aureus* isolated from blood and nasal specimens. J. Clin. Microbiol. 41, 1434–1439.

Becker, K., Friedrich, A.W., Peters, G., von Eiff, C., 2004. Systematic survey on the prevalence of genes coding for staphylococcal enterotoxins SElm, SelO, and SElN. Mol. Nutr. Food Res. 48, 488–495.

Becker, K., Heilmann, C., Peters, G., 2014. Coagulase-negative staphylococci. Clin. Microbiol. Rev. 27, 870–926.

Becker, K., Laham, N.A., Fegeler, W., Proctor, R.A., Peters, G., von Eiff, C., 2006. Fourier-transform infrared spectroscopic analysis is a powerful tool for studying the dynamic changes in *Staphylococcus aureus* small-colony variants. J. Clin. Microbiol. 44, 3274–3278.

Becker, K., Schaumburg, F., Fegeler, C., Friedrich, A.W., Köck, R., 2017. Prevalence of Multiresistant Microorganisms (PMM) Study group, 2017. *Staphylococcus aureus* from the German general population is highly diverse. Int. J. Med. Microbiol. 307, 21–27.

Beenken, K.E., Blevins, J.S., Smeltzer, M.S., 2003. Mutation of *sarA* in *Staphylococcus aureus* limits biofilm formation. Infect. Immun. 71, 4206–4211.

Berends, E.T., Horswill, A.R., Haste, N.M., Monestier, M., Nizet, V., von köckritz-Blickwede, M., 2010. Nuclease expression by *Staphylococcus aureus* facilitates escape from neutrophil extracellular traps. J. Innate Immun. 2, 576–586.

Berube, B.J., Bubeck Wardenburg, J., 2013. *Staphylococcus aureus* α-toxin: nearly a century of intrigue. Toxins (Basel) 5, 1140–1166.

Besier, S., Smaczny, C., von Mallinckrodt, C., Krahl, A., Ackermann, H., Brade, V., Wichelhaus, T.A., 2007. Prevalence and clinical significance of *Staphylococcus aureus* small-colony variants in cystic fibrosis lung disease. J. Clin. Microbiol. 45, 168–172.

Bhakdi, S., Tranum-Jensen, J., 1991. Alpha-toxin of *Staphylococcus aureus*. Microbiol. Rev. 55, 733–751.

Blaiotta, G., Fusco, V., von Eiff, C., Villani, F., Becker, K., 2006. Biotyping of enterotoxigenic *Staphylococcus aureus* by enterotoxin gene cluster (*egc*) polymorphism and *spa* typing analyses. Appl. Environ. Microbiol. 72, 6117–6123.

Bode, L.G., Kluytmans, J.A., Wertheim, H.F., Bogaers, D., Vandenbroucke-Grauls, C.M., Roosendaal, R., Troelstra, A., Box, A.T., Voss, A., van der Tweel, I., van Belkum, A., Verbrugh, H.A., Vos, M.C., 2010. Preventing surgical-site infections in nasal carriers of *Staphylococcus aureus*. N. Engl. J. Med. 362, 9–17.

Bohn, C., Rigoulay, C., Bouloc, P., 2007. No detectable effect of RNA-binding protein Hfq absence in *Staphylococcus aureus*. BMC Microbiol. 7, 10.

Bokarewa, M.I., Jin, T., Tarkowski, A., 2006. *Staphylococcus aureus*: staphylokinase. Int. J. Biochem. Cell Biol. 38, 504–509.

Borderon, E., Horodniceanu, T., 1976. Mutants déficients a colonies naines de *Staphylococcus*: étude de trois souches isolées chez des malades porteurs d'ostéosynthèses. Ann. Microbiol. (Paris) 127, 503–514.

Brosius, J., 2005. Waste not, want not–transcript excess in multicellular eukaryotes. Trends Genet. 21, 287–288.

Bui, L.M., Kidd, S.P., 2015. A full genomic characterization of the development of a stable small colony variant cell-type by a clinical *Staphylococcus aureus* strain. Infect. Genet. Evol. 36, 345–355.

Bukowski, M., Wladyka, B., Dubin, G., 2010. Exfoliative toxins of *Staphylococcus aureus*. Toxins (Basel) 2, 1148–1165.

Cervantes-García, E., García-Gonzalez, R., Reyes-Torres, A., Resendiz-Albor, A.A., Salazar-Schettino, P.M., 2015. *Staphylococcus aureus* small colony variants in diabetic foot infections. Diabet. Foot Ankle 6, 26431.

Chabelskaya, S., Gaillot, O., Felden, B., 2010. A *Staphylococcus aureus* small RNA is required for bacterial virulence and regulates the expression of an immune-evasion molecule. PLoS Pathog. 6, e1000927.

Chambers, H.F., Deleo, F.R., 2009. Waves of resistance: *Staphylococcus aureus* in the antibiotic era. Nat. Rev. Microbiol. 7, 629–641.

Chatterjee, I., Kriegeskorte, A., Fischer, A., Deiwick, S., Theimann, N., Proctor, R.A., Peters, G., Herrmann, M., Kahl, B.C., 2008. In vivo mutations of thymidylate synthase (encoded by *thyA*) are responsible for thymidine dependency in clinical small-colony variants of *Staphylococcus aureus*. J. Bacteriol. 190, 834–842.

Chatterjee, S.S., Chen, L., Joo, H.S., Cheung, G.Y., Kreiswirth, B.N., Otto, M., 2011. Distribution and regulation of the mobile genetic element-encoded phenol-soluble modulin PSM-mec in methicillin-resistant *Staphylococcus aureus*. PLoS One 6, e28781.

Chavakis, T., Wiechmann, K., Preissner, K.T., Herrmann, M., 2005. *Staphylococcus aureus* interactions with the endothelium: the role of bacterial "secretable expanded repertoire adhesive molecules" (SERAM) in disturbing host defense systems. Thromb. Haemost. 94, 278–285.

Cheng, A.G., Mcadow, M., Kim, H.K., Bae, T., Missiakas, D.M., Schneewind, O., 2010. Contribution of coagulases towards *Staphylococcus aureus* disease and protective immunity. PLoS Pathog. 6, e1001036.

Cheung, A.L., Bayer, A.S., Zhang, G., Gresham, H., Xiong, Y.Q., 2004. Regulation of virulence determinants in vitro and in vivo in *Staphylococcus aureus*. FEMS Immunol. Med. Microbiol. 40, 1–9.

Cheung, A.L., Koomey, J.M., Butler, C.A., Projan, S.J., Fischetti, V.A., 1992. Regulation of exoprotein expression in *Staphylococcus aureus* by a locus (*sar*) distinct from agr. Proc. Natl. Acad. Sci. U.S.A. 89, 6462–6466.

Cheung, A.L., Nishina, K.A., Trotonda, M.P., Tamber, S., 2008. The SarA protein family of *Staphylococcus aureus*. Int. J. Biochem. Cell Biol. 40, 355–361.

Cheung, G.Y., Joo, H.S., Chatterjee, S.S., Otto, M., 2014. Phenol-soluble modulins–critical determinants of staphylococcal virulence. FEMS Microbiol. Rev. 38, 698–719.

Clarke, S.R., Mohamed, R., Bian, L., Routh, A.F., Kokai-Kun, J.F., Mond, J.J., Tarkowski, A., Foster, S.J., 2007. The *Staphylococcus aureus* surface protein IsdA mediates resistance to innate defenses of human skin. Cell Host Microbe 1, 199–212.

Cocchiaro, J.L., Gomez, M.I., Risley, A., Solinga, R., Sordelli, D.O., Lee, J.C., 2006. Molecular characterization of the capsule locus from non-typeable *Staphylococcus aureus*. Mol. Microbiol. 59, 948–960.

Cramton, S.E., Gerke, C., Schnell, N.F., Nichols, W.W., Götz, F., 1999. The intercellular adhesion (*ica*) locus is present in *Staphylococcus aureus* and is required for biofilm formation. Infect. Immun. 67, 5427–5433.

Cramton, S.E., Ulrich, M., Götz, F., Döring, G., 2001. Anaerobic conditions induce expression of polysaccharide intercellular adhesin in *Staphylococcus aureus* and Staphylococcus epidermidis. Infect. Immun. 69, 4079–4085.

Crosby, H.A., Kwiecinski, J., Horswill, A.R., 2016. *Staphylococcus aureus* aggregation and coagulation mechanisms, and their function in host-pathogen interactions. Adv. Appl. Microbiol. 96, 1–41.

Crupper, S.S., Gies, A.J., Iandolo, J.J., 1997. Purification and characterization of staphylococcin BacR1, a broad-spectrum bacteriocin. Appl. Environ. Microbiol. 63, 4185–4190.

Cuny, C., Friedrich, A., Kozytska, S., Layer, F., Nübel, U., Ohlsen, K., Strommenger, B., Walther, B., Wieler, L., Witte, W., 2010. Emergence of methicillin-resistant *Staphylococcus aureus* (MRSA) in different animal species. Int. J. Med. Microbiol. 300, 109–117.

Cuny, C., Köck, R., Witte, W., 2013. Livestock associated MRSA (LA-MRSA) and its relevance for humans in Germany. Int. J. Med. Microbiol. 303, 331–337.

de Haas, C.J., Veldkamp, K.E., Peschel, A., Weerkamp, F., van Wamel, W.J., Heezius, E.C., Poppelier, M.J., van Kessel, K.P., van Strijp, J.A., 2004. Chemotaxis inhibitory protein of *Staphylococcus aureus*, a bacterial anti-inflammatory agent. J. Exp. Med. 199, 687–695.

de la Fuente, R., Ballesteros, C., Bautista, V., Medina, A., Orden, J.A., Dominguez-Bernal, G., Vindel, A., 2011. *Staphylococcus aureus* subsp. anaerobius isolates from different countries are clonal in nature. Vet. Microbiol. 150, 198–202.

Deghorain, M., van Melderen, L., 2012. The Staphylococci phages family: an overview. Viruses 4, 3316–3335.

Diederen, B.M., Kluytmans, J.A., 2006. The emergence of infections with community-associated methicillin resistant *Staphylococcus aureus*. J. Infect 52, 157–168.

Diep, B.A., Gill, S.R., Chang, R.F., Phan, T.H., Chen, J.H., Davidson, M.G., Lin, F., Lin, J., Carleton, H.A., Mongodin, E.F., Sensabaugh, G.F., Perdreau-Remington, F., 2006. Complete genome sequence of USA300, an epidemic clone of community-acquired meticillin-resistant *Staphylococcus aureus*. Lancet 367, 731–739.

Dufour, P., Gillet, Y., Bes, M., Lina, G., Vandenesch, F., Floret, D., Etienne, J., Richet, H., 2002. Community-acquired methicillin-resistant *Staphylococcus aureus* infections in France: emergence of a single clone that produces Panton-Valentine leukocidin. Clin. Infect. Dis. 35, 819–824.

Essmann, F., Bantel, H., Totzke, G., Engels, I.H., Sinha, B., Schulze-Osthoff, K., Jänicke, R.U., 2003. *Staphylococcus aureus* α-toxin-induced cell death: predominant necrosis despite apoptotic caspase activation. Cell Death Differ. 10, 1260–1272.

Felden, B., Vandenesch, F., Bouloc, P., Romby, P., 2011. The *Staphylococcus aureus* RNome and its commitment to virulence. PLoS Pathog. 7, e1002006.

Fitzgerald, J.R., Monday, S.R., Foster, T.J., Bohach, G.A., Hartigan, P.J., Meaney, W.J., Smyth, C.J., 2001. Characterization of a putative pathogenicity island from bovine *Staphylococcus aureus* encoding multiple superantigens. J. Bacteriol. 183, 63–70.

Foster, T.J., 2004. The *Staphylococcus aureus* “superbug”. J. Clin. Invest. 114, 1693–1696.

Foster, T.J., Geoghegan, J.A., Ganesh, V.K., Höök, M., 2014. Adhesion, invasion and evasion: the many functions of the surface proteins of *Staphylococcus aureus*. Nat. Rev. Microbiol. 12, 49–62.

Fournier, B., Hooper, D.C., 2000. A new two-component regulatory system involved in adhesion, autolysis, and extracellular proteolytic activity of *Staphylococcus aureus*. J. Bacteriol. 182, 3955–3964.

Gao, W., Chua, K., Davies, J.K., Newton, H.J., Seemann, T., Harrison, P.F., Holmes, N.E., Rhee, H.W., Hong, J.I., Hartland, E.L., Stinear, T.P., Howden, B.P., 2010. Two novel point mutations in clinical *Staphylococcus aureus* reduce linezolid susceptibility and switch on the stringent response to promote persistent infection. PLoS Pathog. 6, e1000944.

Garcia, L.G., Lemaire, S., Kahl, B.C., Becker, K., Proctor, R.A., Denis, O., Tulkens, P.M., van Bambeke, F., 2013. Antibiotic activity against small-colony variants of *Staphylococcus aureus*: review of in vitro, animal and clinical data. J. Antimicrob. Chemother. 68, 1455–1464.

Geissmann, T., Chevalier, C., Cros, M.J., Boisset, S., Fechter, P., Noirot, C., Schrenzel, J., Francois, P., Vandenesch, F., Gaspin, C., Romby, P., 2009. A search for small noncoding RNAs in *Staphylococcus aureus* reveals a conserved sequence motif for regulation. Nucleic Acids Res. 37, 7239–7257.

Gilbert, I., 1931. Dissociation in an encapsulated Staphylococcus. J. Bacteriol. 21, 157–160.

Giraudo, A.T., Cheung, A.L., Nagel, R., 1997. The *sae* locus of *Staphylococcus aureus* controls exoprotein synthesis at the transcriptional level. Arch. Microbiol. 168, 53–58.

Gläser, R., Becker, K., von Eiff, C., Meyer-Hoffert, U., Harder, J., 2014. Decreased susceptibility of *Staphylococcus aureus* small-colony variants toward human antimicrobial peptides. J. Invest. Dermatol. 134, 2347–2350.

Godfrey, H.P., Bugrysheva, J.V., Cabello, F.C., 2002. The role of the stringent response in the pathogenesis of bacterial infections. Trends Microbiol. 10, 349–351.

Götz, F., Perconti, S., Popella, P., Werner, R., Schlag, M., 2014. Epidermin and gallidermin: staphylococcal lantibiotics. Int. J. Med. Microbiol. 304, 63–71.

Gould, I.M., Reilly, J., Bunyan, D., Walker, A., 2010. Costs of healthcare-associated methicillin-resistant *Staphylococcus aureus* and its control. Clin. Microbiol. Infect. 16, 1721–1728.

Groom, A.V., Wolsey, D.H., Naimi, T.S., Smith, K., Johnson, S., Boxrud, D., Moore, K.A., Cheek, J.E., 2001. Community-acquired methicillin-resistant *Staphylococcus aureus* in a rural American Indian community. JAMA 286, 1201–1205.

Guillet, J., Hallier, M., Felden, B., 2013. Emerging functions for the *Staphylococcus aureus* RNome. PLoS Pathog. 9, e1003767.

Hacker, J., Kaper, J.B., 2000. Pathogenicity islands and the evolution of microbes. Annu. Rev. Microbiol. 54, 641–679.

Hartleib, J., Köhler, N., Dickinson, R.B., Chhatwal, G.S., Sixma, J.J., Hartford, O.M., Foster, T.J., Peters, G., Kehrel, B.E., Herrmann, M., 2000. Protein A is the von Willebrand factor binding protein on *Staphylococcus aureus*. Blood 96, 2149–2156.

Haupt, K., Reuter, M., van den Elsen, J., Burman, J., Halbich, S., Richter, J., Skerka, C., Zipfel, P.F., 2008. The *Staphylococcus aureus* protein Sbi acts as a complement inhibitor and forms a tripartite complex with host complement Factor H and C3b. PLoS Pathog. 4, e1000250.

Heilmann, C., 2011. Adhesion mechanisms of staphylococci. Adv. Exp. Med. Biol. 715, 105–123.

Heilmann, C., Niemann, S., Sinha, B., Herrmann, M., Kehrel, B.E., Peters, G., 2004. *Staphylococcus aureus* fibronectin-binding protein (FnBP)-mediated adherence to platelets, and aggregation of platelets induced by FnBPA but not by FnBPB. J. Infect. Dis. 190, 321–329.

Heilmann, C., Schweitzer, O., Gerke, C., Vanittanakom, N., Mack, D., Götz, F., 1996. Molecular basis of intercellular adhesion in the biofilm-forming *Staphylococcus epidermidis*. Mol. Microbiol. 20, 1083–1091.

Herold, B.C., Immergluck, L.C., Maranan, M.C., Lauderdale, D.S., Gaskin, R.E., Boyle-Vavra, S., Leitch, C.D., Daum, R.S., 1998. Community-acquired methicillin-resistant *Staphylococcus aureus* in children with no identified predisposing risk. JAMA 279, 593–598.

Holland, T.L., Arnold, C., Fowler Jr., V.G., 2014. Clinical management of *Staphylococcus aureus* bacteremia: a review. JAMA 312, 1330–1341.

Humphreys, H., Becker, K., Dohmen, P.M., Petrosillo, N., Spencer, M., van Rijen, M., Wechsler-Fordos, A., Pujol, M., Dubouix, A., Garau, J., 2016. *Staphylococcus aureus* and surgical site infections: benefits of screening and decolonization before surgery. J. Hosp. Infect. 94, 295–304.

Hussain, M., Becker, K., von Eiff, C., Peters, G., Herrmann, M., 2001a. Analogs of Eap protein are conserved and prevalent in clinical *Staphylococcus aureus* isolates. Clin. Diagn. Lab. Immunol. 8, 1271–1276.

Hussain, M., Becker, K., von Eiff, C., Schrenzel, J., Peters, G., Herrmann, M., 2001b. Identification and characterization of a novel 38.5-kilodalton cell surface protein of *Staphylococcus aureus* with extended-spectrum binding activity for extracellular matrix and plasma proteins. J. Bacteriol. 183, 6778–6786.

Hymes, J.P., Klaenhammer, T.R., 2016. Stuck in the middle: fibronectin-binding proteins in gram-positive bacteria. Front. Microbiol. 7, 1504.

Jarraud, S., Peyrat, M.A., Lim, A., Tristan, A., Bes, M., Mougel, C., Etienne, J., Vandenesch, F., Bonneville, M., Lina, G., 2001. *egc*, a highly prevalent operon of enterotoxin gene, forms a putative nursery of superantigens in *Staphylococcus aureus*. J. Immunol. 166, 669–677.

Joo, H.S., Cheung, G.Y., Otto, M., 2011. Antimicrobial activity of community-associated methicillin-resistant *Staphylococcus aureus* is caused by phenol-soluble modulin derivatives. J. Biol. Chem. 286, 8933–8940.

Jusko, M., Potempa, J., Kantyka, T., Bielecka, E., Miller, H.K., Kalinska, M., Dubin, G., Garred, P., Shaw, L.N., Blom, A.M., 2014. Staphylococcal proteases aid in evasion of the human complement system. J. Innate Immun. 6, 31–46.

Kahl, B., Herrmann, M., Everding, A.S., Koch, H.G., Becker, K., Harms, E., Proctor, R.A., Peters, G., 1998. Persistent infection with small colony variant strains of *Staphylococcus aureus* in patients with cystic fibrosis. J. Infect. Dis. 177, 1023–1029.

Kahl, B.C., Becker, K., Löffler, B., 2016. Clinical significance and pathogenesis of staphylococcal small colony variants in persistent infections. Clin. Microbiol. Rev. 29, 401–427.

Kalmeijer, M.D., Coertjens, H., van Nieuwland-Bollen, P.M., Bogaers-Hofman, D., de Baere, G.A., Stuurman, A., van Belkum, A., Kluytmans, J.A., 2002. Surgical site infections in orthopedic surgery: the effect of mupirocin nasal ointment in a double-blind, randomized, placebo-controlled study. Clin. Infect. Dis. 35, 353–358.

Karakawa, W.W., Fournier, J.M., Vann, W.F., Arbeit, R., Schneerson, R.S., Robbins, J.B., 1985. Method for the serological typing of the capsular polysaccharides of *Staphylococcus aureus*. J. Clin. Microbiol. 22, 445–447.

Karakawa, W.W., Vann, W.F., 1982. Capsular polysaccharides of *Staphylococcus aureus*. Semin. Pediatr. Infect. Dis. 4, 285–293.

Kaspar, U., Kriegeskorte, A., Schubert, T., Peters, G., Rudack, C., Pieper, D.H., Wos-Oxley, M., Becker, K., 2016. The culturome of the human nose habitats reveals individual bacterial fingerprint patterns. Environ. Microbiol. 18, 2130–2142.

Kloos, W.E., Musselwhite, M.S., 1975. Distribution and persistence of *Staphylococcus* and *Micrococcus* species and other aerobic bacteria on human skin. Appl. Microbiol. 30, 381–385.

Kluytmans, J., van Belkum, A., Verbrugh, H., 1997. Nasal carriage of *Staphylococcus aureus*: epidemiology, underlying mechanisms, and associated risks. Clin. Microbiol. Rev. 10, 505–520.

Kluytmans, J.A., Wertheim, H.F., 2005. Nasal carriage of *Staphylococcus aureus* and prevention of nosocomial infections. Infection 33, 3–8.

Köck, R., Becker, K., Cookson, B., van Gemert-Pijnen, J., Harbarth, S., Kluytmans, J., Mielke, M., Peters, G., Skov, R., Struelens, M., Tacconelli, E., Witte, W., Friedrich, A., 2014. Systematic literature analysis and review of targeted preventive measures to limit healthcare-associated infections by meticillin-resistant *Staphylococcus aureus*. Euro Surveill. 19 pii: 20860.

Köck, R., Becker, K., Cookson, B., van Gemert-Pijnen, J.E., Harbarth, S., Kluytmans, J., Mielke, M., Peters, G., Skov, R.L., Struelens, M.J., Tacconelli, E., Navarro Torné, A., Witte, W., Friedrich, A.W., 2010. Methicillin-resistant *Staphylococcus aureus* (MRSA): burden of disease and control challenges in Europe. Euro Surveill. 15, 19688.

Köck, R., Werner, P., Friedrich, A.W., Fegeler, C., Becker, K., Prevalence of Multiresistant Microorganisms (PMM) Study Group, 2016. Persistence of nasal colonization with human pathogenic bacteria and associated antimicrobial resistance in the German general population. New Microbes New Infect. 9, 24–34.

Kohler, C., von Eiff, C., Peters, G., Proctor, R.A., Hecker, M., Engelmann, S., 2003. Physiological characterization of a heme-deficient mutant of *Staphylococcus aureus* by a proteomic approach. J. Bacteriol. 185, 6928–6937.

Kretschmer, D., Gleske, A.K., Rautenberg, M., Wang, R., Koberle, M., Bohn, E., Schoneberg, T., Rabiet, M.J., Boulay, F., Klebanoff, S.J., van Kessel, K.A., van Strijp, J.A., Otto, M., Peschel, A., 2010. Human formyl peptide receptor 2 senses highly pathogenic *Staphylococcus aureus*. Cell Host Microbe 7, 463–473.

Kriegeskorte, A., Block, D., Drescher, M., Windmüller, N., Mellmann, A., Baum, C., Neumann, C., Lorè, N.I., Bragonzi, A., Liebau, E., Hertel, P., Seggewiss, J., Becker, K., Proctor, R.A., Peters, G., Kahl, B.C., 2014a. Inactivation of *thyA* in *Staphylococcus aureus* attenuates virulence and has a strong impact on metabolism and virulence gene expression. mBio 5 e01447–14.

Kriegeskorte, A., Grubmuller, S., Huber, C., Kahl, B.C., von Eiff, C., Proctor, R.A., Peters, G., Eisenreich, W., Becker, K., 2014b. *Staphylococcus aureus* small colony variants show common metabolic features in central metabolism irrespective of the underlying auxotrophism. Front. Cell Infect. Microbiol. 4, 141.

Kriegeskorte, A., König, S., Sander, G., Pirkl, A., Mahabir, E., Proctor, R.A., von Eiff, C., Peters, G., Becker, K., 2011. Small colony variants of *Staphylococcus aureus* reveal distinct protein profiles. Proteomics 11, 2476–2490.

Kriegeskorte, A., Lore, N.I., Bragonzi, A., Riva, C., Kelkenberg, M., Becker, K., Proctor, R.A., Peters, G., Kahl, B.C., 2015. Thymidine-dependent *Staphylococcus aureus* small-colony variants are induced by trimethoprim-sulfamethoxazole (SXT) and have increased fitness during SXT challenge. Antimicrob. Agents Chemother. 59, 7265–7272.

Kullik, I., Giachino, P., Fuchs, T., 1998. Deletion of the alternative sigma factor σ^B in *Staphylococcus aureus* reveals its function as a global regulator of virulence genes. J. Bacteriol. 180, 4814–4820.

Kwan, T., Liu, J., Dubow, M., Gros, P., Pelletier, J., 2005. The complete genomes and proteomes of 27 *Staphylococcus aureus* bacteriophages. Proc. Natl. Acad. Sci. U.S.A. 102, 5174–5179.

Kwiecinski, J., Jacobsson, G., Karlsson, M., Zhu, X., Wang, W., Bremell, T., Josefsson, E., Jin, T., 2013. Staphylokinase promotes the establishment of *Staphylococcus aureus* skin infections while decreasing disease severity. J. Infect. Dis. 208, 990–999.

Le, K.Y., Otto, M., 2015. Quorum-sensing regulation in staphylococci-an overview. Front. Microbiol. 6, 1174.

Lee, L.Y., Miyamoto, Y.J., Mcintyre, B.W., Höök, M., Mccrea, K.W., Mcdevitt, D., Brown, E.L., 2002. The *Staphylococcus aureus* Map protein is an immunomodulator that interferes with T cell-mediated responses. J. Clin. Invest. 110, 1461–1471.

Li, Z., Stevens, D.L., Hamilton, S.M., Parimon, T., Ma, Y., Kearns, A.M., Ellis, R.W., Bryant, A.E., 2011. Fatal *S. aureus* hemorrhagic pneumonia: genetic analysis of a unique clinical isolate producing both PVL and TSST-1. PLoS One 6, e27246.

Lin, Y.T., Tsai, J.C., Yamamoto, T., Chen, H.J., Hung, W.C., Hsueh, P.R., Teng, L.J., 2016. Emergence of a small colony variant of vancomycin-intermediate *Staphylococcus aureus* in a patient with septic arthritis during long-term treatment with daptomycin. J. Antimicrob. Chemother. 71, 1807–1814.

Lina, G., Bohach, G.A., Nair, S.P., Hiramatsu, K., Jouvin-Marche, E., Mariuzza, R., International Nomenclature Committee for Staphylococcal Superantigens, 2004. Standard nomenclature for the superantigens expressed by *Staphylococcus*. J. Infect. Dis. 189, 2334–2336.

Lina, G., Piémont, Y., Godail-Gamot, F., Bes, M., Peter, M.O., Gauduchon, V., Vandenesch, F., Etienne, J., 1999. Involvement of Panton-Valentine leukocidin-producing *Staphylococcus aureus* in primary skin infections and pneumonia. Clin. Infect. Dis. 29, 1128–1132.

Lindsay, J.A., 2010. Genomic variation and evolution of *Staphylococcus aureus*. Int. J. Med. Microbiol. 300, 98–103.

Lindsay, J.A., Holden, M.T., 2004. *Staphylococcus aureus*: superbug, super genome? Trends Microbiol. 12, 378–385.

Lindsay, J.A., Ruzin, A., Ross, H.F., Kurepina, N., Novick, R.P., 1998. The gene for toxic shock toxin is carried by a family of mobile pathogenicity islands in *Staphylococcus aureus*. Mol. Microbiol. 29, 527–543.

Lobocka, M., Hejnowicz, M.S., Dabrowski, K., Gozdek, A., Kosakowski, J., Witkowska, M., Ulatowska, M.I., Weber-Dabrowska, B., Kwiatek, M., Parasion, S., Gawor, J., Kosowska, H., Glowacka, A., 2012. Genomics of staphylococcal Twort-like phages–potential therapeutics of the post-antibiotic era. Adv. Virus Res. 83, 143–216.

Löffler, B., Tuchscherr, L., Niemann, S., Peters, G., 2014. *Staphylococcus aureus* persistence in non-professional phagocytes. Int. J. Med. Microbiol. 304, 170–176.

Lowy, F.D., 1998. *Staphylococcus aureus* infections. N. Engl. J. Med. 339, 520–532.

Mack, D., Fischer, W., Krokotsch, A., Leopold, K., Hartmann, R., Egge, H., Laufs, R., 1996. The intercellular adhesin involved in biofilm accumulation of *Staphylococcus epidermidis* is a linear β-1,6-linked glucosaminoglycan: purification and structural analysis. J. Bacteriol. 178, 175–183.

Maguire, G.P., Arthur, A.D., Boustead, P.J., Dwyer, B., Currie, B.J., 1996. Emerging epidemic of community-acquired methicillin-resistant *Staphylococcus aureus* infection in the Northern Territory. Med. J. Aust. 164, 721–723.

Malachowa, N., Deleo, F.R., 2010. Mobile genetic elements of *Staphylococcus aureus*. Cell Mol. Life Sci. 67, 3057–3071.

McKenney, D., Pouliot, K.L., Wang, Y., Murthy, V., Ulrich, M., Döring, G., Lee, J.C., Goldmann, D.A., Pier, G.B., 1999. Broadly protective vaccine for *Staphylococcus aureus* based on an in vivo-expressed antigen. Science 284, 1523–1527.

McNamara, P.J., Milligan-Monroe, K.C., Khalili, S., Proctor, R.A., 2000. Identification, cloning, and initial characterization of rot, a locus encoding a regulator of virulence factor expression in *Staphylococcus aureus*. J. Bacteriol. 182, 3197–3203.

Melles, D.C., Gorkink, R.F., Boelens, H.A., Snijders, S.V., Peeters, J.K., Moorhouse, M.J., van der Spek, P.J., van Leeuwen, W.B., Simons, G., Verbrugh, H.A., van Belkum, A., 2004. Natural population dynamics and expansion of pathogenic clones of *Staphylococcus aureus*. J. Clin. Invest. 114, 1732–1740.

Mogensen, T.H., 2009. Pathogen recognition and inflammatory signaling in innate immune defenses. Clin. Microbiol. Rev. 22, 240–273 (Table of Contents).

Monecke, S., Coombs, G., Shore, A.C., Coleman, D.C., Akpaka, P., Borg, M., Chow, H., Ip, M., Jatzwauk, L., Jonas, D., Kadlec, K., Kearns, A., Laurent, F., O'brien, F.G., Pearson, J., Ruppelt, A., Schwarz, S., Scicluna, E., Slickers, P., Tan, H.L., Weber, S., Ehricht, R., 2011. A field guide to pandemic, epidemic and sporadic clones of methicillin-resistant *Staphylococcus aureus*. PLoS One 6, e17936.

Monecke, S., Slickers, P., Ellington, M.J., Kearns, A.M., Ehricht, R., 2007. High diversity of Panton-Valentine leukocidin-positive, methicillin-susceptible isolates of *Staphylococcus aureus* and implications for the evolution of community-associated methicillin-resistant *S. aureus*. Clin. Microbiol. Infect. 13, 1157–1164.

Morgan, M., 2008. Methicillin-resistant *Staphylococcus aureus* and animals: zoonosis or humanosis? J. Antimicrob. Chemother. 62, 1181–1187.

Morikawa, K., Inose, Y., Okamura, H., Maruyama, A., Hayashi, H., Takeyasu, K., Ohta, T., 2003. A new staphylococcal sigma factor in the conserved gene cassette: functional significance and implication for the evolutionary processes. Genes Cells 8, 699–712.

Morikawa, K., Takemura, A.J., Inose, Y., Tsai, M., Nguyen Thi, L.T., Ohta, T., Msadek, T., 2012. Expression of a cryptic secondary sigma factor gene unveils natural competence for DNA transformation in *Staphylococcus aureus*. PLoS Pathog. 8, e1003003.

Much, H., 1908. Über eine Vorstufe des Fibrinfermentes in Kulturen von Staphylokokkus aureus. Biochem. Zeitschr. 14, 143–155.

Muenks, C.E., Hogan, P.G., Wang, J.W., Eisenstein, K.A., Burnham, C.A., Fritz, S.A., 2016. Diversity of *Staphylococcus aureus* strains colonizing various niches of the human body. J. Infect. 72, 698–705.

Nair, R., Perencevich, E.N., Blevins, A.E., Goto, M., Nelson, R.E., Schweizer, M.L., 2016. Clinical effectiveness of mupirocin for preventing *Staphylococcus aureus* infections in nonsurgical settings: a meta-analysis. Clin. Infect. Dis. 62, 618–630.

Navaratna, M.A., Sahl, H.G., Tagg, J.R., 1998. Two-component anti-*Staphylococcus aureus* lantibiotic activity produced by *Staphylococcus aureus* C55. Appl. Environ. Microbiol. 64, 4803–4808.

Navarre, W.W., Schneewind, O., 1994. Proteolytic cleavage and cell wall anchoring at the LPXTG motif of surface proteins in gram-positive bacteria. Mol. Microbiol. 14, 115–121.

Netz, D.J., Pohl, R., Beck-Sickinger, A.G., Selmer, T., Pierik, A.J., Bastos mdo, C., Sahl, H.G., 2002. Biochemical characterisation and genetic analysis of aureocin A53, a new, a typical bacteriocin from *Staphylococcus aureus*. J. Mol. Biol. 319, 745–756.

Novick, R.P., Christie, G.E., Penades, J.R., 2010. The phage-related chromosomal islands of Gram-positive bacteria. Nat. Rev. Microbiol. 8, 541–551.

Novick, R.P., Ross, H.F., Projan, S.J., Kornblum, J., Kreiswirth, B., Moghazeh, S., 1993. Synthesis of staphylococcal virulence factors is controlled by a regulatory RNA molecule. EMBO J. 12, 3967–3975.

Novick, R.P., Subedi, A., 2007. The SaPIs: mobile pathogenicity islands of *Staphylococcus*. Chem. Immunol. Allergy 93, 42–57.

Nygaard, T.K., Pallister, K.B., Dumont, A.L., Dewald, M., Watkins, R.L., Pallister, E.Q., Malone, C., Griffith, S., Horswill, A.R., Torres, V.J., Voyich, J.M., 2012. Alpha-toxin induces programmed cell death of human T cells, B cells, and monocytes during USA300 infection. PLoS One 7, e36532.

O'Brien, C.N., Guidry, A.J., Fattom, A., Shepherd, S., Douglass, L.W., Westhoff, D.C., 2000. Production of antibodies to *Staphylococcus aureus* serotypes 5, 8, and 336 using poly(DL-lactide-co-glycolide) microspheres. J. Dairy Sci. 83, 1758–1766.

O'Neill, E., Pozzi, C., Houston, P., Humphreys, H., Robinson, D.A., Loughman, A., Foster, T.J., O'gara, J.P., 2008. A novel *Staphylococcus aureus* biofilm phenotype mediated by the fibronectin-binding proteins, FnBPA and FnBPB. J. Bacteriol. 190, 3835–3850.

Otto, M., 2010. *Staphylococcus* colonization of the skin and antimicrobial peptides. Expert Rev. Dermatol. 5, 183–195.

Otto, M., 2014. *Staphylococcus aureus* toxins. Curr. Opin. Microbiol. 17, 32–37.

Pané-Farré, J., Jonas, B., Förstner, K., Engelmann, S., Hecker, M., 2006. The σ^B regulon in *Staphylococcus aureus* and its regulation. Int. J. Med. Microbiol. 296, 237–258.

Panton, P.N., Valentine, F.C., 1932. Staphylococcal toxin. Lancet 219, 506–508.

Pantosti, A., 2012. Methicillin-resistant *Staphylococcus aureus* associated with animals and its relevance to human health. Front. Microbiol. 3, 127.

Pantůček, R., Doškař, J., RŮŽičková, V., Kašpárek, P., Oráčová, E., Kvardová, V., Rosypal, S., 2004. Identification of bacteriophage types and their carriage in *Staphylococcus aureus*. Arch. Virol. 149, 1689–1703.

Paterson, G.K., Harrison, E.M., Murray, G.G., Welch, J.J., Warland, J.H., Holden, M.T., Morgan, F.J., Ba, X., Koop, G., Harris, S.R., Maskell, D.J., Peacock, S.J., Herrtage, M.E., Parkhill, J., Holmes, M.A., 2015. Capturing the cloud of diversity reveals complexity and heterogeneity of MRSA carriage, infection and transmission. Nat. Commun. 6, 6560.

Patti, J.M., Allen, B.L., Mcgavin, M.J., Höök, M., 1994. MSCRAMM-mediated adherence of microorganisms to host tissues. Annu. Rev. Microbiol. 48, 585–617.

Peacock, S.J., Moore, C.E., Justice, A., Kantzanou, M., Story, L., Mackie, K., O'neill, G., Day, N.P., 2002. Virulent combinations of adhesin and toxin genes in natural populations of *Staphylococcus aureus*. Infect. Immun. 70, 4987–4996.

Perl, T.M., Cullen, J.J., Wenzel, R.P., Zimmerman, M.B., Pfaller, M.A., Sheppard, D., Twombley, J., French, P.P., Herwaldt, L.A., Mupirocin and the Risk of Staphylococcus aureus Study Team, 2002. Intranasal mupirocin to prevent postoperative *Staphylococcus aureus* infections. N. Engl. J. Med. 346, 1871–1877.

Peschel, A., 2002. How do bacteria resist human antimicrobial peptides? Trends Microbiol. 10, 179–186.

Peschel, A., Sahl, H.G., 2006. The co-evolution of host cationic antimicrobial peptides and microbial resistance. Nat. Rev. Microbiol. 4, 529–536.

Peton, V., Le Loir, Y., 2014. *Staphylococcus aureus* in veterinary medicine. Infect. Genet. Evol. 21, 602–615.

Pichon, C., Felden, B., 2005. Small RNA genes expressed from *Staphylococcus aureus* genomic and pathogenicity islands with specific expression among pathogenic strains. Proc. Natl. Acad. Sci. U.S.A. 102, 14249–14254.

Prat, C., Bestebroer, J., de Haas, C.J., van Strijp, J.A., van Kessel, K.P., 2006. A new staphylococcal anti-inflammatory protein that antagonizes the formyl peptide receptor-like 1. J. Immunol. 177, 8017–8026.

Proctor, R.A., Kriegeskorte, A., Kahl, B.C., Becker, K., Loffler, B., Peters, G., 2014. *Staphylococcus aureus* Small Colony Variants (SCVs): a road map for the metabolic pathways involved in persistent infections. Front. Cell Infect. Microbiol. 4, 99.

Proctor, R.A., van Langevelde, P., Kristjansson, M., Maslow, J.N., Arbeit, R.D., 1995. Persistent and relapsing infections associated with small-colony variants of *Staphylococcus aureus*. Clin. Infect. Dis. 20, 95–102.

Proctor, R.A., von Eiff, C., Kahl, B.C., Becker, K., Mcnamara, P., Herrmann, M., Peters, G., 2006. Small colony variants: a pathogenic form of bacteria that facilitates persistent and recurrent infections. Nat. Rev. Microbiol. 4, 295–305.

Qin, L., Mccausland, J.W., Cheung, G.Y., Otto, M., 2016. PSM-mec - a virulence determinant that connects transcriptional regulation, virulence, and antibiotic resistance in staphylococci. Front. Microbiol. 7, 1293.

Queck, S.Y., Khan, B.A., Wang, R., Bach, T.H., Kretschmer, D., Chen, L., Kreiswirth, B.N., Peschel, A., Deleo, F.R., Otto, M., 2009. Mobile genetic element-encoded cytolysin connects virulence to methicillin resistance in MRSA. PLoS Pathog. 5, e1000533.

Quie, P.G., 1969. Microcolonies (G-variants) of *Staphylococcus aureus*. Yale J. Biol. Med. 41, 394–403.

Rasigade, J.P., Vandenesch, F., 2014. *Staphylococcus aureus*: a pathogen with still unresolved issues. Infect. Genet. Evol. 21, 510–514.

Richardson, A.R., Dunman, P.M., Fang, F.C., 2006. The nitrosative stress response of *Staphylococcus aureus* is required for resistance to innate immunity. Mol. Microbiol. 61, 927–939.

Richardson, A.R., Libby, S.J., Fang, F.C., 2008. A nitric oxide-inducible lactate dehydrogenase enables *Staphylococcus aureus* to resist innate immunity. Science 319, 1672–1676.

Robbins, J.B., Schneerson, R., Horwith, G., Naso, R., Fattom, A., 2004. *Staphylococcus aureus* types 5 and 8 capsular polysaccharide-protein conjugate vaccines. Am. Heart J. 147, 593–598.

Rolauffs, B., Bernhardt, T.M., von Eiff, C., Hart, M.L., Bettin, D., 2002. Osteopetrosis, femoral fracture, and chronic osteomyelitis caused by *Staphylococcus aureus* small colony variants (SCV) treated by girdlestone resection–6-year follow-up. Arch. Orthop. Trauma Surg. 122, 547–550.

Ruzin, A., Lindsay, J., Novick, R.P., 2001. Molecular genetics of SaPI1-a mobile pathogenicity island in *Staphylococcus aureus*. Mol. Microbiol. 41, 365–377.

Ruzin, A., Severin, A., Ritacco, F., Tabei, K., Singh, G., Bradford, P.A., Siegel, M.M., Projan, S.J., Shlaes, D.M., 2002. Further evidence that a cell wall precursor [C_{55}-MurNAc-(peptide)-GlcNAc] serves as an acceptor in a sorting reaction. J. Bacteriol. 184, 2141–2147.

Sahl, H.G., Bierbaum, G., 1998. Lantibiotics: biosynthesis and biological activities of uniquely modified peptides from gram-positive bacteria. Annu. Rev. Microbiol. 52, 41–79.

Saïd-Salim, B., Dunman, P.M., Mcaleese, F.M., Macapagal, D., Murphy, E., Mcnamara, P.J., Arvidson, S., Foster, T.J., Projan, S.J., Kreiswirth, B.N., 2003. Global regulation of *Staphylococcus aureus* genes by Rot. J. Bacteriol. 185, 610–619.

Salgado, D.R., Bozza, F.A., Pinto, M., Sampaio, J., 2001. Outbreak with small colony variants of methicillin-resistant *Staphylococcus aureus* in an ICU. In: 41st Interscience Conference on Antimicrobial Agents and Chemotherapy, 16–19, 12, 2001, Chicago, III.

Sato'o, Y., Omoe, K., Ono, H.K., Nakane, A., Hu, D.L., 2013. A novel comprehensive analysis method for *Staphylococcus aureus* pathogenicity islands. Microbiol. Immunol. 57, 91–99.

Schaumburg, F., Köck, R., Mellmann, A., Richter, L., Hasenberg, F., Kriegeskorte, A., Friedrich, A.W., Gatermann, S., Peters, G., von Eiff, C., Becker, K., 2012. Population dynamics among methicillin-resistant *Staphylococcus aureus* isolates in Germany during a 6-year period. J. Clin. Microbiol. 50, 3186–3192.

Schaumburg, F., Pauly, M., Anoh, E., Mossoun, A., Wiersma, L., Schubert, G., Flammen, A., Alabi, A.S., Muyembe-Tamfum, J.J., Grobusch, M.P., Karhemere, S., Akoua-Koffi, C., Couacy-Hymann, E., Kremsner, P.G., Mellmann, A., Becker, K., Leendertz, F.H., Peters, G., 2015. *Staphylococcus aureus* complex from animals and humans in three remote African regions. Clin. Microbiol. Infect. 21 (345), e1–e8.

Schijffelen, M.J., Boel, C.H., van Strijp, J.A., Fluit, A.C., 2010. Whole genome analysis of a livestock-associated methicillin-resistant *Staphylococcus aureus* ST398 isolate from a case of human endocarditis. BMC Genom. 11, 376.

Schwartz, K., Syed, A.K., Stephenson, R.E., Rickard, A.H., Boles, B.R., 2012. Functional amyloids composed of phenol soluble modulins stabilize *Staphylococcus aureus* biofilms. PLoS Pathog. 8, e1002744.

Seggewiß, J., Becker, K., Kotte, O., Eisenacher, M., Yazdi, M.R., Fischer, A., Mcnamara, P., Al Laham, N., Proctor, R., Peters, G., Heinemann, M., von Eiff, C., 2006. Reporter metabolite analysis of transcriptional profiles of a *Staphylococcus aureus* strain with normal phenotype and its isogenic hemB mutant displaying the small-colony-variant phenotype. J. Bacteriol. 188, 7765–7777.

Seifert, H., Wisplinghoff, H., Schnabel, P., von Eiff, C., 2003. Small colony variants of *Staphylococcus aureus* and pacemaker-related infection. Emerg. Infect. Dis. 9, 1316–1318.

Sendi, P., Rohrbach, M., Graber, P., Frei, R., Ochsner, P.E., Zimmerli, W., 2006. *Staphylococcus aureus* small colony variants in prosthetic joint infection. Clin. Infect. Dis. 43, 961–967.

Sinha, B., Fraunholz, M., 2010. *Staphylococcus aureus* host cell invasion and post-invasion events. Int. J. Med. Microbiol. 300, 170–175.

Spanu, T., Romano, L., D'inzeo, T., Masucci, L., Albanese, A., Papacci, F., Marchese, E., Sanguinetti, M., Fadda, G., 2005. Recurrent ventriculoperitoneal shunt infection caused by small-colony variants of *Staphylococcus aureus*. Clin. Infect. Dis. 41, e48–52.

Spaulding, A.R., Salgado-Pabon, W., Kohler, P.L., Horswill, A.R., Leung, D.Y., Schlievert, P.M., 2013. Staphylococcal and streptococcal superantigen exotoxins. Clin. Microbiol. Rev. 26, 422–447.

Sunderkötter, C., Becker, K., 2015. Frequent bacterial skin and soft tissue infections: diagnostic signs and treatment. J. Dtsch. Dermatol. Ges. 13, 501–524 quiz 525–526.

Sung, J.M., Lloyd, D.H., Lindsay, J.A., 2008. *Staphylococcus aureus* host specificity: comparative genomics of human versus animal isolates by multi-strain microarray. Microbiology 154, 1949–1959.

Surewaard, B.G., de Haas, C.J., Vervoort, F., Rigby, K.M., Deleo, F.R., Otto, M., van Strijp, J.A., Nijland, R., 2013. Staphylococcal alpha-phenol soluble modulins contribute to neutrophil lysis after phagocytosis. Cell Microbiol. 15, 1427–1437.

Suzuki, Y., Omoe, K., Hu, D.L., Sato'o, Y., Ono, H.K., Monma, C., Arai, T., Konishi, N., Kato, R., Hirai, A., Nakama, A., Kai, A., Kamata, Y., 2014. Molecular epidemiological characterization of *Staphylococcus aureus* isolates originating from food poisoning outbreaks that occurred in Tokyo, Japan. Microbiol. Immunol. 58, 570–580.

Thakker, M., Park, J.S., Carey, V., Lee, J.C., 1998. *Staphylococcus aureus* serotype 5 capsular polysaccharide is antiphagocytic and enhances bacterial virulence in a murine bacteremia model. Infect. Immun. 66, 5183–5189.

Theocharis, A.D., Skandalis, S.S., Gialeli, C., Karamanos, N.K., 2016. Extracellular matrix structure. Adv. Drug Deliv. Rev. 97, 4–27.

Tomasini, A., Francois, P., Howden, B.P., Fechter, P., Romby, P., Caldelari, I., 2014. The importance of regulatory RNAs in *Staphylococcus aureus*. Infect. Genet. Evol. 21, 616–626.

Tong, S.Y., Davis, J.S., Eichenberger, E., Holland, T.L., Fowler Jr., V.G., 2015. *Staphylococcus aureus* infections: epidemiology, pathophysiology, clinical manifestations, and management. Clin. Microbiol. Rev. 28, 603–661.

Tsompanidou, E., Denham, E.L., Becher, D., de Jong, A., Buist, G., van Oosten, M., Manson, W.L., Back, J.W., van Dijl, J.M., Dreisbach, A., 2013a. Distinct roles of phenol-soluble modulins in spreading of *Staphylococcus aureus* on wet surfaces. Appl. Environ. Microbiol. 79, 886–895.

Tsompanidou, E., Denham, E.L., van Dijl, J.M., 2013b. Phenol-soluble modulins, hellhounds from the staphylococcal virulence-factor pandemonium. Trends Microbiol. 21, 313–315.

Tuchscherr, L., Medina, E., Hussain, M., Volker, W., Heitmann, V., Niemann, S., Holzinger, D., Roth, J., Proctor, R.A., Becker, K., Peters, G., Löffler, B., 2011. *Staphylococcus aureus* phenotype switching: an effective bacterial strategy to escape host immune response and establish a chronic infection. EMBO Mol. Med. 3, 129–141.

Ubeda, C., Maiques, E., Barry, P., Matthews, A., Tormo, M.A., Lasa, I., Novick, R.P., Penades, J.R., 2008. SaPI mutations affecting replication and transfer and enabling autonomous replication in the absence of helper phage. Mol. Microbiol. 67, 493–503.

Ubeda, C., Tormo, M.A., Cucarella, C., Trotonda, P., Foster, T.J., Lasa, I., Penades, J.R., 2003. Sip, an integrase protein with excision, circularization and integration activities, defines a new family of mobile *Staphylococcus aureus* pathogenicity islands. Mol. Microbiol. 49, 193–210.

Valle, J., Toledo-Arana, A., Berasain, C., Ghigo, J.M., Amorena, B., Penadés, J.R., Lasa, I., 2003. SarA and not σ^B is essential for biofilm development by *Staphylococcus aureus*. Mol. Microbiol. 48, 1075–1087.

van Belkum, A., Verkaik, N.J., de Vogel, C.P., Boelens, H.A., Verveer, J., Nouwen, J.L., Verbrugh, H.A., Wertheim, H.F., 2009. Reclassification of *Staphylococcus aureus* nasal carriage types. J. Infect. Dis. 199, 1820–1826.

Vesga, O., Groeschel, M.C., Otten, M.F., Brar, D.W., Vann, J.M., Proctor, R.A., 1996. *Staphylococcus aureus* small colony variants are induced by the endothelial cell intracellular milieu. J. Infect. Dis. 173, 739–742.

Viana, D., Blanco, J., Tormo-Más, M.Á., Selva, L., Guinane, C.M., Baselga, R., Corpa, J., Lasa, Í., Novick, R.P., Fitzgerald, J.R., Penadés, J.R., 2010. Adaptation of *Staphylococcus aureus* to ruminant and equine hosts involves SaPI-carried variants of von Willebrand factor-binding protein. Mol. Microbiol. 77, 1583–1594.

von Eiff, C., Becker, K., 2007. Small-colony variants (SCVs) of staphylococci: a role in foreign body-associated infections. Int. J. Artif. Organs 30, 778–785.

von Eiff, C., Becker, K., Machka, K., Stammer, H., Peters, G., 2001a. Nasal carriage as a source of *Staphylococcus aureus* bacteremia. N. Engl. J. Med. 344, 11–16.

von Eiff, C., Becker, K., Metze, D., Lubritz, G., Hockmann, J., Schwarz, T., Peters, G., 2001b. Intracellular persistence of *Staphylococcus aureus* small-colony variants within keratinocytes: a cause for antibiotic treatment failure in a patient with Darier's disease. Clin. Infect. Dis. 32, 1643–1647.

von Eiff, C., Bettin, D., Proctor, R.A., Rolauffs, B., Lindner, N., Winkelmann, W., Peters, G., 1997. Recovery of small colony variants of *Staphylococcus aureus* following gentamicin bead placement for osteomyelitis. Clin. Infect. Dis. 25, 1250–1251.

von Eiff, C., Jansen, B., Kohnen, W., Becker, K., 2005. Infections associated with medical devices: pathogenesis, management and prophylaxis. Drugs 65, 179–214.

von Eiff, C., Mcnamara, P., Becker, K., Bates, D., Lei, X.H., Ziman, M., Bochner, B.R., Peters, G., Proctor, R.A., 2006a. Phenotype microarray profiling of *Staphylococcus aureus menD* and *hemB* mutants with the small-colony-variant phenotype. J. Bacteriol. 188, 687–693.

von Eiff, C., Peters, G., Becker, K., 2006b. The small colony variant (SCV) concept – the role of staphylococcal SCVs in persistent infections. Injury 2 (37 Suppl.), S26–S33.

von Eiff, C., Taylor, K.L., Mellmann, A., Fattom, A.I., Friedrich, A.W., Peters, G., Becker, K., 2007. Distribution of capsular and surface polysaccharide serotypes of *Staphylococcus aureus*. Diagn. Microbiol. Infect. Dis. 58, 297–302.

Vuong, C., Voyich, J.M., Fischer, E.R., Braughton, K.R., Whitney, A.R., Deleo, F.R., Otto, M., 2004. Polysaccharide intercellular adhesin (PIA) protects *Staphylococcus epidermidis* against major components of the human innate immune system. Cell Microbiol. 6, 269–275.

Walker, J.N., Crosby, H.A., Spaulding, A.R., Salgado-Pabon, W., Malone, C.L., Rosenthal, C.B., Schlievert, P.M., Boyd, J.M., Horswill, A.R., 2013. The *Staphylococcus aureus* ArlRS two-component system is a novel regulator of agglutination and pathogenesis. PLoS Pathog. 9, e1003819.

Wang, R., Braughton, K.R., Kretschmer, D., Bach, T.H., Queck, S.Y., Li, M., Kennedy, A.D., Dorward, D.W., Klebanoff, S.J., Peschel, A., Deleo, F.R., Otto, M., 2007. Identification of novel cytolytic peptides as key virulence determinants for community-associated MRSA. Nat. Med. 13, 1510–1514.

Weidenmaier, C., Kokai-Kun, J.F., Kristian, S.A., Chanturiya, T., Kalbacher, H., Gross, M., Nicholson, G., Neumeister, B., Mond, J.J., Peschel, A., 2004. Role of teichoic acids in *Staphylococcus aureus* nasal colonization, a major risk factor in nosocomial infections. Nat. Med. 10, 243–245.

Werbick, C., Becker, K., Mellmann, A., Juuti, K.M., von Eiff, C., Peters, G., Kuusela, P.I., Friedrich, A.W., Sinha, B., 2007. Staphylococcal chromosomal cassette *mec* type I, *spa* type, and expression of Pls are determinants of reduced cellular invasiveness of methicillin-resistant *Staphylococcus aureus* isolates. J. Infect. Dis. 195, 1678–1685.

Wertheim, H.F., Melles, D.C., Vos, M.C., van Leeuwen, W., van Belkum, A., Verbrugh, H.A., Nouwen, J.L., 2005. The role of nasal carriage in *Staphylococcus aureus* infections. Lancet Infect. Dis. 5, 751–762.

Wertheim, H.F., Vos, M.C., Ott, A., van Belkum, A., Voss, A., Kluytmans, J.A., Van Keulen, P.H., Vandenbroucke-Grauls, C.M., Meester, M.H., Verbrugh, H.A., 2004. Risk and outcome of nosocomial *Staphylococcus aureus* bacteraemia in nasal carriers versus non-carriers. Lancet 364, 703–705.

Wertheim, H.F., Walsh, E., Choudhurry, R., Melles, D.C., Boelens, H.A., Miajlovic, H., Verbrugh, H.A., Foster, T., van Belkum, A., 2008. Key role for clumping factor B in *Staphylococcus aureus* nasal colonization of humans. PLoS Med. 5, e17.

Williams, R.E., 1963. Healthy carriage of *Staphylococcus aureus*: its prevalence and importance. Bacteriol. Rev. 27, 56–71.

Wladyka, B., Wielebska, K., Wloka, M., Bochenska, O., Dubin, G., Dubin, A., Mak, P., 2013. Isolation, biochemical characterization, and cloning of a bacteriocin from the poultry-associated *Staphylococcus aureus* strain CH-91. Appl. Microbiol. Biotechnol. 97, 7229–7239.

Wolter, D.J., Emerson, J.C., Mcnamara, S., Buccat, A.M., Qin, X., Cochrane, E., Houston, L.S., Rogers, G.B., Marsh, P., Prehar, K., Pope, C.E., Blackledge, M., Deziel, E., Bruce, K.D., Ramsey, B.W., Gibson, R.L., Burns, J.L., Hoffman, L.R., 2013. *Staphylococcus aureus* small-colony variants are independently associated with worse lung disease in children with cystic fibrosis. Clin. Infect. Dis. 57, 384–391.

Xia, G., Kohler, T., Peschel, A., 2010. The wall teichoic acid and lipoteichoic acid polymers of *Staphylococcus aureus*. Int. J. Med. Microbiol. 300, 148–154.

Xia, G., Wolz, C., 2014. Phages of *Staphylococcus aureus* and their impact on host evolution. Infect. Genet. Evol. 21, 593–601.

Yarwood, J.M., Mccormick, J.K., Schlievert, P.M., 2001. Identification of a novel two-component regulatory system that acts in global regulation of virulence factors of *Staphylococcus aureus*. J. Bacteriol. 183, 1113–1123.

Yoong, P., Torres, V.J., 2013. The effects of *Staphylococcus aureus* leukotoxins on the host: cell lysis and beyond. Curr. Opin. Microbiol. 16, 63–69.

Yu, V.L., Goetz, A., Wagener, M., Smith, P.B., Rihs, J.D., Hanchett, J., Zuravleff, J.J., 1986. *Staphylococcus aureus* nasal carriage and infection in patients on hemodialysis. Efficacy of antibiotic prophylaxis. N. Engl. J. Med. 315, 91–96.

Ziebandt, A.K., Kusch, H., Degner, M., Jaglitz, S., Sibbald, M.J., Arends, J.P., Chlebowicz, M.A., Albrecht, D., Pantuček, R., Doškar, J., Ziebuhr, W., Bröker, B.M., Hecker, M., van Dijl, J.M., Engelmann, S., 2010. Proteomics uncovers extreme heterogeneity in the *Staphylococcus aureus* exoproteome due to genomic plasticity and variant gene regulation. Proteomics 10, 1634–1644.

Zubair, S., Fischer, A., Liljander, A., Meens, J., Hegerman, J., Gourlé, H., Bishop, R.P., Roebbelen, I., Younan, M., Mustafa, M.I., Mushtaq, M., Bongcam-Rudloff, E., Jores, J., 2015. Complete genome sequence of *Staphylococcus aureus*, strain ILRI_Eymole1/1, isolated from a Kenyan dromedary camel. Stand Genom. Sci. 10, 109.

CHAPTER

STAPHYLOCOCCUS AUREUS ENTEROTOXINS

3

Dong-Liang Hu[1,3], Lizhe Wang[2], Rendong Fang[4], Masashi Okamura[1], Hisaya K. Ono[1]

[1]Kitasato University School of Veterinary Medicine, Towada, Aomori, Japan; [2]Zhuhai Foodawa Food Technology Ltd., Zhuhai, P. R. China; [3]College of Veterinary Medicine, Jilin University, Changchun, Jilin, P. R. China; [4]Southwest University, Chongqing, P. R. China

1. INTRODUCTION

Staphylococcus aureus causes a broad range of infections, food poisoning, and toxic shock syndrome (TSS). For the majority of diseases caused by *S. aureus*, the pathogenesis is multifactorial, i.e., is related to a number of virulence factors. However, there are also correlations between strains isolated from particular diseases and expression of particular virulence determinants, which suggests their role in a particular disease. Virulence factors of *S. aureus* include pathogenic antigens and surface proteins, secreted enzymes, membrane-damaging toxins, toxic shock syndrome toxins (TSSTs), and proteins that invade or avoid immune system (see Chapter 2 for more details). Additionally, staphylococcal enterotoxins (SEs) and SE-like toxins (SEls) are considered to be major virulence factors of *S. aureus* in particular in the context of food safety. These toxins have become to be a superfamily with 23 types including SEs and SEls.

The genes of SEs and SEls are encoded by diverse accessory genetic elements, and the molecular of the toxins share certain structural and biological properties. SEs and SEls are also heat-stable and multifunctional proteins that trigger food poisoning outbreaks, pyrogenicity, superantigenicity, and the capacity to induce lethal hypersensitivity to endotoxin.

SEs are important toxins, that stimulate nausea, vomiting, and abdomen clumps in human and animals. SEls are newly identified toxins with very similar physicochemical properties compared with SEs but which do not induce any emetic activity or which have not yet been tested in a vomiting experiment. Both of SEs and SEls have superantigenic activity and belong to a large family of bacterial superantigen (SAg) exotoxins.

2. SUPERFAMILY OF STAPHYLOCOCCAL ENTEROTOXINS AND STAPHYLOCOCCAL ENTEROTOXIN-LIKE TOXINS

S. aureus produces a large variety ("superfamily") of SEs that are important toxins of food poisoning representing superantigenic toxins. SEs and SEls share common phylogenetic relationships, structure, function, and sequence homology. These toxins are simple proteins composed of approximately 168–261 amino acids and demonstrate molecular size of 19–29 kD (Table 3.1). Since the first

Staphylococcus aureus. http://dx.doi.org/10.1016/B978-0-12-809671-0.00003-6

Table 3.1 Biological Characteristics of Staphylococcal Enterotoxins (SEs) and Staphylococcal Enterotoxin-Like Toxins (SEls)

				Emetic Activity[a]		
SEs and SEls	**Mature Length (aa)**	**Molecular Weight (kDa)**	**Superantigenic Activity**	**Monkey[b]**	**House Musk Shrew[c]**	**Genetic Elements**
SEA	233	27.1	+	5 (ED_{50})	0.3	Prophage
SEB	239	28.4	+	5 (ED_{50})	10	Chromosome, SaPI3, plasmid (pZA10)
SEC1	239	27.5	+	5 (ED_{50})	NE[d]	SaPI
SEC2	239	27.6	+	5 (ED_{50})	1000	SaPI
SEC3	239	27.6	+	5 (ED_{50})	NE	SaPI
SED	233	26.9	+	5 (ED_{50})	40	Plasmid (pIB485)
SEE	230	26.4	+	10–20 (ED_{50})	10	Prophage (Hypothetical location)
SEG	233	27.0	+	160–320	200	*egc*1, *egc*2, *egc*3, *egc*4
SEH	217	25.1	+	30	1000	Transposon (MGEmw2/ mssa476 seh/Δseo)
SEI	218	24.9	+	300–600	1	*egc*1, *egc*2, *egc*3
SE1J	245	28.6	+	NE	NE	Plasmid (pIB485, pF5)
SEK	219	25.3	+	100 (2/6)	NE	Prophages, SaPI1, SaPI3, SaPI5, SaPIbov1
SEL	216	24.7	+	100 (1/6)	NE	SaPIn1, SaPIm1, SaPImw2, SaPIbov1
SEM	217	24.8	+	100 (1/7)	NE	*egc*1, *egc*2
SEN	227	26.1	+	100 (1/8)	NE	*egc*1, *egc*2, *egc*3, *egc*4
SEO	232	26.8	+	100 (2/6)	NE	*egc*1, *egc*2, *egc*3, *egc*4, *Transposon*
SEP	233	26.7	+	100 (3/6)	50	Prophage (Sa3n)
SEQ	216	25.2	+	100 (2/6)	NE	SaPI1, SaPI3, SaPI5, Prophage
SER	233	27.0	+	<100	<1000	Plasmid (pIB485, pF5)
SES	257	26.2	+	<100	20	Plasmid (pF5)
SET	216	22.6	+	<100	1000	Plasmid (pF5)
SE1U	261	27.2	+	NE	NE	*egc*2, *egc*3
SE1V	239	27.6	+	NE	NE	*egc*4
SE1W	256	26.7	+	NE	NE	*egc*4
SE1X	168	19.3	+	NE	NE	Chromosome
SE1Y	221	22.5	+/--	NE	500	Chromosome

+, positive reaction.
[a]μg/animal.
[b]Oral administration.
[c]Intraperitoneal administration.
[d]Not examined.

characterization of staphylococcal enterotoxin A (SEA) in 1959 by Casman and Bergdoll, five SEs, named SEA to SEE, have been known from the differences in the antigenicity of SEs (Balaban and Rasooly, 2000; Bergdoll et al., 1959; Casman, 1960). The staphylococcal enterotoxin C (SEC) was further divided into three antigenic subtypes, SEC1, SEC2, and SEC3 (Balaban and Rasooly, 2000). Todd et al. (1978) had reported a toxin that induces TSS caused by *S. aureus* infection, and the toxin had been named staphylococcal enterotoxin F (SEF). However, SEF does not have emetic activity, later it was renamed TSST-1. From 1990s, the presence of additional SEs began to be revealed, and after 2001, many new types of SEs and SEls were discovered, one after another (Omoe et al., 2003; Ono et al., 2008, 2015; Thomas et al., 2007; Thomas et al., 2006; Wilson et al., 2011). To date, the presence of total 23 different SEs and SEls has been reported and described (Table 3.1). These toxins include SEA to SEE, SEG to SET, and staphylococcal enterotoxin-like toxin U (SElU) to SElY. The International Nomenclature Committee for Staphylococcal Superantigens (INCSS) proposed the standard nomenclature for the toxins that have been newly discovered (Thomas et al., 2007; Lina et al., 2004). INCSS naming convention is to emphasize the relevance of the food poisoning (emetic activity). To name the SE, it is obliged to prove emetic activity by oral administration to a primate model. If it shows negative emetic potential in the vomiting experiments in a primate model or the vomiting experiments have not yet been carried out, this toxin should be named "SEl," even if the SAg is considered to be closely related to SE structure.

Omoe et al. (2013) assessed the emetic potentials of some newly discovered SEls, including SElK, SElL, SElM, SElN, SElO, SElP, and SElQ, using a monkey-feeding assay. It was shown that all of the SEls tested induced emetic responses in the monkeys at a dose of 100 μg/kg, although the numbers of affected monkeys were smaller than the numbers that were affected after consuming SEA or SEB. These results indicated that the newly identified SEls are emetic toxins allowing for its renaming as SEK, SEL, SEM, SEN, SEO, SEP, and SEQ, respectively, according to the INCSS naming convention (Lina et al., 2004). More recently, Ono et al. (2015) reported a novel staphylococcal emetic toxin, SElY, which had strong emetic activity in a small emetic animal model using house musk shrews.

3. GENE LOCATIONS OF STAPHYLOCOCCAL ENTEROTOXINS AND STAPHYLOCOCCAL ENTEROTOXIN-LIKE TOXINS

Genes encoding for SEs and SEls are harbored by numerous mobile elements, including *S. aureus* pathogenicity islands (SaPIs), genomic islands (υSa), prophages, and plasmids.

3.1 *STAPHYLOCOCCUS AUREUS* PATHOGENICITY ISLANDS

SaPIs are mobile pathogenicity islands, which range in size from 15 to 17 kilobase (kb), with the exceptions of SaPIbov2 (27 kb) and a highly degenerated SaPI (3.14 kb) present in some sequenced genomes (Lindsay et al., 1998; Tallent et al., 2007). The complete nucleotide sequence is known for 20 SaPIs, and some of them carry genes encoding one or more SEs (Fig. 3.1). *sek* and *seq* are found together with *tst* in SaPI1; *sel* and *sec* are found in SaPIbov1; *seb*, *seq*, and *sek* have been reported in SaPI3 (Novick and Subedi, 2007). Induction of a SaPI is likely to originate an increase in the copy number of the toxin genes and therefore to an increase in toxin production (Sumby and Waldor, 2003). Sato'o et al. (2013) designed a series of primers corresponding to sequences flanking six SaPI insertion sites in *S. aureus*

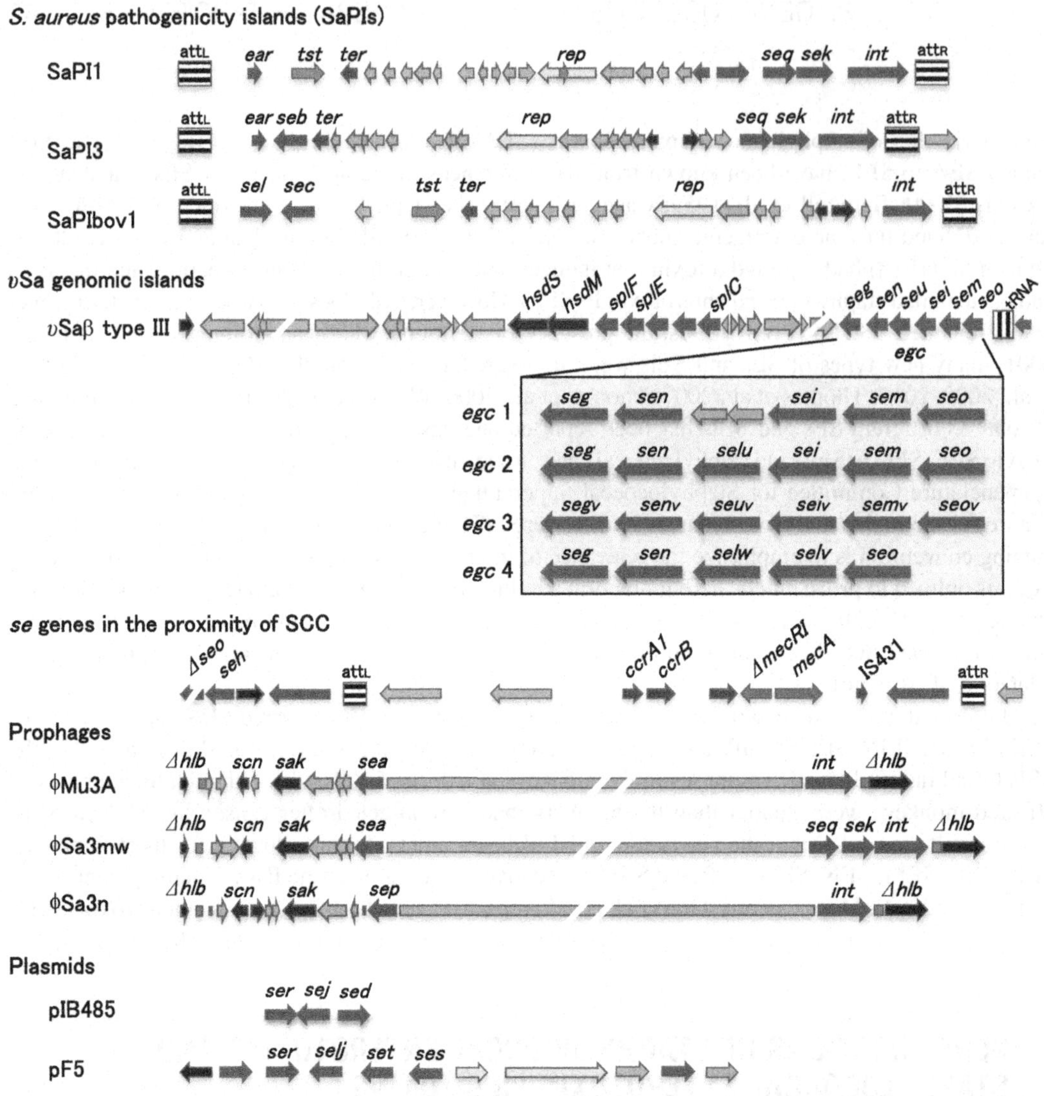

FIGURE 3.1

Staphylococcal enterotoxin (SE) and SE-like toxin genes carried by *Staphylococcus aureus* pathogenicity islands, υSa genomic islands, prophages, and plasmids.

Based on sequencing data and modified from Novick, R.P., Subedi, A., 2007. The SaPIs: mobile pathogenicity islands of Staphylococcus. *Chem. Immunol. Allergy 93, 42–57; Baba, T., Bae, T., Schneewind, O., Takeuchi, F., Hiramatsu, K., 2008. Genome sequence of* Staphylococcus aureus *strain Newman and comparative analysis of staphylococcal genomes: polymorphism and evolution of two major pathogenicity islands. J. Bacteriol. 190, 300–310; Thomas, D.Y., Jarraud, S., Lemercier, B., Cozon, G., Echasserieau, K., Etienne, J., Gougeon, M.L., Lina, G., Vandenesch, F., 2006. Staphylocccal enterotoxin-like toxins U2 and V, two new staphylococcal superantigens arising from recombination within the enterotoxin gene cluster. Infect. Immun. 74, 4724–4734; Collery, M.M., Smyth, D.S., Tumilty, J.J., Twohig, J.M., Smyth, C.J., 2009. Associations between enterotoxin gene cluster types* egc*1, * egc*2 and* egc*3, agr types, enterotoxin and enterotoxin-like gene profiles, and molecular typing characteristics of human nasal carriage and animal isolates of* Staphylococcus aureus. *J. Med. Microbiol. 58, 13–25; Ono, H.K., Omoe, K., Imanishi, K., Iwakabe, Y., Hu, D.-L., Kato, H., Saito, N., Nakane, A., Uchiyama, T., Shinagawa, K., 2008. Identification and characterization of two novel staphylococcal enterotoxins, types S and T. Infect. Immun. 76, 4999–5005; Ono, H.K., Sato'o, Y., Narita, K., Naito, I., Hirose, S., Hisatsune, J., Asano, K., Hu, D.-L., Omoe, K., Sugai, M., Nakane, A., 2015. Identification and characterization of a novel staphylococcal emetic toxin. Appl. Environ. Microbiol. 8, 7034–7040.*

genome and established a long and accurate (LA)-PCR analysis method. LA-PCR products of 13–17 kbp were observed in strains with *seb*, *sek*, or *seq* genes. Nucleotide sequencing analysis revealed seven novel SaPIs: *seb*-harboring SaPIivm10, SaPishikawa11, SaPIivm60, SaPIno10 and SaPIhirosaki4, *sek*- and *seq*-harboring SaPIj11, and nonsuperantigen-harboring SaPIhhms2. These SaPIs have mosaic structures containing components of known SaPIs and other unknown genes. Strains carrying different SaPIs were found to have significantly different production of SAg toxins (Sato'o et al., 2013, 2014).

3.2 GENOMIC ISLANDS (υSA)

The υSa genomic islands are nonphage and non-SCC (staphylococcal chromosome) genomic islands that are exclusively present in *S. aureus*. These islands are inserted at specific loci in the chromosome and are associated with either intact or remnant DNA recombinases (Baba et al., 2002, 2008). Two major υSa genomic islands, namely υSaα and υSaβ are present in all *S. aureus* genomes. Both υSaα and υSaβ contain clusters of genes encoding virulence factors, of which υSaβ carries—among others—the enterotoxin gene cluster (*egc*), which includes a variable number of *se* or *sel* genes forming an operon (Fig. 3.1).

3.3 ENTEROTOXIN GENE CLUSTER

The *egc* is organized as an operon. The dynamic evolution of this cluster is reflected in the number of variants (Fig. 3.1). The first *egc* (*egc*1) consists of five *se* genes (*seg*, *sei*, *sem*, *sen*, and *seo*) and two pseudogenes (φ*ent1* and φ*ent2*) (Monday and Bohach, 2001). The second *egc* variant (*egc*2) contains an additional *sel* gene (*selu*) (Letertre et al., 2003). Allelic variants of each of the *egc*2 genes compose the *egc*3 cluster (Letertre et al., 2003; Collery et al., 2009), and two new *sel* genes (*selv*, *selw*) are present in *egc*4 (Thomas et al., 2006). Incomplete *egc* clusters, lacking one or more genes of the classical *egc*1, as well as variants carrying insertion sequences within *seg*, *sen*, or *sei*, have also been described (Thomas et al., 2006; Omoe et al., 2005). This locus probably plays the role of a nursery for *se* genes.

3.4 STAPHYLOCOCCAL CASSETTE CHROMOSOME

The *seh* gene has been found in close proximity of the non-*mec*A containing SCC element harbored by the SCC*mec* type IV of *S. aureus* MW2. The *seh* gene is flanked by a truncated *seo* gene and a putative transposase gene (Fig. 3.1). Acquisition of the *seh* element could have stabilized the integration of SCC*mec* type IV, which is unable to excise (Noto and Archer, 2006).

3.5 PROPHAGE

SEs genes can be carried by prophages (Betley and Mekalanos, 1985; Coleman et al., 1989). The phages carrying *se* genes (*sea*, *sek*, *sep*, and *seq*) belong to the Siphoviridae family. Three *se*/*sel* genes (*sea*, *selk*, and *selq*) are present together in ϕSa3ms and ϕSa3mw, whereas a single *se*/*sel* gene (*sea* or *sep*) is carried by ϕMu3A, ϕSa3a, and other prophages (Fig. 3.1). Integration of these phages into the *S. aureus* chromosome occurs by a site-specific recombination event between the *att*P site in the phage genome and the *att*B site located within the β-hemolysin gene in the bacterial chromosome (Coleman et al., 1989).

3.6 PLASMID

Plasmids are efficient vehicles for the spread of resistance and virulence determinants through horizontal gene transfer. In *S. aureus*, two kinds of plasmids, pIB485 and pF5, carrying *se/sel* genes have been characterized (Omoe et al., 2003; Ono et al., 2008; Couch et al., 1988). The pIB485, a 27.6 kb plasmid, in which first *sed, selj*, and latter *ser* were identified (Ono et al., 2008; Goerke et al., 2009). Enterotoxin SER was discovered by Omoe et al. (2003) in *S. aureus* strains associated with a food poisoning outbreak that occurred in Fukuoka, Japan, in 1997, and the *ser* gene was shown to be located on a family of closely related plasmids, termed Fukuoka plasmids, pF5 and pF5-like (Fig. 3.1). These plasmids have similar restriction profiles and carry *selj* along with *ser*. Two novel SE genes (*ses* and *set*) have also been detected on the pF5 (Omoe et al., 2003; Ono et al., 2008). Interestingly, the *ser* gene, together with *sed* and *selj*, has also been found in pIB485-like plasmids from laboratory strains, food poisoning outbreak isolates, and healthy human isolates in Japan (Omoe et al., 2003).

4. MOLECULAR STRUCTURES OF STAPHYLOCOCCAL ENTEROTOXINS AND STAPHYLOCOCCAL ENTEROTOXIN-LIKE TOXINS

SEs and SEls are globular, single-chain proteins, with molecular weights ranging from 19 to 29 kDa. These toxins are structurally homologous, even though the primary sequences of the proteins are diverse. Examination of the aligned sequences shows that there are several linear stretches of amino acids that are more similar among the SEs, SEls, and other SAg toxins (McCormick et al., 2001). According to the homology of nucleotide and amino-acid sequences of the superantigenic toxins, they can be classified into five groups (Fig. 3.2).

Group I toxins include TSST-1, staphylococcal superantigen-like protein (SSL7), and the recently described SElX (Wilson et al., 2011; Mitchell et al., 2000; Brosnahan and Schlievert, 2011; Acharya et al., 1994; Prasad et al., 1993). They have unique primary amino-acid sequences and lack the emetic cystine loop of SEs (McCormick et al., 2001). Group I contain only a low-affinity major histocompatibility class (MHC) II binding site in their O/B folds that interacts with the α-chains of MHC II molecules (Fig. 3.3) (Kim et al., 1994).

Group II toxins (SEB group) include SEB, SEC, SEG, SElU, and SElW (SElU2) (Fitzgerald et al., 2003; Hovde et al., 1994). These toxins contain the core SAg structure plus a cystine loop that has a varying 10- to 19-amino-acid sequence separating the cysteine residues (Hovde et al., 1994; Schlievert et al., 2000). Group II SAgs also contain one MHC II site, α-chain MHC II binding site, and this interaction does not depend on interaction with the antigenic peptide within the MHC II peptide-binding groove (Letertre et al., 2003; Jardetzky et al., 1994). The Vβ-TCR (T cell receptor) binding site of the group II SAg toxin is located on the top front of the molecules (Letertre et al., 2003; Leder et al., 1998; Li et al., 1998).

Group III toxins (SEA group) include SEA, SED, SEE, SEH, SElJ, SEN, and SEP (Fig. 3.2). Among the different SE types, SEA, SEE, SEP, and SElJ share the highest amino-acid sequence homology, ranging from 65% to 83%. They contain a cystine loop like the Group II SAg, but the loop is always nine amino acids long (Fitzgerald et al., 2003; Spaulding et al., 2013). Importantly, the Group III SAg toxins contain the low-affinity α-chain MHC II binding site and the high-affinity site, referred

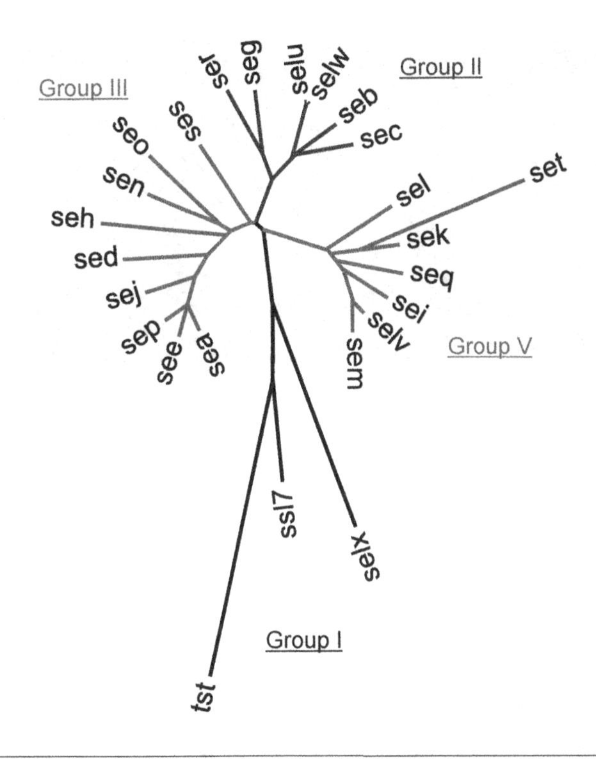

FIGURE 3.2

Grouping of Staphylococcal enterotoxins and SE-like toxins based on amino-acid sequence comparisons.

Modified from Wilson, G.J., Seo, K.S., Cartwright, R.A., Connelley, T., Chuang-Smith, O.N., Merriman, J.A., Guinane, C.M., Park, J.Y., Bohach, G.A., Schlievert, P.M., Morrison, W.I., Fitzgerald, J.R., 2011. A novel core genome-encoded superantigen contributes to lethality of community-associated MRSA necrotizing pneumonia. PLoS Pathog. 7, e1002271. Group I toxins are shown in black; Group II toxins are shown in blue (drak gray in print versions); Group III toxins are shown in red (light gray in print versions) and the Group V are shown in green (gray in print versions). Because Group IV toxins are produced by group A streptococci *but not* S. aureus, *this group is not shown in Fig. 3.2.*

to as a Zn^{2+}-dependent β-chains MHC II binding site, in the β-grasp domains (Fig. 3.3) (Fitzgerald et al., 2003; Petersson et al., 2002). The presence of two MHC II sites allows the SEs to cross bridge MHC II molecules on antigen presenting cells (APCs) and increase SAg activity (Kozono et al., 1995). The presence of the Zn^{2+}-dependent high-affinity site on these SAgs makes them 10- to 100-fold more active overall in causing cytokine production from T cells and APCs than other SAgs that have only a low-affinity MHC II binding site.

The Group IV superantigenic toxins are produced by *Streptococci* but not *S. aureus*. These SAgs contain both low (α-chain)- and high (β-chain)-affinity MHC II binding sites, similar to the site found in SEA. Group IV SAgs lack the cystine loop required for emetic activity. This group will not be further discussed.

Group V (SEI group) is the last described group and contains many of the recently discovered SAgs, i.e., SEI, SEK, SEL, SEM, SEQ, SET, and SElV (Su and Wong, 1995; Orwin et al., 2001, 2002, 2003).

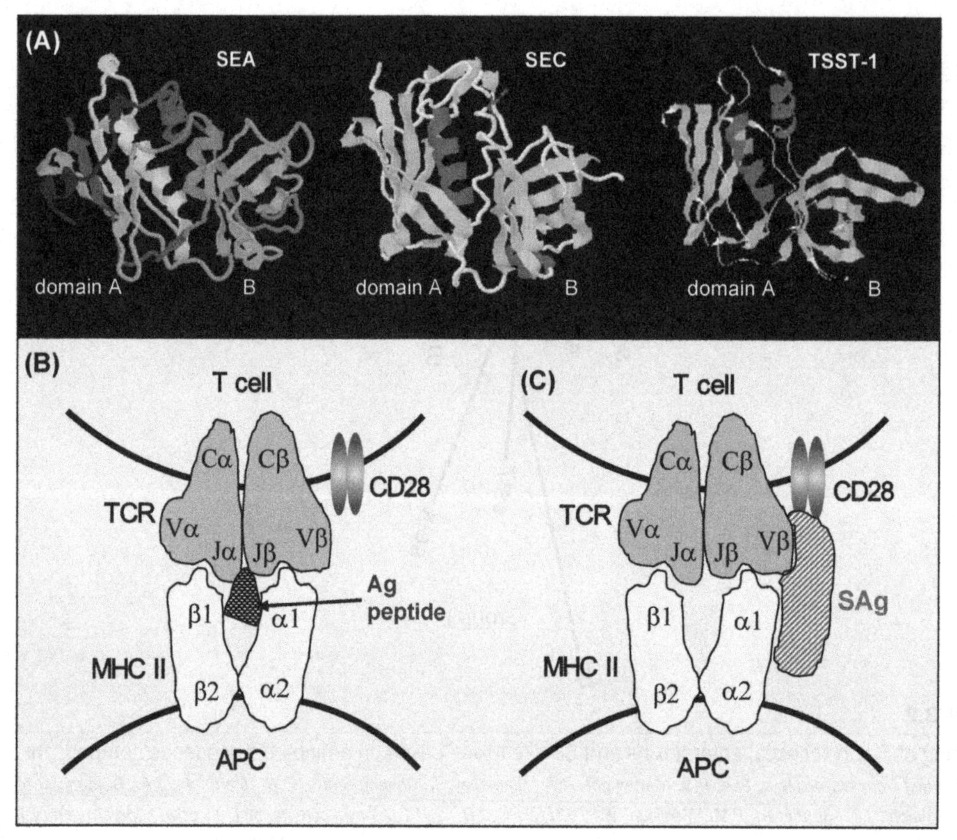

FIGURE 3.3

Three-dimensional structure of staphylococcal enterotoxin A (SEA), staphylococcal enterotoxin C (SEC), and toxic shock syndrome toxin 1 (TSST-1) as representative of the superantigenic toxins (A). Diagram of normal antigen (B) and superantigen (C) bind to T cell receptor, CD28 molecules, and major histocompatibility class (MHC) II of antigen presenting cell (APC).

Prepared and modified from Hu, D.-L., Imai, A., Ono, K., Sasaki, S., Nakane, A., Sugii, S., Shinagawa, K., 1998. Epitope analysis of staphylococcal enterotoxin A using different synthetic peptides. J. Vet. Med. Sci. 60, 993–996; Hu, D.-L., Nakane, A., 2014. Mechanisms of staphylococcal enterotoxin-induced emesis. Eur. J. Pharm. 722, 95–107.

Group V toxins contain both the low- and high-affinity MHC II binding sites, but lack the cystine loop, which appears to be critical for the specificity of the interaction of the SAgs with their respective Vβ-TCRs (Gunther et al., 2007).

Although these toxins have different homology in the amino acids sequences, the secondary structure is a mixture of α-helix and β-sheet components, and the three-dimensional structures show very similar conformation (Swaminathan et al., 1992, 1995). The canonical structure consists of an A domain, a B domain, and an α-helix spans the center of the structure and connects the A, B domains (Fig. 3.3). Three-dimensional structures of SEA, SEC, and TSST-1 have verified this assertion. They contain two unequal domains. The domain A contains both the amino and carboxyl termini, as well as a β-grasp motif. The amino-terminal residues drape over the edge of the β-sheet in a loosely attached structure. The interfaces between the A and B domains are marked by a set of α-helices, which form a long groove in the back side of the molecule and a shallow cavity at the top (Fig. 3.3A). The internal β-barrel region is richly hydrophobic, and the external surface is covered by a number of hydrophilic residues. The characteristic SE disulfide bond is located at the end of domain B, opposing the α-helical cap.

5. SUPERANTIGENIC ACTIVITY OF STAPHYLOCOCCAL ENTEROTOXINS AND STAPHYLOCOCCAL ENTEROTOXIN-LIKE TOXINS

A number of studies have been undertaken to understand the structural basis for superantigenic activity. Crystallographic and structural studies revealed their common molecular structure and binding motifs (Kotzin et al., 1993; Papageorgiou and Acharya, 2000), unlike conventional antigens, these toxins bypass normal processing by APC and induce a large proportion (5%–30%) of T cells to proliferate and subsequently stimulate massive cytokine release (Fig. 3.3B and C). The excessive release of cytokines, such as tumor necrosis factor α (TNF-α), interleukin 1 (IL-1), IL-2, and gamma interferon (IFN-γ), that mediate the toxic effects of these SAg toxins (Jupin et al., 1988; Trede et al., 1991; Faulkner et al., 2005).

The groove formed between the conserved N- and C-terminal domains of SEs represents an important interaction site for the TCR Vβ chain (Leder et al., 1998; Seth et al., 1994; Moza et al., 2007; Krakauer, 1994). Each SE or SEl binds to a distinct repertoire of Vβ-bearing T cells, revealing a unique biological "fingerprint". A direct binding of SEB to the T cell costimulatory receptor CD28 was reported recently (Arad et al., 2012). Several approaches, such as synthetic peptides corresponding to regions of SEA (Hu et al., 1998, 1999, 2005, Pontzer et al., 1991; Griggs et al., 1992), proteolytic digestion fragments of the toxins (Spero and Morlock, 1978; Ezepchuk and Noskov, 1986; Bohach et al., 1989; Binek et al., 1992), monoclonal antibodies specific for identified epitopes (Hu et al., 1998, 1999; Bohach et al., 1989), and recombinant mutant SEA, SEB, and SEC (Hu et al., 2006, 2009; Kappler et al., 1992; Irwin et al., 1992) have been used to determine the regions of SEs involved in the mitogenic activity of these toxins. Based on several mutagenic studies, a general mode of TCR binding has been elucidated. The binding of SEA, SEB, and SEC to the TCR was shown to occur through the shallow cavity at the top of the molecule.

SEs have evolved several distinct modes of interaction with the MHC class II molecules. The information on the SE-MHC interaction was provided by the crystal structure of SE complexed with MHC class II. SEA contains two MHC class II binding sites (Fig. 3.4). The zinc-dependent site located in domain A is the major interaction region, and several important residues (H187, H225, and D227) for binding to MHC class II were identified by mutagenesis (Hu et al., 2009, 2007; Schad et al., 1995). The second (minor) binding site of SEA is F47 located in domain B, which is not zinc-dependent (Fig. 3.4A). The cooperation between the two binding sites may be responsible for the high affinity of SEA for MHC class II. SED has also the potential to form a zinc-dependent binding site similar to that in SEA (Fraser, 1993; Hudson et al., 1993). However, no such site is predicted by the structure of SEB and SEC. SEB had a considerably higher affinity for the MHC class II molecule, which may well explain the difference in activity. Additional works have shown that mutation at the residue N23 of SEC2 and at residue 135 of TSST-1 results in greatly impaired SEC2 and TSST-1 superantigenic activity (Fig. 3.4B and C) (Hu et al., 2005, 2003).

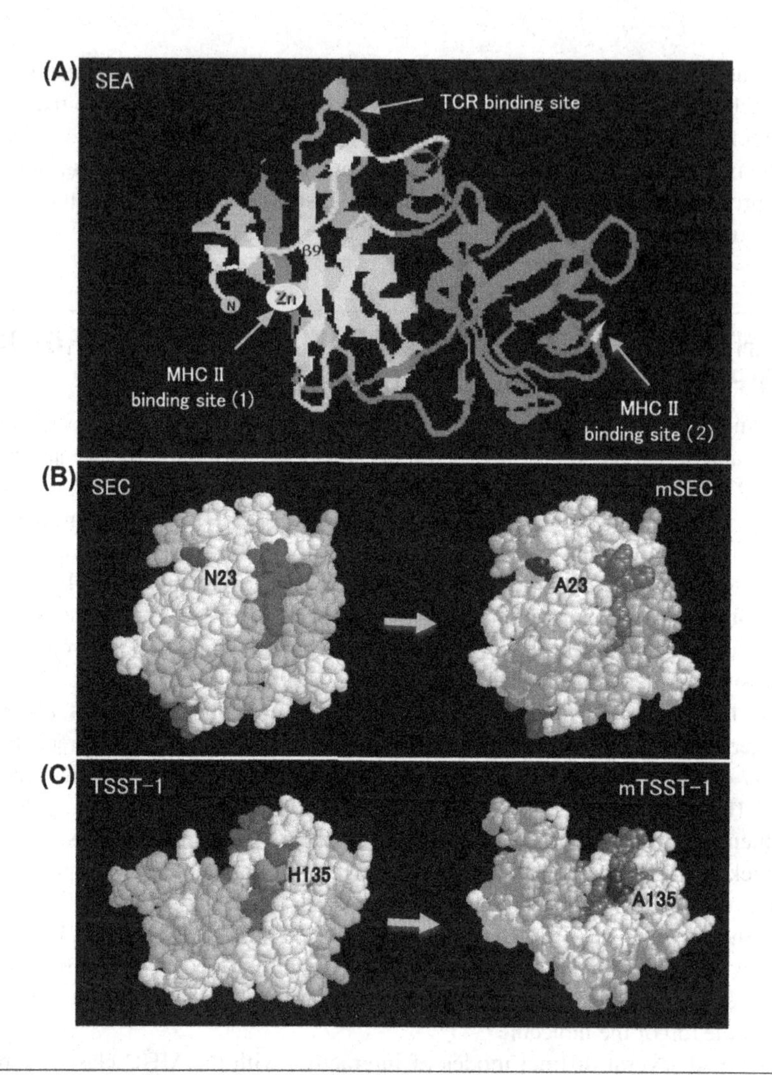

FIGURE 3.4

The important binding sites of staphylococcal enterotoxin A (SEA) to T cell receptor (TCR), CD28 of T cells and major histocompatibility class (MHC) II of antigen presenting cell (APC) (A). The binding sites of staphylococcal enterotoxin C (SEC) (B, left) and toxic shock syndrome toxin 1 (TSST-1) (C, left) to TCR, and the mutant sites of the SEC (B, right) and TSST-1 (C, right) on the three-dimensional structure of the toxins, respectively.

These pictures were created according to the sequences of the toxins and the results from Hu, D.-L., Cui, J.C., Omoe, K., Sashinami, H., Yokomizo, Y., Shinagawa, K., Nakane, A., 2005. A mutant of staphylococcal enterotoxin C devoid of bacterial superantigenic activiy elicits Th2 immune response for protection against Staphylococcus aureus *infection. Infect. Immun. 73, 174–180; Hu, D.-L., Omoe, K., Sashinami, H., Shinagawa, K., Nakane, A., 2009. Immunization with nontoxic mutant staphylococcal enterotoxin A, D227A, protects against enterotoxin-induced emesis in house musk shrews. J. Infect. Dis. 199, 302–310; Hu, D.-L., Omoe, K., Sasaki, S., Yokomizo, Y., Sashinami, H., Sakuraba, H., Shinagawa, K., Nakane, A., 2003. Vaccination with nontoxic TSST-1 protects* Staphylococcus aureus *infection. J. Infect. Dis. 188, 743–752.*

6. BIOLOGICAL CHARACTERISTICS AND EMETIC ACTIVITY OF STAPHYLOCOCCAL ENTEROTOXINS AND STAPHYLOCOCCAL ENTEROTOXIN-LIKE TOXINS

SEs and SEls share common biochemical and structural properties and are remarkably resistant to heat. The potency of these toxins can only be gradually decreased by prolonged boiling or autoclaving. They are highly stable and resistant to most proteolytic enzymes, and thus retain their activity in the digestive tract after ingestion. Several groups have compared the respective integrity and toxicity of SEA and TSST-1 after treatment with heat, pepsin, and trypsin in relation to the condition of food cooking or luminal location in stomach and intestine. The superantigenic and emetic activities of SEA showed marked resistance to heat treatment, pepsin, and trypsin digestion. Although SEA was degraded to smaller fragments after treatment with heat or pepsin, it retained significant superantigenic and emetic activity, indicating that toxin-contaminated foods, even they were cooked, may also induce food poisoning (Evans et al., 1983; Li et al., 2011).

Although the clinical manifestations are well known, the physiopathology of the symptoms and the mechanisms of SEs-induced emesis are only partially understood. The lack of progress in studying the mechanisms of the emetic activity of SEs can be partially attributed to the lack of convenient and appropriate animal models. Monkeys have been considered to be the primary animal models, but the use in investigating SEs is severely restricted by the high cost, the availability of the animals, and ethical considerations. Other animals such as mouse, rat, rabbit, and cat are less susceptible to SEs or their responses to SEs are not specific (Hu et al., 1999; Stiles and Denniston, 1971).

House musk shrew, *Suncus murinus*, has been described as a small emetic animal model for SEA (Hu et al., 1999). The emetic activity of SEA in house musk shrews was found to be dose-dependent and the SEA-induced emesis in the animals could be neutralized by anti-SEA antibody (Hu et al., 1998). Hu et al. further investigated the emetic response of the animals to other classic and several newly discovered SEls. After intraperitoneal administration, all of the tested SEs and SEls caused vomiting responses in the house musk shrews (Ono et al., 2015; Omoe et al., 2013; Hu et al., 2003). It is noteworthy that different types of toxins had different emetic activities in house musk shrews. It seems that the receptors exist in the gut of the house musk shrew, and that the differences in affinities among different types of SEs and their receptors might explain the different emetic activities of different types of SEs. Ono et al. (2012) investigated the behavior of SEA in the gastrointestinal tract in vivo, suggested that the submucosal mast cells in the gastrointestinal tract are one of the target cells of SEA and that serotonin (5-HT) released from submucosal mast cells plays an important role in SEA-induced emesis. The emesis in the house musk shrew is inhibited by the serotonin (5-HT) synthesis inhibitor and the 5-HT_3 receptor antagonist, showing the important role of 5-HT in SEA-induced emesis (Hu et al., 2007). In addition, SEA-induced emesis is blocked by surgical vagotomy in house musk shrews and primates (Hu et al., 2007; Sugiyama and Hayama, 1965), suggesting that 5-HT released from submucosal mast cells may bind to the 5-HT_3 receptor expressed on enteric nerves in the gastrointestinal tract and thereby induce the depolarization of these nerves.

SEA did not behave as a bacterial cytotoxin to intestinal epithelial cells in morphologic feature (Buxser and Bonventre, 1981), but could cross the intestinal epithelium in immunologically intact form and participate in the initiation, exaggeration, or reactivation of enteric inflammatory disease (McKay and Singh, 1997; McKay et al., 2000). Hu et al. (2005) demonstrated that SEA could induce an increase in intracellular calcium ($[Ca^{2+}]_i$) in intestinal epithelial cells, and the increase is originated from intracellular calcium stores (Hu and Nakane, 2014). These results may have important implications for the pathogenesis of food poisoning.

Recently, many new SEs and SEls have been reported, although the role of the toxins in food poisoning remains unclear. The emetic responses induced by other classic SEs including SEB, SEC2, SED, and SEE, and some nonclassic SEs such as SEG, SEH, and SEI in the house musk shrew were investigated. SEA, SEE, and SEI showed higher emetic activity in the house musk shrew than the other SEs. SEB, SEC2, SED, SEG, and SEH also induced emetic responses in this animal model but relatively high doses were required (Hu et al., 2003). More recently, the emetic potentials of SEK, SEL, SEM, SEN, SEO, SEP, and SEQ were assessed using a monkey-feeding assay. All the SEs that were tested induced emetic reactions in monkeys at a dose of 100 μg/kg, although the numbers of affected monkeys were significantly smaller than the numbers that were affected after consuming SEA or SEB (Omoe et al., 2013). These results suggest that the new SEs may also play some role in staphylococcal food poisoning.

7. CONCLUSION

The large family of SEs and SEls continues to grow. The most interesting question that remains in this field is why do *S. aureus* possess such a large, genetically and antigenically distinct, extremely potent, and seemingly redundant group of these toxins? SEs and SEls function both as potent gastrointestinal toxins as well as superantigenic toxins. It is still unknown how these toxins enter the body via the intestine, induce emetic responses in humans and animals, and what is the receptor of target cells in intestine and/or neuron system for SE-induced food poisoning? It is clear that these remarkable toxins represent a highly unique and well-adapted virulence factor, although the evolutionary function of these toxins in the life cycle of *S. aureus* still remains unclear. Continued efforts into understanding the complex biology and the relationship of the different functions of SEs and SEls superfamily will undoubtedly answer many of these questions (Williams et al., 2000).

REFERENCES

Arad, G., Levy, R., Nasie, I., Hillman, D., Rotfogel, Z., Barash, U., Supper, E., Shpilka, T., Minis, A., Kaempfer, R., 2012. Binding of superantigen toxins into CD28 homodimer interface is essential for induction of cytokine genes that mediate lethal shock. PLoS Biol. 9, e1001149.

Acharya, K.R., Passalacquam, E.F., Jones, E.Y., Harlos, K., Stuart, D.I., Brehm, R.D., Tranter, H.S., 1994. Structural basis of superantigen action inferred from crystal structure of toxic-shock syndrome toxin-1. Nature 367, 94–97.

Balaban, N., Rasooly, A., 2000. Staphylococcal enterotoxins. Int. J. Food Microbiol. 61, 1–10.

Bergdoll, M.S., Surgalla, M.J., Dack, G.M., 1959. Staphylococcal enterotoxin: identification of a specific precipitating antibody with enterotoxin neutralizing property. J. Immunol. 83, 334–338.

Baba, T., Takeuchi, F., Kuroda, M., Yuzawa, H., Aoki, K., Oguchi, A., Nagai, Y., Iwama, N., Asano, K., Naimi, T., Kuroda, H., Cui, L., Yamamoto, K., Hiramatsu, K., 2002. Genome and virulence determinants of high virulence community-acquired MRSA. Lancet 359, 1819–1827.

Baba, T., Bae, T., Schneewind, O., Takeuchi, F., Hiramatsu, K., 2008. Genome sequence of *Staphylococcus aureus* strain Newman and comparative analysis of staphylococcal genomes: polymorphism and evolution of two major pathogenicity islands. J. Bacteriol. 190, 300–310.

Betley, M.J., Mekalanos, J.J., 1985. Staphylococcal enterotoxin A is encoded by phage. Science 229, 185–187.

Bohach, G.A., Handley, J.P., Schlievert, P.M., 1989. Biological and immunological properties of the carboxyl terminus of staphylococcal enterotoxin C1. Infect. Immun. 57, 23–28.

Binek, M., Newcomb, J.R., Rogers, C.M., Rogers, T.J., 1992. Localisation of the mitogenic epitope of staphylococcal enterotoxin B. J. Med. Microbiol. 36, 156–163.

Brosnahan, A.J., Schlievert, P.M., 2011. Gram positive bacterial superantigen outside-in signaling causes toxic shock syndrome. FEBS J. 278, 4649–4667.

Buxser, S., Bonventre, P.F., 1981. Staphylococcal enterotoxins fail to disrupt membrane integrity or synthetic functions of Henle 407 intestinal cells. Infect. Immun. 31, 929–934.

Casman, E.P., 1960. Further serological studies of staphylococcal enterotoxin. J. Bacteriol. 79, 849–856.

Collery, M.M., Smyth, D.S., Tumilty, J.J., Twohig, J.M., Smyth, C.J., 2009. Associations between enterotoxin gene cluster types *egc1*, *egc2* and *egc3*, *agr* types, enterotoxin and enterotoxin-like gene profiles, and molecular typing characteristics of human nasal carriage and animal isolates of *Staphylococcus aureus*. J. Med. Microbiol. 58, 13–25.

Coleman, D.C., Sullivan, D.J., Russell, R.J., Arbuthnott, J.P., Carey, B.F., Pomeroy, H.M., 1989. *Staphylococcus aureus* bacteriophages mediating the simultaneous lysogenic conversion of beta-lysin, staphylokinase and enterotoxin A: molecular mechanism of triple conversion. J. Gen. Microbiol. 135, 1679–1697.

Couch, J.L., Soltis, M.T., Betley, M.J., 1988. Cloning and nucleotide sequence of the type E staphylococcal enterotoxin gene. J. Bacteriol. 170, 2954–2960.

Ezepchuk, Y.V., Noskov, A.N., 1986. NH2 terminal localization of that part of the staphylococcal enterotoxins polypept ide chain responsible for binding with membrane receptor and mitogenic effect. Int. J. Biochem. 18, 485–488.

Evans, J.B., Ananaba, G.A., Pate, C.A., Bergdoll, M.S., 1983. Enterotoxin production by atypical *Staphylococcus aureus* from poultry. J. Appl. Bacteriol. 54, 257–261.

Fitzgerald, J.R., Reid, S.D., Ruotsalainen, E., Tripp, T.J., Liu, M., Cole, R., Kuusela, P., Schlievert, P.M., Järvinen, A., Musser, J.M., 2003. Genome diversification in *Staphylococcus aureus*: molecular evolution of a highly variable chromosomal region encoding the staphylococcal exotoxin-like family of proteins. Infect. Immun. 71, 2827–2838.

Faulkner, L., Cooper, A., Fantino, C., Altmann, D.M., Sriskandan, S., 2005. The mechanism of superantigen-mediated toxic shock: not a simple Th1 cytokine storm. J. Immunol. 175, 6870–6877.

Fraser, J.D., 1993. T-cell recognition of superantigens. Res. Immunol. 144, 173–174.

Goerke, C., Pantucek, R., Holtfreter, S., Schulte, B., Zink, M., Grumann, D., Bröker, B.M., Doskar, J., Wolz, C., 2009. Diversity of prophages in dominant *Staphylococcus aureus* clonal lineages. J. Bacteriol. 191, 3462–3468.

Gunther, S., Varma, A.K., Moza, B., Kasper, K.J., Wyatt, A.W., Zhu, P., Rahman, A.K., Li, Y., Mariuzza, R.A., McCormick, J.K., Sundberg, E.J., 2007. A novel loop domain in superantigens extends their T cell receptor recognition site. J. Mol. Biol. 371, 210–221.

Griggs, N.D., Pontzer, C.H., Jarpe, M.A., Johnson, H.M., 1992. Mapping of multiple binding domains of the superantigen staphylococcal enterotoxin A for HLA. J. Immunol. 148, 2516–2521.

Hovde, C.J., Marr, J.C., Hoffmann, M.L., Hackett, S.P., Chi, Y.I., Crum, K.K., Stevens, D.L., Stauffacher, C.V., Bohach, G.A., 1994. Investigation of the role of the disulphide bond in the activity and structure of staphylococcal enterotoxin C1. Mol. Microbiol. 13, 897–909.

Hu, D.-L., Imai, A., Ono, K., Sasaki, S., Nakane, A., Sugii, S., Shinagawa, K., 1998. Epitope analysis of staphylococcal enterotoxin A using different synthetic peptides. J. Vet. Med. Sci. 60, 993–996.

Hu, D.-L., Omoe, K., Nakane, A., Sugii, S., Ono, K., Sasaki, S., Shinagawa, K., 1999. Studies on the functional site on staphylococcal enterotoxin A responsible for production of murine gamma interferon. FEMS Immunol. Med. Microbiol. 25, 237–244.

Hu, D.-L., Cui, J.C., Omoe, K., Sashinami, H., Yokomizo, Y., Shinagawa, K., Nakane, A., 2005. A mutant of staphylococcal enterotoxin C devoid of bacterial superantigenic activiy elicits Th2 immune response for protection against *Staphylococcus aureus* infection. Infect. Immun. 73, 174–180.

Hu, D.-L., Omoe, K., Narita, K., Cui, J.C., Shinagawa, K., Nakane, A., 2006. Intranasal vaccination with a double mutant of staphylococcal enterotoxin C provides protection against *Staphylococcus aureus* infection. Microbes Infect. 8, 2841–2848.

Hu, D.-L., Omoe, K., Sashinami, H., Shinagawa, K., Nakane, A., 2009. Immunization with nontoxic mutant staphylococcal enterotoxin A, D227A, protects against enterotoxin-induced emesis in house musk shrews. J. Infect. Dis. 199, 302–310.

Hu, D.-L., Zhu, G., Mori, F., Omoe, K., Kaneko, S., Wakabayashi, K., Shinagawa, K., Nakane, A., 2007. Staphylococcal enterotoxin induces emesis through increasing serotonin release in intestine and it is downregulated by cannabinoid receptor 1. Cell. Microbiol. 9 (9), 2267–2277.

Hudson, K.R., Robinson, H., Fraser, J.D., 1993. Two adjacent residues in staphylococcal enterotoxins A and E determine T cell receptor V beta specificity. J. Exp. Med. 177, 175–184.

Hu, D.-L., Omoe, K., Sasaki, S., Yokomizo, Y., Sashinami, H., Sakuraba, H., Shinagawa, K., Nakane, A., 2003. Vaccination with nontoxic TSST-1 protects *Staphylococcus aureus* infection. J. Infect. Dis. 188, 743–752.

Hu, D.-L., Omoe, K., Shimura, H., Ono, K., Sugii, S., Shinagawa, K., 1999. Emesis in the shrew mouse (*Suncus murinus*) induced by peroral and intraperitoneal administration of staphylococcal enterotoxin A. J. Food Prot. 62, 1350–1353.

Hu, D.-L., Omoe, K., Shimoda, Y., Nakane, A., Shinagawa, K., 2003. Induction of emetic response to staphylococcal enterotoxins in the house musk shrew (*Suncus murinus*). Infect. Immun. 71, 567–570.

Hu, D.-L., Suga, S., Omoe, K., Abe, Y., Shinagawa, K., Wakui, M., Nakane, A., 2005. Staphylococcal enterotoxin A modulates intracellular Ca^{2+} signal pathway in human intestinal epithelial cells. FEBS Lett. 579, 4407–4412.

Hu, D.-L., Nakane, A., 2014. Mechanisms of staphylococcal enterotoxin-induced emesis. Eur. J. Pharm. 722, 95–107.

Irwin, M.J., Hudson, K.R., Fraser, J.D., Gascoigne, N.R., 1992. Enterotoxin residues determining T cell receptor V beta binding specificity. Nature 359, 841–843.

Jardetzky, T.S., Brown, J.H., Gorga, J.C., Stern, L.J., Urban, R.G., Chi, Y.I., Stauffacher, C., Strominger, J.L., Wiley, D.C., 1994. Three-dimensional structure of a human class II histocompatibility molecule complexed with superantigen. Nature 368, 711–718.

Jupin, C., Anderson, S., Damais, C., Alouf, J.E., Parant, M., 1988. Toxic shock syndrome toxin 1 as an inducer of human tumor necrosis factors and gamma interferon. J. Exp. Med. 167, 752–761.

Kim, J., Urban, R.G., Strominger, J.L., Wiley, D.C., 1994. Toxic shock syndrome toxin-1 complexed with a class II major histocompatibility molecule HLA-DR1. Science 266, 1870–1874.

Kotzin, B.L., Leung, D.Y.M., Kappler, J., Marrack, P., 1993. Superantigens and their potential role in human disease. Adv. Immunol. 54, 99–166.

Kozono, H., Parker, D., White, J., Marrack, P., Kappler, J., 1995. Multiple binding sites for bacterial superantigens on soluble class II MHC molecules. Immunity 3, 187–196.

Krakauer, T., 1994. Co-stimulatory receptors for the superantigen staphyloccoccal enterotoxin B on human vascular endothelial cells and T cells. J. Leukoc. Biol. 56, 458–463.

Kappler, J.W., Herman, A., Clements, J., Marrack, P., 1992. Mutations defining functional regions of the superantigen staphylococcal enterotoxin B. J. Exp. Med. 175, 387–396.

Lina, G., Bohach, G.A., Nair, S.P., Hiramatsu, K., Jouvin-Marche, E., Mariuzza, R., 2004. Standard nomenclature for the superantigens expressed by Staphylococcus. J. Infect. Dis. 189, 2334–2336.

Lindsay, J.A., Ruzin, A., Ross, H.F., Kurepina, N., Novick, R.P., 1998. The gene for toxic shock toxin is carried by a family of mobile pathogenicity islands in *Staphylococcus aureus*. Mol. Microbiol. 29, 527–543.

Letertre, C., Perelle, S., Dilasser, F., Fach, P., 2003. Identification of a new putative enterotoxin SEU encoded by the *egc* cluster of *Staphylococcus aureus*. J. Appl. Microbiol. 95, 38–43.

Leder, L., Llera, A., Lavoie, P.M., Lebedeva, M.I., Li, H., Sekaly, R.P., Bohach, G.A., Gahr, P.J., Schlievert, P.M., Karjalainen, K., Mariuzza, R.A., 1998. A mutational analysis of the binding of staphylococcal enterotoxins B and C3 to the T cell receptor beta chain and major histocompatibility complex class II. J. Exp. Med. 187, 823–833.

Li, H., Llera, A., Tsuchiya, D., Leder, L., Ysern, X., Schlievert, P.M., Karjalainen, K., Mariuzza, R.A., 1998. Three-dimensional structure of the complex between a T cell receptor beta chain and the superantigen staphylococcal enterotoxin B. Immunity 9, 807–816.

Li, S.J., Hu, D.-L., Maina, E.K., Shinagawa, K., Omoe, K., Nakane, A., 2011. Superantigenic activity of toxic shock syndrome toxin-1 is resistant to heating and digestive enzymes. J. Appl. Microbiol. 110, 729–736.

Monday, S.R., Bohach, G.A., 2001. Genes encoding staphylococcal enterotoxins G and I are linked and separated by DNA related to other staphylococcal enterotoxins. J. Nat. Toxins 10, 1–8.

McCormick, J.K., Yarwood, J.M., Schlievert, P.M., 2001. Toxic shock syndrome and bacterial superantigens: an update. Annu. Rev. Microbiol. 55, 77–104.

Mitchell, D.T., Levitt, D.G., Schlievert, P.M., Ohlendorf, D.H., 2000. Structural evidence for the evolution of pyrogenic toxin superantigens. J. Mol. Evol. 51, 520–531.

Moza, B., Varma, A.K., Buonpane, R.A., Zhu, P., Herfst, C.A., Nicholson, M.J., Wilbuer, A.K., Seth, N.P., Wucherpfennig, K.W., McCormick, J.K., Kranz, D.M., Sundberg, E.G., 2007. Structural basis of T-cell specificity and activation by the bacterial superantigen TSST-1. EMBO J. 26, 1187–1197.

McKay, D.M., Singh, P.K., 1997. Superantigen activation of immune cells evokes epithelial (T84) transport and barrier abnormalities via IFN-gamma and TNF alpha: inhibition of increased permeability, but not diminished secretory responses by TGF-beta2. J. Immunol. 159, 2382–2390.

McKay, D.M., Botelho, F., Ceponis, P.J., Richards, C.D., 2000. Superantigen immune stimulation activates epithelial STAT-1 and PI3-K: PI3-K regulation of permeability. Am. J. Physiol. Gastrointest. Liver Physiol. 279, G1094–G1103.

Novick, R.P., Subedi, A., 2007. The SaPIs: mobile pathogenicity islands of *Staphylococcus*. Chem. Immunol. Allergy 93, 42–57.

Noto, M.J., Archer, G.L., 2006. A subset of *Staphylococcus aureus* strains harboring staphylococcal cassette chromosome mec (SCCmec) type IV is deficient in CcrAB-mediated SCCmec excision. Antimicrob. Agents Chemother. 50, 2782–2788.

Omoe, K., Hu, D.-L., Takahashi-Omoe, H., Nakane, A., Shinagaw, K., 2003. Identification and characterization of a new staphylococcal enterotoxin-related putative toxin encoded by two kinds of plasmids. Infect. Immun. 71, 6088–6094.

Ono, H.K., Omoe, K., Imanishi, K., Iwakabe, Y., Hu, D.-L., Kato, H., Saito, N., Nakane, A., Uchiyama, T., Shinagawa, K., 2008. Identification and characterization of two novel staphylococcal enterotoxins, types S and T. Infect. Immun. 76, 4999–5005.

Ono, H.K., Sato'o, Y., Narita, K., Naito, I., Hirose, S., Hisatsune, J., Asano, K., Hu, D.-L., Omoe, K., Sugai, M., Nakane, A., 2015. Identification and characterization of a novel staphylococcal emetic toxin. Appl. Environ. Microbiol. 8, 7034–7040.

Omoe, K., Hu, D.-L., Ono, H.K., Shimizu, S., Takahashi-Omoe, H., Nakane, A., Uchiyama, T., Shinagawa, K., Imanishi, K., 2013. Emetic potentials of newly identified staphylococcal enterotoxin-like toxins. Infect. Immun. 81, 3627–3631.

Omoe, K., Hu, D.-L., Takahashi-Omoe, H., Nakane, A., Shinagawa, K., 2005. Comprehensive analysis of classical and newly described staphylococcal superantigenic toxin genes in *Staphylococcus aureus* isolates. FEMS Microbiol. Lett. 246, 191–198.

Orwin, P.M., Leung, D.Y., Donahue, H.L., Novick, R.P., Schlievert, P.M., 2001. Biochemical and biological properties of staphylococcal enterotoxin K. Infect. Immun. 69, 360–366.

Orwin, P.M., Leung, D.Y., Tripp, T.J., Bohach, G.A., Earhart, C.A., Ohlendorf, D.H., Schlievert, P.M., 2002. Characterization of a novel staphylococcal enterotoxin-like superantigen, a member of the group V subfamily of pyrogenic toxins. Biochemistry 41, 14033–14040.

Orwin, P.M., Fitzgerald, J.R., Leung, D.Y., Gutierrez, J.A., Bohach, G.A., Schlievert, P.M., 2003. Characterization of *Staphylococcus aureus* enterotoxin L. Infect. Immun. 71, 2916–2919.

Ono, H., Nishizawa, M., Yamamoto, Y., Hu, D.-L., Nakane, A., Shinagawa, K., Omoe, K., 2012. Submucosal mast cells in the gastrointestinal tract are a target of staphylococcal enterotoxin type A. FEMS Immun. Med. Microbiol. 64, 392–402.

Prasad, G.S., Earhart, C.A., Murray, D.L., Novick, R.P., Schlievert, P.M., Ohlendorf, D.H., 1993. Structure of toxic shock syndrome toxin 1. Biochemistry 32, 13761–13766.

Petersson, K., Thunnissen, M., Forsberg, G., Walse, B., 2002. Crystal structure of a SEA variant in complex with MHC class II reveals the ability of SEA to crosslink MHC molecules. Structure 10, 1619–1626.

Papageorgiou, A.C., Acharya, K.R., 2000. Microbial superantigens: from structure to function. Trends Microbiol. 8, 369–375.

Pontzer, C.H., Russell, J.K., Johnson, H.M., 1991. Structural basis for differential binding of staphylococcal enterotoxin A and toxic shock syndrome toxin 1 to class II major histocompatibility molecules. Proc. Natl. Acad. Sci. U.S.A. 88, 125–128.

Sumby, P., Waldor, M.K., 2003. Transcription of the toxin genes present within the staphylococcal phage phiSa3ms is intimately linked with the phage's life cycle. J. Bacteriol. 185, 6841–6851.

Sato'o, Y., Omoe, K., Ono, H.K., Nakane, A., Hu, D.-L., 2013. A novel comprehensive analysis method for *Staphylococcus aureus* pathogenicity islands. Microbiol. Immunol. 57, 91–99.

Sato'o, Y., Omoe, K., Naito, I., Ono, H.K., Nakane, A., Sugai, M., Yamagishi, N., Hu, D.-L., 2014. Molecular epidemiology and identification of a *Staphylococcus aureus* clone causing food poisoning outbreaks in Japan. J. Clin. Microbiol. 52, 2637–2640.

Schlievert, P.M., Jablonski, L.M., Roggiani, M., Sadler, I., Callantine, S., Mitchell, D.T., Ohlendorf, D.H., Bohach, G.A., 2000. Pyrogenic toxin superantigen site specificity in toxic shock syndrome and food poisoning in animals. Infect. Immun. 68, 3630–3634.

Spaulding, A.R., Salgado-Pabón, W., Kohler, P.L., Horswill, A.R., Leung, D.Y., Schlievert, P.M., 2013. Staphylococcal and streptococcal superantigen exotoxins. Clin. Microbiol. Rev. 26 (3), 422–447.

Su, Y.C., Wong, A.C., 1995. Identification and purification of a new staphylococcal enterotoxin, H. Appl. Environ. Microbiol. 61, 1438–1443.

Swaminathan, S., Furey, W., Pletcher, J., Sax, M., 1992. Crystal structure of staphylococcal enterotoxin B, a superantigen. Nature 359, 801–806.

Swaminathan, S., Furey, W., Pletcher, J., Sax, M., 1995. Residues defining V beta specificity in staphylococcal enterotoxins. Nat. Struct. Biol. 8, 680–686.

Seth, A., Stern, L.J., Ottenhoff, T.H., Engel, I., Owen, M.J., Lamb, J.R., Klausner, R.D., Wiley, D.C., 1994. Binary and ternary complexes between T-cell receptor, class II MHC and superantigen in vitro. Nature 369, 324–327.

Spero, L., Morlock, B.A., 1978. Biological activities of the peptides of staphylococcal enterotoxin C formed by limited tryptic hydrolysis. J. Biol. Chem. 253, 8787–8791.

Schad, E.M., Zaitseva, I., Zaitsev, V.N., Dohlsten, M., Kalland, T., Schlievert, P.M., Ohlendorf, D.H., Svensson, L.A., 1995. Crystal structure of the superantigen staphylococcal enterotoxin type A. EMBO J. 14, 3292–3301.

Stiles, J.W., Denniston, J.C., 1971. Response of the rhesus monkey, Macaca mulatta, to continuously infused staphylococcal enterotoxin B. Lab. Invest. 25, 617–625.

Sugiyama, H., Hayama, T., 1965. Abdominal viscera as site of emetic action for staphylococcal enterotoxin in monkey. J. Infect. Dis. 115, 330–336.

Todd, J., Fishaut, M., Kapral, F., Welch, T., 1978. Toxic-shock syndrome associated with phage-group-I Staphylococci. Lancet. 2, 1116–1118.

Thomas, D., Chou, S., Dauwalder, O., Lina, G., 2007. Diversity in *Staphylococcus aureus* enterotoxins. Chem. Immunol. Allergy 93, 24–41.

Thomas, D.Y., Jarraud, S., Lemercier, B., Cozon, G., Echasserieau, K., Etienne, J., Gougeon, M.L., Lina, G., Vandenesch, F., 2006. Staphylococcal enterotoxin-like toxins U2 and V, two new staphylococcal superantigens arising from recombination within the enterotoxin gene cluster. Infect. Immun. 74, 4724–4734.

Tallent, S.M., Langston, T.B., Moran, R.G., Christie, G.E., 2007. Transducing particles of *Staphylococcus aureus* pathogenicity island SaPI1 are comprised of helper phage-encoded proteins. J. Bacteriol. 189, 7520–7524.

Trede, N.S., Geha, R.S., Chatila, T., 1991. Transcriptional activation of IL-1 beta and tumor necrosis factor-alpha genes by MHC class II ligands. J. Immunol. 146, 2310–2315.

Wilson, G.J., Seo, K.S., Cartwright, R.A., Connelley, T., Chuang-Smith, O.N., Merriman, J.A., Guinane, C.M., Park, J.Y., Bohach, G.A., Schlievert, P.M., Morrison, W.I., Fitzgerald, J.R., 2011. A novel core genome-encoded superantigen contributes to lethality of community-associated MRSA necrotizing pneumonia. PLoS Pathog. 7, e1002271.

Williams, R.J., Ward, J.M., Henderson, B., Poole, S., O'Hara, B.P., Wilson, M., Nair, S.P., 2000. Identification of a novel gene cluster encoding staphylococcal exotoxin-like proteins: characterization of the prototypic gene and its protein product, SET1. Infect. Immun. 68, 4407–4415.

CHAPTER

ANTIMICROBIAL RESISTANCE PROPERTIES OF *STAPHYLOCOCCUS AUREUS*

4

Andrea T. Feßler[3], Jun Li[1,2], Kristina Kadlec[1], Yang Wang[2], Stefan Schwarz[3]

[1]Friedrich-Loeffler-Institut (FLI), Neustadt-Mariensee, Germany; [2]College of Veterinary Medicine, China Agricultural University, Beijing, P. R. China; [3]Freie Universität Berlin, Berlin, Germany

1. INTRODUCTION

During the last decades, several articles and book chapters focussed on the genetic basis of antimicrobial resistance in *Staphylococcus aureus*. Some articles included staphylococci of human and animal origin (Lyon and Skurray, 1987; Jensen and Lyon, 2009; Schwarz et al., 2011; Wendland et al., 2013a), whereas others focussed on either staphylococci of animal origin in general (Aarestrup and Schwarz, 2006) or on specific animal-associated staphylococci such as bovine *S. aureus*, among others, (Werckenthin et al., 2001) or livestock-associated methicillin-resistant *Staphylococcus aureus* (LA-MRSA) (Kadlec et al., 2012b). *S. aureus* isolates do not live in genetic isolation on the skin or the mucosal surfaces of humans and animals but are in close contact with other bacteria in polymicrobial environments of the same host. In addition, *S. aureus* may also be exchanged between humans and/or animals either by direct contact, inhalation of *S. aureus*–containing aerosols and dust, contact with *S. aureus*-containing excretions, or via handling or ingesting *S. aureus*–contaminated food (Schwarz et al., 2017). Previous work on the resistome of *S. aureus* revealed that (1) *S. aureus* can act as donor and recipient of resistance genes, (2) many resistance genes found in *S. aureus* are located on mobile genetic elements, and (3) the acquisition of novel resistance genes by *S. aureus* is a continuous process that results from its interaction with other bacteria (Kadlec et al., 2012b; Wendlandt et al., 2013a).

This chapter is based on some previous reviews (Wendlandt et al., 2013a; Schwarz et al., 2011, 2014, 2016) and summarizes the latest information on resistance genes and resistance-mediating mutations detected in *S. aureus*. Preference is given to those resistance genes and mutations for which nucleotide sequence data have been deposited in the databases.

2. β-LACTAM RESISTANCE

Two different resistance mechanisms mainly account for β-lactam resistance in *S. aureus*: (1) enzymatic inactivation by a *blaZ*-encoded β-lactamase, which confers resistance to penicillins except isoxazolyl-penicillins, or (2) target site replacement by the gene products of the *mecA* or *mecC* (formerly known as *mecA*$_{LGA251}$) genes conferring resistance to virtually all β-lactams and

Staphylococcus aureus. http://dx.doi.org/10.1016/B978-0-12-809671-0.00004-8

Table 4.1 Resistance Genes Detected in *Staphylococcus aureus*

Resistance to …	Mechanism	Enzyme	Gene	Localization	Mobile Genetic Element	Accession Number or References
β-Lactams	Enzymatic inactivation	β-Lactamase	*blaZ*	Tn, P::Tn, C::Tn	Tn*552*	X52734
β-Lactams	Target site replacement	Alternative penicillin-binding protein	*mecA*	C	SCC*mec*	AB033763
β-Lactams	Target site replacement	Alternative penicillin-binding protein	*mecC* ($mecA_{LGA251}$)	C	SCC*mec*	FR823292 FR821779
Tetracyclines (except minocycline and glycylcyclines)	Active efflux	MFS exporter	*tet*(K)	P, C::P	pT181	J01764.1 S67449
Tetracyclines (except minocycline and glycylcyclines)	Active efflux	MFS exporter	*tet*(L)	P	pKKS2187	FM207105.1
Tetracyclines (except minocycline and glycylcyclines)	Active efflux	MFS exporter	*tet*(38)	C	Unknown	AY825285
Tetracyclines (including minocycline but excluding glycylcyclines)	Target site protection	Ribosome-protective protein	*tet*(M)	Tn, C::Tn, P::Tn	Tn*916*	M21136.1
Tetracyclines (including minocycline but excluding glycylcyclines)	Target site protection	Ribosome-protective protein	*tet*(S)	Unknown	Unknown	Guran and Kahya (2015)
Tetracyclines (including minocycline but excluding glycylcyclines)	Target site protection	Ribosome-protective protein	*tet*(W)	Unknown	Unknown	Guran and Kahya (2015)
Nonfluorinated phenicols	Enzymatic inactivation	Acetyltransferase	cat_{pC194}	P	pC194	NC_002013

Nonfluorinated phenicols	Enzymatic inactivation	Acetyltransferase	*cat*$_{pC221}$	P	pC221	NC_006977
Nonfluorinated phenicols	Enzymatic inactivation	Acetyltransferase	*cat*$_{pC223}$	P	pC223	NC_005243
All phenicols	Active efflux	MFS exporter	*fexA*	Tn, P::Tn, C::Tn	Tn*558*	AM086211.1
Aminoglycosides (gentamicin, kanamycin, tobramycin, amikacin)	Enzymatic inactivation	Acetyltransferase and phosphotransferase	*aacA-aphD*	Tn, P::Tn, C::Tn	Tn*4001*	GU565967
Aminoglycosides (kanamycin, neomycin, tobramycin)	Enzymatic inactivation	Adenyltransferase	*aadD*	P, C::P	pUB110 pKKS825	M37273.1 FN377602.2
Aminoglycosides (streptomycin)	Enzymatic inactivation	Adenyltransferase	*aadE*	Tn, P::Tn, C::Tn	Tn*5405*	AB699882.1
Aminoglycosides (streptomycin)	Enzymatic inactivation	Adenyltransferase	*str*	P	pS194 pS0385-2	NC_005564 AM990994.1
Aminoglycosides (kanamycin, neomycin, amikacin)	Enzymatic inactivation	Phosphotransferase	*aphA3*	Tn, P::Tn, C::Tn	Tn*5405*	AB699882.1
Aminocyclitols/aminoglycosides (apramycin, decreased susceptibility to gentamicin)	Enzymatic inactivation	Acetyltransferase	*apmA*	P	pAFS11 pKKS49	FN806789.3 HE611647.1
Streptothricins	Enzymatic inactivation	Acetyltransferase	*sat4*	Tn, P::Tn	Tn*5405*	AB699882.1
Aminocyclitols (spectinomycin)	Enzymatic inactivation	Adenyltransferase	*spc*	Tn, P::Tn, C::Tn	Tn*554*	X03216.1
Aminocyclitols (spectinomycin)	Enzymatic inactivation	Adenyltransferase	*spd*	P	pDJ91S	KC895984
Aminocyclitols (spectinomycin)	Enzymatic inactivation	Adenyltransferase	*spw*	P, C::P	pV7037	JX560992.1
Macrolides	Enzymatic inactivation	Esterase	*ere*(A)	Unknown	Unknown	Li et al. (2015b)
Macrolides	Enzymatic inactivation	Esterase	*ere*(B)	Unknown	Unknown	Schmitz et al., 2002

Continued

Table 4.1 Resistance Genes Detected in *Staphylococcus aureus*—cont'd

Resistance to ...	Mechanism	Enzyme	Gene	Localization	Mobile Genetic Element	Accession Number or References
Macrolides, lincosamides, streptogramin B	Target site modification	rRNA methylase	*erm*(A)	Tn, P::Tn, C::Tn	Tn*554*	X03216.1
Macrolides, lincosamides, streptogramin B	Target site modification	rRNA methylase	*erm*(B)	Tn, P::Tn, C::Tn	Tn*917*	FN806789.3
Macrolides, lincosamides, streptogramin B	Target site modification	rRNA methylase	*erm*(C)	P	pE194, pNE131	V01278.1 M12730.1
Macrolides, lincosamides, streptogramin B	Target site modification	rRNA methylase	*erm*(F)	Unknown	unknown	Luna et al. (2002)
Macrolides, lincosamides, streptogramin B	Target site modification	rRNA methylase	*erm*(T)	P, C::P	pKKS25	FN390947.1
Macrolides, lincosamides, streptogramin B	Target site modification	rRNA methylase	*erm*(Y)	P	pMS97	AB179623
Macrolides, lincosamides, streptogramin B	Target site modification	rRNA methylase	*erm*(33)	Unknown	Unknown	Li et al. (2015a)
Macrolides, streptogramin B	Active efflux	ABC transporter	*msr*(A)	C, P	pMS97 pEP2104	AB179623 NG_048001.1
Macrolides	Enzymatic inactivation	Phosphotransferase	*mph*(C)	C, P	pMS97	AB179623 AB013298.1
Macrolides	Active efflux	MFS exporter	*mef*(A)	Unknown	Unknown	Luna et al. (2002)
Lincosamides	Enzymatic inactivation	Nucleotidyltransferase	*lnu*(A)	P	pBMSa1	AY541446 JQ861958.1
Lincosamides	Enzymatic inactivation	Nucleotidyltransferase	*lnu*(B)	P, C::P	pV7037	JX560992.1 JQ861959.1

Streptogramin A	Enzymatic inactivation	Acetyltransferase	*vat*(A)	P, C		L07778.1
Streptogramin A	Enzymatic inactivation	Acetyltransferase	*vat*(B)	P	pIP1156	U19459.1
Streptogramin A	Enzymatic inactivation	Acetyltransferase	*vat*(C)	P	pIP1714	AF015628.1
Streptogramin B	Enzymatic inactivation	Hydrolase	*vgb*(A)	P	pIP524	M20129.1
Streptogramin B	Enzymatic inactivation	Hydrolase	*vgb*(B)	P	pIP1714	AF015628.1
Lincosamides, pleuromutilins, streptogramin A	Active efflux	ABC transporter	*vga*(A)	Tn, P::Tn	Tn*5406*	M90056.1 FN806791.1
Lincosamides, pleuromutilins, streptogramin A	Active efflux	ABC transporter	*vga*(A)$_{LC}$	Unknown	Unknown	Qin et al. (2011)
Lincosamides, pleuromutilins, streptogramin A	Active efflux	ABC transporter	*vga*(A)$_{V}$	C, P	Unknown	AF186237
Lincosamides, pleuromutilins, streptogramin A	Active efflux	ABC transporter	*vga*(B)	P	pIP1633	U82085.2
Lincosamides, pleuromutilins, streptogramin A	Active efflux	ABC transporter	*vga*(C)	P	pKKS825	FN377602.2
Lincosamides, pleuromutilins, streptogramin A	Active efflux	ABC transporter	*vga*(E)	Tn, C::Tn	Tn*6133*	FR772051.1
Lincosamides	Active efflux	ABC transporter	*lsa*(B)	P	pSAMS13-0401	KU510528.1
Lincosamides, pleuromutilins, streptogramin A	Active efflux	ABC transporter	*lsa*(E)	P, C::P	pV7037	JX560992.1
All phenicols, lincosamides, oxazolidinones, pleuromutilins, streptogramin A	Target site modification	rRNA methylase	*cfr*	P, C	pSCFS1	AJ579365.1

Continued

Table 4.1 Resistance Genes Detected in *Staphylococcus aureus*—cont'd

Resistance to …	Mechanism	Enzyme	Gene	Localization	Mobile Genetic Element	Accession Number or References
Trimethoprim	Target replacement	Trimethoprim-resistant dihydrofolate reductase	*dfrA* (*dfrS1*)	Tn, P::Tn, C::Tn	Tn*4003*	GU565967.1
Trimethoprim	Target replacement	Trimethoprim-resistant dihydrofolate reductase	*dfrD*	C	Unknown	KX149097.1
Trimethoprim	Target replacement	Trimethoprim-resistant dihydrofolate reductase	*dfrG*	C	Unknown	NG_047756.1
Trimethoprim	Target replacement	Trimethoprim-resistant dihydrofolate reductase	*dfrK*	Tn, P, C::Tn	Tn*559*, pKKS2187	FN390947
Vancomycin	Target modification	Several proteins involved in synthesis of D-Ala-D-Lac depsipeptide low-affinity peptidoglycan precursors	*van*(A) gene cluster	Tn, P::Tn	Tn*1546*	AE017171
Bleomycin	Enzymatic inactivation	Bleomycin-binding protein	*ble*	P, C::P	pUB110	NC_001384.1
Fusidic acid	Target site protection	Ribosome-protective protein	*fusB*	P	p11819-97 pUB101	NC_017350.1 NC_005127.1
Fusidic acid	Target site protection	Ribosome-protective protein	*fusC*	P, C	SCC*mec*	HE980450.1
Mupirocin	Target replacement	Mupirocin-insensitive isoleucyl-tRNA synthase	*ileS2* (*mupA*, *ileS*)	P	pPR5	HQ625438.1

Modified from references Schwarz et al. (2014); Wendlandt et al. (2013a).

being the genetic basis of methicillin resistance in *S. aureus* (MRSA) (Table 4.1). The *blaZ-blaI-blaR1* operon has been identified on transposon Tn*552* (Rowland and Dyke, 1989). This transposon or parts thereof were detected on plasmids and in the chromosomal DNA of *S. aureus* from humans and animals (Lyon and Skurray, 1987; Jensen and Lyon, 2009; Wendland et al., 2013a). The *blaZ* operon was also present as additional β-lactam resistance operon in many MRSA from animals, such as pigs (Kadlec et al., 2009; Argudín et al., 2011; Overesch et al., 2011), cattle (Feβler et al., 2010), horses (Walther et al., 2009), donkeys (Gharsa et al., 2012a), sheep (Gharsa et al., 2012b), chickens, and turkeys (Monecke et al., 2012b).

The *mecA* and *mecC* genes are also located on a mobile genetic element. They are part of a mobile genetic island called the "Staphylococcal Cassette Chromosome *mec*" (SCC*mec*). Currently, at least 13 different major types of SCC*mec* elements plus various subtypes have been described in MRSA from humans and animals (IWG-SCC, 2009). The two *mec* genes (*mecA* and *mecC*) encode alternative penicillin-binding proteins, which exhibit a largely reduced affinity to almost all β-lactam antibiotics currently available for therapeutic applications in human and veterinary medicine. In contrast to the wealth of data about *mecA* genes (Peacock and Patterson, 2015), less information is currently available about the recently identified *mecC* gene (Becker et al., 2014). The *mecC* gene was initially found in *S. aureus* from humans and cattle (García-Álvarez et al., 2011; Shore et al., 2011). Further screening for this gene identified it in *S. aureus* from cattle in France (Laurent et al., 2012) and a cat in Norway (Medhus et al., 2013). A study from the United Kingdom identified it in *S. aureus* isolates from a dog, brown rats, a rabbit, a harbor seal, sheep, and a chaffinch (Paterson et al., 2012). In Germany, Walther et al. (2012) identified it in *S. aureus* from cats, dogs, and a guinea pig. The *mecC* gene was also present in *S. aureus* isolates from a wild bird (Robb et al., 2012), a hedgehog (Monecke et al., 2012a), and horses (Haenni et al., 2015). Numerous publications described the presence of MRSA strains with the same characteristics carrying either *mecA* (Feßler et al., 2010; Grøntvedt et al., 2016; Smith et al., 2013; van den Eede et al., 2013) or *mecC* (Benito et al., 2016; Drougka et al., 2016; Harrison et al., 2013) among humans and animals, suggesting a zoonotic transmission of these isolates.

3. TETRACYCLINE RESISTANCE

Resistance to tetracyclines in *S. aureus* is commonly mediated either by active efflux or ribosome-protective proteins. The mostly plasmid-borne genes *tet*(K) and *tet*(L) encode membrane-associated efflux proteins of the Major Facilitator Superfamily (MFS) (Table 4.1). The Tet(K) and Tet(L) proteins exhibit 14 transmembrane domains (Roberts, 1996). They can export tetracyclines except minocycline, whereas the most common ribosome-protective protein, Tet(M), also confers minocycline resistance. None of these genes mediates resistance to the glycylcycline tigecycline. The *tet*(K) gene is often located on small plasmids of about 4.5 kb, with pT181 being the prototype *tet*(K)-carrying plasmid (Khan and Novick, 1983). pT181-like plasmids have also been identified to be integrated via closely related insertion sequences of the types IS*257* or IS*431* into larger plasmids of *S. aureus* (Needham et al., 1994) or into type III or type V SCC*mec* cassettes of MRSA (Kondo et al., 2007; Li et al., 2011). In contrast to *tet*(K), the *tet*(L) gene is rarely seen in *S. aureus* (Schwarz et al., 1998), except in livestock-associated MRSA from pigs, cattle, and poultry (Kadlec et al., 2009; Feßler et al., 2010; Argudín et al., 2011; Monecke et al., 2012b). The *tet*(L) gene has also been detected mainly on plasmids

(Schwarz et al., 2014). In contrast to *tet*(K)-carrying plasmids, those carrying *tet*(L) are more diverse in size (5.5–>40 kb) and often harbor one or more additional resistance genes (Schwarz et al., 2014; Kadlec and Schwarz, 2009b, 2010a). The analysis of the whole genome sequence of *S. aureus* strain ISP794 identified the presence of another tetracycline efflux protein, designated Tet(38), which has about 46% similarity to the Tet(K) protein (Truong-Bolduc et al., 2005).

The *tet*(M) genes are commonly located on conjugative transposons of enterococcal origin, such as Tn*916* or Tn*1545* (Roberts, 1996), and have been found in MRSA and methicillin-susceptible *S. aureus* (MSSA) from humans (Lyon and Skurray, 1987; Jensen and Lyon, 2009). The *tet*(M) gene is frequently found not only in LA-MRSA from food-producing animals (Kadlec et al., 2009; Feßler et al., 2010; Argudín et al., 2011; Monecke et al., 2012b) but also in MSSA from chickens and turkeys (Monecke et al., 2012b). Recent studies of LA-MRSA from pigs, cattle, chickens, and ducks showed that such strains—regardless of their origin—often carried the genes *tet*(M), *tet*(K), and *tet*(L) in various combinations, with two or all three *tet* genes being present in the same isolate (Kadlec et al., 2009; Feßler et al., 2010; Argudín et al., 2011; Monecke et al., 2012b). Two additional *tet* genes, *tet*(S) and *tet*(W), both coding for ribosome-protective proteins, have been identified in *S. aureus* from ground meat (Guran and Kahya, 2015).

4. PHENICOL RESISTANCE

Resistance to phenicols in *S. aureus* can be due to enzymatic inactivation, active efflux, or target site modification. Enzymatic inactivation via acetylation is mainly mediated by one of three types of chloramphenicol acetyltransferases (Schwarz et al., 2004) (Table 4.1). The corresponding *cat* genes have been named according to the *S. aureus* plasmids on which they have been identified first: cat_{pC221}, cat_{pC223}, and cat_{pC194} (Horinouchi and Weisblum, 1982; Brenner and Shaw, 1985; Projan et al., 1985; Smith and Thomas, 2004). These *cat* genes code for Cat monomers of 215, 215, and 216 aa, respectively, which form the trimeric mature Cat enzyme (Schwarz et al., 2004). In *S. aureus* of animal origin, the cat_{pC221} and cat_{pC223} genes were detected on a variety of plasmids in bovine *S. aureus* (Cardoso and Schwarz, 1992a,b). Plasmids carrying the cat_{pC194} gene have so far only been found in *S. aureus* of human origin (Schwarz et al., 2004).

In contrast, the 475-aa FexA protein of the MFS, which consists of 14 transmembrane domains, exports not only nonfluorinated phenicols but also fluorinated phenicols such as florfenicol (Kehrenberg and Schwarz, 2004) (Table 4.1). The *fexA* gene is located on a small nonconjugative transposon, designated Tn*558* (Kehrenberg and Schwarz, 2005). It has also been detected in MRSA from pigs (Kadlec et al., 2009; Kehrenberg et al., 2009; Argudín et al., 2011; Wang et al., 2012b), cattle (Feßler et al., 2010), and a horse (Kehrenberg and Schwarz, 2006).

Phenicol resistance may also be based on target site modification via the gene *cfr*; for this, please see the subchapter on oxazolidinone resistance.

5. MACROLIDE–LINCOSAMIDE–STREPTOGRAMIN RESISTANCE

Resistance to macrolides, lincosamides, and streptogramin antibiotics in *S. aureus* can be due to a number of resistance genes, which, in part, specify different resistance phenotypes.

5.1 COMBINED RESISTANCE TO MACROLIDES, LINCOSAMIDES, AND STREPTOGRAMIN B

Macrolides, lincosamides, and streptogramin B (MLS_B) resistance in *S. aureus* is based on the presence of one or more *erm* genes of the classes A, B, C, F, T, Y, and 33 (Table 4.1). These genes code for methylases that modify the target site(s) A2058/A2059 in 23S rRNA and thereby inhibit the binding of MLS_B compounds. The *erm*(A) and *erm*(B) genes are commonly located on small nonconjugative transposons Tn*554* and Tn*917*/Tn*551*, respectively (Lyon and Skurray, 1987; Schwarz et al., 2011, 2014). Tn*554* has also been frequently found to be integrated into SCC*mec* II elements (IWG-SCC, 2009). Most commonly, Tn*554* site-specifically integrates into a chromosomal site called att*554* (Murphy et al., 1991). However, if this site is occupied or deleted, Tn*554* or parts thereof are also found on large plasmids (Townsend et al., 1986). A single case described the presence of the *erm*(A) gene together with the multiresistance gene *cfr* on a small 7054 bp plasmid in an *S. aureus* isolate from bovine mastitis (Wang et al., 2012b). The *erm*(B) gene was present in canine and porcine *S. aureus* (Eady et al., 1993; Lüthje and Schwarz, 2007). Moreover, the *erm*(B) gene has also been detected in LA-MRSA from pigs (Kadlec et al., 2009) and cattle (Feßler et al., 2010). The *erm*(A) and *erm*(B) genes have been detected in a wide variety of *S. aureus* (including MRSA) from humans (Monecke et al., 2011). The gene *erm*(C) is commonly located on small plasmids ranging in size between 2.3 and 4.4 kb, which most often do not carry additional resistance genes (Schwarz et al., 2014). Such plasmids have been identified in *S. aureus* (including MRSA) from humans and various animal species (Monecke et al., 2011; Lyon and Skurray, 1987; Schwarz et al., 2011; Wendlandt et al., 2013a). Small *erm*(C)-carrying plasmids can integrate into larger plasmids via insertion sequences, such as IS*257* (Diep et al., 2006) or the closely related IS*Sau10* (Gómez-Sanz et al., 2013a). The gene *erm*(T) was first identified in *S. aureus* in 2010 in a porcine MRSA CC398 isolate (Kadlec and Schwarz, 2010a). Since then, it has been identified in MRSA from cattle (Feßler et al., 2010), chickens, and turkeys (Monecke et al., 2012b). In most cases, the *erm*(T) gene was plasmid-borne and the corresponding plasmids harbored additional resistance genes. In MSSA of Sequence Type (ST)398 (*spa*-type t571), however, the *erm*(T) gene was located in the chromosomal DNA (Vandendriessche et al., 2011). An *erm*(T)-carrying plasmid, pUR3912, was found to be able to integrate in and excise from the chromosomal DNA of *S. aureus* by recombination via IS*431* (Gómez-Sanz et al., 2013b). The *erm*(Y) gene was found on the 33,347-bp plasmid pMS97 from *S. aureus* together with another two macrolide resistance genes *msr*(A) and *mph*(C) (Matsuoka et al., 2003). An *erm*(33) gene, which represents an in vivo recombination product between *erm*(A) and *erm*(C), was recently found in two MRSA ST9-t899 isolates from pigs (Li et al., 2015b). The gene *erm*(F) was detected in five *S. aureus* isolates from healthy Portuguese children (Luna et al., 2002).

5.2 COMBINED RESISTANCE TO MACROLIDES AND STREPTOGRAMIN B

The *msr*(A) gene codes for a 488-aa ABC transporter protein, which confers resistance to macrolides and streptogramin B antibiotics (Ross et al., 1990) (Table 4.1). The *msr*(A) gene has been detected in *S. aureus* isolates of poultry (Nawaz et al., 2000) and dog origin (Lüthje and Schwarz, 2007). In human *S. aureus*, the *msr*(A) gene has been also detected in community-associated MRSA USA300 isolates (Kennedy et al., 2010; Glaser et al., 2016). Moreover, the *msr*(A) gene was found on a plasmid designated $pETB_{TY825}$ from a patient with impetigo, where the *msr*(A) gene was colocated with the resistance genes *aacA-aphD* (aminoglycoside resistance) and *blaZ* (penicillin resistance), as well as the gene encoding the exfoliative toxin B (Hisatsune et al., 2013).

5.3 RESISTANCE TO MACROLIDES ONLY

The gene *mph*(C) codes for a macrolide phosphotransferase of 299 aa, which confers only low-level resistance to macrolides (Matsuoka et al., 1998; Schnellmann et al., 2006; Lüthje and Schwarz, 2006) (Table 4.1). However, the *mph*(C) gene is often linked to *msr*(A), thereby conferring higher levels of macrolide resistance (Lüthje and Schwarz, 2006). In addition to *mph*(C), the macrolide esterase gene *ere*(B) has been identified in five *S. aureus* of human origin (Schmitz et al., 2002), and the gene *ere*(A) was found in 14 of 40 erythromycin-resistant *S. aureus* from bovine mastitis (Li et al., 2015b). The gene *mef*(A), which codes for a 405-aa efflux protein of the MFS, was detected in a single *S. aureus* isolate of human origin (Luna et al., 2002).

5.4 RESISTANCE TO LINCOSAMIDES ONLY

Resistance to lincosamides in *S. aureus* can be due to any of the genes *lnu*(A) or *lnu*(B). The gene *lnu*(A) encodes a lincosamide nucleotidyltransferase of 161 aa (Brisson-Noël and Courvalin, 1986) (Table 4.1). This gene is often located on small plasmids. The *lnu*(A) gene has been identified in *S. aureus*, including MRSA from dairy cattle (Loeza-Lara et al., 2004), pigs (Argudín et al., 2011; Lozano et al., 2012b), and a turkey (Monecke et al., 2012b). The gene *lnu*(B) encodes a lincosamide nucleotidyltransferase of 267 aa (Table 4.1). This gene was identified as part of a multiresistance gene cluster not only on a plasmid and in the chromosomal DNA of porcine MRSA ST9 isolates from China (Li et al., 2013; Wendlandt et al., 2013c, 2014b) but also in human MRSA ST398 and MSSA ST9 isolates from Spain and Germany (Lozano et al., 2012b; Wendlandt et al., 2014b). The ABC transporter gene *lsa*(B), which confers only reduced susceptibility to lincosamides (Schwarz et al., 2016), was recently detected in a single MRSA ST22 isolate of human origin (Shore et al., 2016).

5.5 RESISTANCE TO STREPTOGRAMIN A ONLY

Resistance to streptogramin A antibiotics in *S. aureus* is mediated by *vat*(A), *vat*(B), or *vat*(C) genes, which code for acetyltransferases of 219, 212, or 212 aa, respectively (Schwarz et al., 2011, 2014). These genes are commonly plasmid-borne and are preferentially found in *S. aureus* of human origin (Allignet and El Solh, 1995, 1999; Allignet et al., 1998; Yu et al., 2014). The gene *vga*(B), which codes for an ABC transporter of 552 aa (Allignet and El Solh, 1997), also confers streptogramin A resistance. This gene has been so far only seen in *S. aureus* of human origin.

5.6 RESISTANCE TO STREPTOGRAMIN B ONLY

The resistance genes *vgb*(A) and *vgb*(B) encode streptogramin B lyases, also known as streptogramin B lactonases or streptogramin B lactone hydrolases, of 299 and 295 aa, respectively (Roberts et al., 1999; Schwarz et al., 2011). These genes have so far only been identified in *S. aureus* of human origin (Allignet and El Solh, 1999; Allignet et al., 1998).

5.7 COMBINED RESISTANCE TO LINCOSAMIDES, PLEUROMUTILINS, AND STREPTOGRAMIN A

Combined resistance to lincosamides, pleuromutilins, and streptogramin A in *S. aureus* is mediated by the genes *vga*(A), *vga*(C), and *vga*(E), which code for ABC transporter proteins of 522, 522, and 524 aa, respectively (Haroche et al., 2000; Gentry et al., 2008; Kadlec and Schwarz, 2009a; Lozano et al., 2012a; Schwendener and Perreten, 2011; Hauschild et al., 2012b). The *vga*(A) genes are widely disseminated in *S. aureus* from humans and have also been detected in *S. aureus* (mostly MRSA) from pigs (Kadlec et al., 2009, 2012b; Overesch et al., 2011), cattle (Feßler et al., 2010), and turkeys (Monecke et al., 2012b). Moreover, the gene *vga*(C) was detected in MRSA from pigs (Kadlec and Schwarz, 2009a; Kadlec et al., 2010) and cattle (Feßler et al., 2010), whereas the gene *vga*(E) was identified in MRSA from pigs (Schwendener and Perreten, 2011), cattle (Hauschild et al., 2012b), and from chickens and turkeys (Hauschild et al., 2012b; Monecke et al., 2012b).

The *lsa*(E) gene, which also mediates combined resistance to pleuromutilins, lincosamides, and streptogramin A, has been described in human MRSA/MSSA ST398 and MSSA ST9 (Wendlandt et al., 2013a; Sarrou et al., 2016). This gene codes for an ABC transporter protein of 494 aa. The *lsa*(E) gene, together with the genes *lnu*(B), *aadE*, and *spw*, forms the core component of a multiresistance gene cluster, which most likely originates from *Enterococcus faecalis* (Li et al., 2013; Wendlandt et al., 2014b).

6. AMINOGLYCOSIDE RESISTANCE

Resistance to aminoglycosides is based on a number of inactivating enzymes, which differ in their specific substrate spectra (Ramirez and Tolmasky, 2010) (Table 4.1). In *S. aureus* of human and animal origin, the following genes for aminoglycoside-modifying enzymes have been detected so far.

The gene *aacA-aphD* [also known as *aac(6′)-Ie–aph(2″)-Ia*] encodes a bifunctional enzyme of 479 aa that shows acetyltransferase and phosphotransferase activity and confers resistance to gentamicin, kanamycin, tobramycin, and, when overexpressed, also to amikacin (Rouch et al., 1987). This gene is located on transposon Tn*4001*, which consists of a central resistance gene region that is bracketed by two IS*256* elements located in opposite orientations (Byrne et al., 1989). The *aacA-aphD* gene is widespread in staphylococci of animal origin, including *S. aureus* (mostly MRSA) from pigs (Schwarz et al., 2008; Kadlec et al., 2009; Argudín et al., 2011; Overesch et al., 2011), cattle (Turutoglu et al., 2009; Feßler et al., 2010), horses (Cuny et al., 2006; Walther et al., 2009; Sieber et al., 2011), chickens, and turkeys (Monecke et al., 2012b).

The gene *aadD* [also known as *ant(4′)-Ia*] encodes a 256-aa adenyltransferase, which confers resistance to kanamycin, neomycin, and tobramycin. It was initially identified on plasmid pUB110 (McKenzie et al., 1987) in human *S. aureus* isolates. This plasmid is also often integrated in type II SCC*mec* elements via IS*431* (IWG-SCC, 2009). In *S. aureus* of animal origin, it has been detected, often on plasmids, in livestock-associated MRSA/MSSA from pigs (Kadlec and Schwarz, 2009a; Argudín et al., 2011), cattle (Feßler et al., 2010), turkeys (Monecke et al., 2012b), and horses (Walther et al., 2009).

The gene *aphA3* [also known as *aph(3′)-IIIa*] encodes a phosphotransferase of 264 aa, which mediates resistance to kanamycin, neomycin, and amikacin (Derbise et al., 1996). In *S. aureus* of animal origin, this gene was found in MRSA from pigs (Argudín et al., 2011), cattle (Turutoglu et al., 2009; Feßler et al., 2010), and horses (Walther et al., 2009).

The gene *aadE* [also known as *ant(6)-Ia*] encodes a 302-aa adenyltransferase, which confers streptomycin resistance. Both genes, *aadE* and *aphA3*, are together with the streptothricin resistance gene *sat4* part of transposon Tn*5405* (Derbise et al., 1996). The *aadE* gene seems to occur mainly in human *S. aureus*. In *S. aureus* from animal sources, this gene was mainly seen as one of the resistance genes in a multiresistance gene cluster that harbors the resistance genes *lsa*(E) and *lnu*(B), among others (Lozano et al., 2012b; Li et al., 2013; Wendlandt et al., 2013c, 2014b, 2015).

The gene *str* [also known as *aad(6)*] encodes a 282-aa adenyltransferase (Projan et al., 1985), which mediates streptomycin resistance. This gene is often located on small plasmids of which pS194 is considered as the prototype plasmid. The *str* gene has also been identified in porcine *S. aureus* including MRSA (Stegmann and Perreten, 2010; Overesch et al., 2011).

7. AMINOCYCLITOL RESISTANCE

The first spectinomycin resistance gene identified in *S. aureus* was the gene *spc* [also known as *aad(9)-Ia*], which encodes a 260-aa adenyltransferase (Murphy, 1985) (Table 4.1). Similar to the MLS_B resistance gene *erm*(A), the gene *spc* is part of the transposon Tn*554*. The gene *spc* is not only widespread in human *S. aureus* isolates but has also been detected in MRSA from pigs (Kadlec et al., 2009; Overesch et al., 2011), cattle (Feßler et al., 2010), as well as chickens and turkeys (Monecke et al., 2012b). A second spectinomycin resistance gene, *spw*, encoding an adenyltransferase of 269 aa, was found as part of a multiresistance gene cluster in MRSA ST398 and MRSA/MSSA ST9 isolates from humans, animals, and food of animal origin (Wendlandt et al., 2013b,c, 2014b). The Spw protein had only 64.7% identity to the Tn*554*-associated Spc protein (Wendlandt et al., 2013b). In 2014, a third spectinomycin resistance gene, designated *spd*, which encodes an adenyltransferase of 259-aa, was detected (Jamrozy et al., 2014). The Spd protein shared only 45% sequence identity with the Spw protein. The *spd* gene was initially detected in MRSA ST398 from humans and various animal species (Jamrozy et al., 2014) and soon thereafter also in porcine MSSA ST433 (Wendlandt et al., 2014a).

The gene *apmA* is the first and, so far, the only apramycin resistance gene found in staphylococci (Table 4.1). It encodes a 274-aa acetyltransferase, which confers resistance to apramycin and decreased susceptibility to gentamicin (Feßler et al., 2011). It was detected either on large multiresistance plasmids in bovine or porcine LA-MRSA of CC398 (Feßler et al., 2011), or, in a single case, on a small plasmid that conferred only apramycin resistance in a porcine LA-MRSA of CC398 (Kadlec et al., 2012a). Recently, the complete sequence of the *apmA*-carrying plasmid pAFS11 was published, which showed that this plasmid carried several antimicrobial resistance genes in combination with cadmium and copper resistance genes and a novel *ica*-like gene cluster (Feßler et al., 2017).

8. STREPTOTHRICIN RESISTANCE

Resistance against streptothricins in *S. aureus* is mediated by the gene *sat4*, which codes for a 176-aa acetyltransferase protein (Derbise et al., 1996) (Table 4.1). The *sat4* gene is also part of Tn*5405* and as such, commonly detected in *S. aureus* isolates that also harbor the genes *aphA3* and *aadE*. It has also been found in *S. aureus* from healthy human carriers in Spain (Argudín et al., 2014). In staphylococci of animal origin, this gene has been detected in MRSA of Clonal Complex (CC)8 (ST254) from horses in Germany (Walther et al., 2009).

9. OXAZOLIDINONE RESISTANCE

Oxazolidinone resistance in *S. aureus* are based on mutations either in the 23S rRNA and/or in the genes for the ribosomal proteins L2, L3, and L22 (Xiong et al., 2000; Meka et al., 2004; Locke et al., 2009; Román et al., 2013). Some of these alterations also confer cross-resistance to pleuromutilins (Miller et al., 2008). Transferable oxazolidinone resistance in *S. aureus* is based on the expression of the gene *cfr*. This gene codes for an rRNA methylase of 349 aa, which methylates the adenine at position 2503 in the 23S rRNA (Kehrenberg et al., 2005). Because this adenine residue is located in the overlapping binding region of phenicols, lincosamides, oxazolidinones, pleuromutilins, and streptogramin A antibiotics, methylation of A2503 interferes with the binding and correct positioning of these antimicrobial agents. Thus, Cfr confers a penta-resistance phenotype, which includes the aforementioned classes of antimicrobial agents (Long et al., 2006) (Table 4.1). The *cfr* gene is usually located on plasmids, most of which also harbor additional resistance genes, which facilitate the coselection of *cfr* (Schwarz et al., 2014). Numerous reports described the presence of *cfr* in human *S. aureus* (Shen et al., 2013), including its presence in a Panton–Valentine leukocidin-positive ST8 MRSA IVa (USA300) isolate (Shore et al., 2010) and in pandemic ST22 MRSA isolates (Shore et al., 2016). In *S. aureus* of animal origin, the *cfr* gene has been detected in MRSA and MSSA from pigs (Kehrenberg and Schwarz, 2006; Kehrenberg et al., 2009; Argudín et al., 2011; Wang et al., 2012a), cattle (Wang et al., 2012b), and a horse (Kehrenberg and Schwarz, 2006). Recently, the *cfr* gene was found to be integrated into a type IVb SCC*mec* cassette in an ST9-t899 MRSA of porcine origin (Li et al., 2015a).

10. TRIMETHOPRIM RESISTANCE

Resistance to trimethoprim in *S. aureus* is based on any of the four genes, *dfrA* (also known as *dfrS1*), *dfrD*, *dfrG*, or *dfrK* (Kadlec et al., 2012b), which encode trimethoprim-resistant dihydrofolate reductases of 161, 166, 165, or 163-aa, respectively (Table 4.1). The Tn*4003*-associated gene *dfrA* is most widespread among staphylococci from humans (Lyon and Skurray, 1987). In staphylococci of animal origin, it has rarely been found. There are only few reports about its occurrence in *S. aureus* (mostly MRSA) from pigs (Argudín et al., 2011) and horses (Walther et al., 2009; Sieber et al., 2011). The gene *dfrD* has been detected very rarely in *S. aureus*. There is only a single report that describes its occurrence together with *dfrA* in a single porcine MRSA isolate (Argudín et al., 2011). In contrast, the gene *dfrG* seems to be an emerging trimethoprim resistance gene in *S. aureus* causing infections and colonization among humans from Africa (Nurjadi et al., 2014). This gene has also been detected in MRSA from pigs (Kadlec et al., 2009; Argudín et al., 2011; Overesch et al., 2011). The gene *dfrK* has initially been found on multiresistance plasmids from porcine MRSA ST398 where it was linked to the tetracycline resistance gene *tet*(L) (Kadlec and Schwarz, 2009a,b, 2010a). Later on, the *dfrK* gene was also identified as part of the transposon Tn*559* in a porcine MSSA CC398 isolate (Kadlec and Schwarz, 2010b). The *dfrK* gene has been found mainly in MRSA from food-producing animals (Kadlec et al., 2009; Argudín et al., 2011; Feßler et al., 2010; Monecke et al., 2012b).

11. SULFONAMIDE RESISTANCE

No specific sulfonamide resistance genes have been detected so far in *S. aureus*. Sulfonamide resistance in *S. aureus* is believed to result either from an increased production of *p*-aminobenzoic acid (Landy et al., 1943) or from mutations in the chromosomal gene that encodes the enzyme dihydropteroate synthase (Hampele et al., 1997). Such mutations may reduce the affinity of this enzyme for sulfonamides.

12. FUSIDIC ACID RESISTANCE

Resistance to fusidic acid is based on the expression of the genes *fusB* (*far1*) or *fusC*, which encode cytoplasmatic proteins of 213 or 212 aa (O'Neill et al., 2007), or on single point mutations in the *fusA* gene (Table 4.1). For the FusB protein it has been shown that it binds to the staphylococcal elongation factor G (EF-G) and thereby protects the translation system from inhibition by fusidic acid (O'Neill and Chopra, 2006). In contrast to human medicine, fusidic acid resistance has rarely been detected in animal staphylococci. The *fusB* gene has been found in MRSA isolates from sheep (Gharsa et al., 2012b) and the *fusC* gene in *S. aureus* from donkeys (Gharsa et al., 2012a). Loeffler et al. (2008) identified six MRSA isolates from cats and dogs with high minimal inhibitory concentrations (MICs) of fusidic acid (≥512 mg/L), but no information is available about the corresponding resistance genes or mutations.

13. MUPIROCIN RESISTANCE

Resistance to mupirocin in *S. aureus* is commonly due to a mupirocin-insensitive isoleucyl-tRNA synthase of 1024 aa, encoded by the gene *ileS*, also known as *ileS2* or *mupA* (Hodgson et al., 1994; Needham et al., 1994) (Table 4.1). This gene is often found on plasmids in staphylococci from humans. Although a small number of mupirocin-resistant MRSA/MSSA from dogs have been reported (Manian, 2003; Fulham et al., 2011), no information about the corresponding resistance genes or mutations is available. Wipf and Perreten (2016) recently described mupirocin resistance–mediating mutations, N213D and V588F, in the bacterial isoleucyl-t-RNA-synthetase of a *S. aureus* ST5-t002 isolate from a dog.

14. FLUOROQUINOLONE RESISTANCE

In contrast to Enterobacteriaceae, no plasmid-mediated quinolone resistance genes have so far been detected in *S. aureus*. (Fluoro)quinolone resistance in *S. aureus* is mainly based on mutations in the quinolone resistance–determining regions of the topoisomerase genes *gyrA*, *gyrB*, *grlA*, and *grlB*. The corresponding amino acid substitutions, so far found in *S. aureus*, are summarized in Table 4.2. Several studies reported such mutations in naturally occurring *S. aureus* from humans (Takahashi et al., 1998; Iihara et al., 2006; Nakaminami et al., 2014), animals, and food of animal origin (Iihara et al., 2006; Hauschild et al., 2012a; Li et al., 2016; Wipf and Perreten, 2016).

Table 4.2 Amino Acid Substitutions Associated With (fluoro)quinolone Resistance in *Staphylococcus aureus*

Amino Acid Substitution in																			
GyrA				GyrB					GrlA				GrlB						
81	84	85	88	437	456	458	477	517	79	80	84	108	422	443	444	451	471		
Ser	Ser	Ser	Glu	Asp	Pro	Arg	Glu	Arg	Asp	Ser	Glu	Ser	Glu	Asp	Arg	Pro	Glu	Origin	References
	Leu		Lys							Phe	Lys							Human	Takahashi et al. (1998)
	Leu									Phe	Lys							Human	Takahashi et al. (1998)
	Leu		Lys		Ser					Phe	Lys							Human	Takahashi et al. (1998)
	Leu				Ser					Phe	Lys							Human	Takahashi et al. (1998)
	Leu									Phe						Ser		Human	Horii et al. (2003)
	Leu									Tyr	Lys							Human	Horii et al. (2003)
	Leu								Val	Tyr								Human	Horii et al. (2003)
			Lys							Phe								Human	Horii et al. (2003)
	Leu									Tyr								Human	Horii et al. (2003)
			Lys							Tyr								Human	Horii et al. (2003)
			Lys						Val	Phe								Human	Horii et al. (2003)
	Leu	Pro								Phe	Lys							Human	Iihara et al. (2006)
	Leu	Pro								Phe		Asn					Lys[b]	Human	Iihara et al. (2006)
	Leu		Gly							Tyr	Lys							Human	Iihara et al. (2006)
	Leu									Phe								Human/ A[a]	Iihara et al. (2006), Wipf and Perreten (2016)
	Leu										Lys			Glu	Ser			Human	Iihara et al. (2006)
	Leu										Lys							Human	Iihara et al. (2006)
												Asn						Human	Iihara et al. (2006)
	Leu									Phe							Asp	Poultry	Hauschild et al. (2012a)
	Leu									Phe			Asp					B[a]	Hauschild et al. (2012a)
	Leu							Lys		Phe								Turkey meat	Hauschild et al. (2012a)
	Ala									Phe			Asp					Pig	Hauschild et al. (2012a)
										Phe			Asp					C[a]	Hauschild et al. (2012a)

Continued

Table 4.2 Amino Acid Substitutions Associated With (fluoro)quinolone Resistance in *Staphylococcus aureus*—cont'd

Amino Acid Substitution in																			
GyrA				GyrB					GrlA				GrlB						
81	84	85	88	437	456	458	477	517	79	80	84	108	422	443	444	451	471		
Ser	Ser	Ser	Glu	Asp	Pro	Arg	Glu	Arg	Asp	Ser	Glu	Ser	Glu	Asp	Arg	Pro	Glu	Origin	References
										Tyr			Asp					Pig/ Poultry	Hauschild et al. (2012a)
													Asp					Pig	Hauschild et al. (2012a)
			Asp							Leu	Asp							Poultry	Hauschild et al. (2012a)
			Gly							Tyr								Human	Nakaminami et al. (2014)
			Gly							Phe								Human	Nakaminami et al. (2014)
	Leu		Arg							Tyr	Lys							Human	Nakaminami et al. (2014)
	Leu		Gly				Asp			Tyr	Lys							Human	Nakaminami et al. (2014)
	Ala									Phe								Pig	Li et al. (2016)
Pro	Leu									Phe								Dog	Wipf and Perreten (2016)

[a] *A, turkey meat/chicken product/pig/poultry/cattle/cat/dog; B, turkey product/pig/poultry; C, chicken meat/turkey product/pig.*

[b] *According to the reference, this mutation is a silent mutation.*

In addition to mutations mediating fluoroquinolone resistance, efflux-mediated fluoroquinolone resistance via NorA is also known in *S. aureus* (Kaatz et al., 1993). NorA-mediated fluoroquinolone resistance can occur in the absence of topoisomerase mutations and in some *S. aureus* isolates may be the result of a mutation in the promoter region causing increased expression of *norA* (Kaatz and Seo, 1997). The results of further studies suggest that besides NorA, there are likely other multidrug transporters, which contribute to fluoroquinolone resistance in *S. aureus* (Kaatz et al., 2002).

15. RIFAMPICIN RESISTANCE

Rifampicin interacts with the β subunit of the bacterial RNA polymerase encoded by the *rpoB* gene. Several studies have shown that rifampicin resistance can quickly arise from mutations in the *rpoB* gene. So far, numerous mutations in the *rpoB* gene have been detected in clinical *S. aureus* isolates from humans or laboratory strains (Aubry-Damon et al., 1998; Wichelhaus et al., 1999, 2002; O'Neill et al., 2006; Mick et al., 2010; Villar et al., 2011; Zhou et al., 2012). In contrast, little information is available about rifampicin-resistant *S. aureus* from animals (Li et al., 2016). Within the *rpoB* gene, the observed mutations clustered in three different regions as shown in Table 4.3. It became apparent that not all mutations had the same effect on the rifampicin MICs, and that occasionally isolates with two or three *rpoB* mutations exhibited higher MICs than isolates with a single *rpoB* mutation (Aubry-Damon et al., 1998; Villar et al., 2011; Li et al., 2016).

16. GLYCOPEPTIDE RESISTANCE

Vancomycin resistance is rarely encountered in *S. aureus*. All so far published reports about vancomycin resistance in *S. aureus* are from isolates of human origin. Vancomycin resistance is based on the presence of the *vanA* gene cluster, which commonly resides on nonconjugative transposons of the type Tn*1545*. The *vanA* gene cluster codes for several proteins involved in the synthesis of D-Ala-D-Lac depsipeptide peptidoglycan precursors, which exhibit a low affinity to glycopeptides. The first clinical highly vancomycin-resistant *S. aureus* (VRSA) isolate was detected in a human patient in 2002. This isolate harbored the 57,889-bp conjugative multiresistance plasmid pLW043, which consisted of a pSK41-like *S. aureus* resistance plasmid (with the resistance genes *dfrA*, *aacA-aphD*, *blaZ*, and *qacC*) with an insertion of a Tn*1546*-like transposon carrying the *vanA* gene cluster (Weigel et al., 2003) (Table 4.1). Detailed analysis of the Tn*1546* elements so far found in VRSA isolates revealed alterations, which, however, did not affect the structural genes required for vancomycin resistance (Périchon and Courvalin, 2009). So far, only the localization on plasmids was described for the transposons with the *vanA* gene cluster (Périchon and Courvalin, 2009; Walters et al., 2015).

The bleomycin resistance gene *ble* was located together with the *aadD* gene on the small plasmid pUB110 (McKenzie et al., 1987). Moreover, this gene was also found in the chromosomal DNA from a human MRSA isolate in Japan, where it clustered with the kanamycin resistance gene *aadD* pointing toward an integration of these genes from pUB110 in the chromosomal DNA in association with IS*431 mec* (Bhuiyan et al., 1995; Sugiyama et al., 1996). The presence of integrated pUB110 has been used for the subtyping of SCC*mec* elements (Kondo et al., 2007).

Table 4.3 Amino Acid Substitutions Associated With Rifampicin Resistance in *Staphylococcus aureus*

Amino Acid Substitution in RpoB															
Cluster I											Cluster II		Cluster III		
463	464	465	466	468	471	473	477	481	484	486	527	529	550		
Ser	Ser	Gln	Leu	Gln	Asp	Ala	Ala	His	Arg	Ser	Ile	Ser	Asp	Origin	References
					Tyr									Human/Pig	Aubry-Damon et al. (1998) and Li et al. (2016)
							Asp							Human	Aubry-Damon et al. (1998)
								Tyr						Human	Aubry-Damon et al. (1998)
									His					Human	Aubry-Damon et al. (1998)
													Gly	Human	Aubry-Damon et al. (1998)
								Asn						Human/Pig	Wichelhaus et al. (1999) and Li et al. (2016)
			Ser					Asn						Human	Wichelhaus et al. (1999)
								Asn				Leu		Human/Pig	Wichelhaus et al. (1999) and Li et al. (2016)
								Asn			Met			Human	Wichelhaus et al. (1999)
		Arg						Asn				Leu		Human	Wichelhaus et al. (1999)
						Thr	Thr	Asn						Human	Wichelhaus et al. (1999)
				Lys										Human	Wichelhaus et al. (1999)
				Ile										Human	O'Neill et al. (2006)
				Leu										Human	O'Neill et al. (2006)
										Leu				Human	O'Neill et al. (2006)
					Tyr	Glu								Human/Pig	O'Neill et al. (2006) and Li et al. (2016)
Asn	Pro													Human	O'Neill et al. (2006)
					Gly				Cys					Human	O'Neill et al. (2006)
				Lys				Asn						Human	Mick et al. (2010)
								Asn			Leu			Human	Mick et al. (2010)
							Thr	Asn						Human	Mick et al. (2010)
							Asp	Asn						Human	Zhou et al. (2012)
			Ser				Asp	Asn						Human	Zhou et al. (2012)

17. CONCLUSION

The data presented in this chapter show that *S. aureus* isolates from humans and animals harbor a wide variety of antimicrobial resistance genes and resistance-mediating mutations. Some of the resistance genes are widely disseminated in MRSA and MSSA of human and animal origin, whereas others are preferentially present in isolates of specific origins. The selection pressure as imposed by the use of antimicrobial agents seems to play an important role in this regard. Thus, resistances to antimicrobial agents, such as tetracyclines, β-lactams, macrolides, trimethoprim, or aminoglycosides, which are used for a long time in human and veterinary medicine, are found in *S. aureus* isolates of both human and animal origin. In turn, resistances to antimicrobial agents, such as vancomycin, mupirocin, streptogramins, or linezolid, which are rarely or not at all used in animals, are found mainly in human staphylococci. An exception is the multiresistance gene *cfr*, which confers cross-resistance to five classes of antimicrobial agents, some of which are important for therapeutic applications in human and/or veterinary medicine. The fact that many resistance genes found in *S. aureus* are located on mobile genetic elements, such as plasmids and transposons, supports the spread of these resistance genes across strain, species, and even genus boundaries. In this regard, *S. aureus* can act as donor and recipient of resistance genes and thereby plays an important role in the dissemination of resistance genes within the Gram-positive resistance gene pool. In this regard, it should be noted that many resistance plasmids of *S. aureus* harbor more than one resistance gene, which favors the coselection and persistence of resistance genes even in the absence of a direct selective pressure. Because the selective pressure is a major driving force in the development and spread of resistance properties, foremost attempts should emphasize the reduction of antimicrobial use in human and veterinary medicine to the necessary minimum and carefully take into consideration alternative approaches to prevent and control bacterial infections.

ACKNOWLEDGMENTS

We apologize in advance to all the investigators whose research could not be appropriately cited owing to space limitations.

The work on antimicrobial resistance genes in staphylococci is financially supported by grants from the German Federal Ministry of Education and Research (BMBF) through the German Aerospace Center (DLR), grant numbers 01KI1014D (MedVet-Staph) and 01KI1301D (MedVet-Staph 2), and the National Natural Science Foundation of China (31472237).

REFERENCES

Aarestrup, F.M., Schwarz, S., 2006. Antimicrobial resistance in staphylococci and streptococci of animal origin. In: Aarestrup, F.M. (Ed.), Antimicrobial Resistance in Bacteria of Animal Origin. ASM Press, Washington, DC, pp. 187–212.

Allignet, J., El Solh, N., 1995. Diversity among the gram-positive acetyltransferases inactivating streptogramin A and structurally related compounds and characterization of a new staphylococcal determinant, *vatB*. Antimicrob. Agents Chemother. 39 (9), 2027–2036.

Allignet, J., El Solh, N., 1997. Characterization of a new staphylococcal gene, *vgaB*, encoding a putative ABC transporter conferring resistance to streptogramin A and related compounds. Gene 202 (1–2), 133–138.

Allignet, J., El Solh, N., 1999. Comparative analysis of staphylococcal plasmids carrying three streptogramin-resistance genes: *vat-vgb-vga*. Plasmid 42 (2), 134–138.

Allignet, J., Liassine, N., El Solh, N., 1998. Characterization of a staphylococcal plasmid related to pUB110 and carrying two novel genes, *vatC* and *vgbB*, encoding resistance to streptogramins A and B and similar antibiotics. Antimicrob. Agents Chemother. 42 (7), 1794–1798.

Argudín, M.A., Mendoza, M.C., Martín, M.C., Rodicio, M.R., 2014. Molecular basis of antimicrobial drug resistance in *Staphylococcus aureus* isolates recovered from young healthy carriers in Spain. Microb. Pathog. 74, 8–14.

Argudín, M.A., Tenhagen, B.A., Fetsch, A., Sachsenröder, J., Käsbohrer, A., Schroeter, A., Hammerl, J.A., Hertwig, S., Helmuth, R., Bräunig, J., Mendoza, M.C., Appel, B., Rodicio, M.R., Guerra, B., 2011. Virulence and resistance determinants of German *Staphylococcus aureus* ST398 isolates from nonhuman sources. Appl. Environ. Microbiol. 77 (9), 3052–3060.

Aubry-Damon, H., Soussy, C.J., Courvalin, P., 1998. Characterization of mutations in the *rpoB* gene that confer rifampin resistance in *Staphylococcus aureus*. Antimicrob. Agents Chemother. 42 (10), 2590–2594.

Becker, K., Ballhausen, B., Köck, R., Kriegeskorte, A., 2014. Methicillin resistance in *Staphylococcus* isolates: the "*mec* alphabet" with specific consideration of *mecC*, a *mec* homolog associated with zoonotic *S. aureus* lineages. Int. J. Med. Microbiol. 304 (7), 794–804.

Benito, D., Gómez, P., Aspiroz, C., Zarazaga, M., Lozano, C., Torres, C., 2016. Molecular characterization of *Staphylococcus aureus* isolated from humans related to a livestock farm in Spain, with detection of MRSA-CC130 carrying *mecC* gene: a zoonotic case? Enferm. Infecc. Microbiol. Clin. 34 (5), 280–285.

Bhuiyan, M.Z., Ueda, K., Inouye, Y., Sugiyama, M., 1995. Molecular cloning and expression in *Escherichia coli* of bleomycin-resistance gene from a methicillin-resistant *Staphylococcus aureus* and its association with IS*431 mec*. Appl. Microbiol. Biotechnol. 43 (1), 65–69.

Brenner, D.G., Shaw, W.V., 1985. The use of synthetic oligonucleotides with universal templates for rapid DNA sequencing: results with staphylococcal replicon pC221. EMBO J. 4 (2), 561–568.

Brisson-Noël, A., Courvalin, P., 1986. Nucleotide sequence of gene *linA* encoding resistance to lincosamides in *Staphylococcus haemolyticus*. Gene 43 (3), 247–253.

Byrne, M.E., Rouch, D.A., Skurray, R.A., 1989. Nucleotide sequence analysis of IS*256* from the *Staphylococcus aureus* gentamicin-tobramycin-kanamycin-resistance transposon Tn*4001*. Gene 81 (2), 361–367.

Cardoso, M., Schwarz, S., 1992a. Chloramphenicol resistance plasmids in *Staphylococcus aureus* isolated from bovine subclinical mastitis. Vet. Microbiol. 30 (2–3), 223–232.

Cardoso, M., Schwarz, S., 1992b. Nucleotide sequence and structural relationships of a chloramphenicol acetyltransferase encoded by the plasmid pSCS6 from *Staphylococcus aureus*. J. Appl. Bacteriol. 72 (4), 289–293.

Cuny, C., Kuemmerle, J., Stanek, C., Willey, B., Strommenger, B., Witte, W., 2006. Emergence of MRSA infections in horses in a veterinary hospital: strain characterisation and comparison with MRSA from humans. Euro. Surveill. 11 (1), 44–47.

Derbise, A., Dyke, K.G., el Solh, N., 1996. Characterization of a *Staphylococcus aureus* transposon, Tn*5405*, located within Tn*5404* and carrying the aminoglycoside resistance genes, *aphA-3* and *aadE*. Plasmid 35 (3), 174–188.

Diep, B.A., Gill, S.R., Chang, R.F., Phan, T.H., Chen, J.H., Davidson, M.G., Lin, F., Lin, J., Carleton, H.A., Mongodin, E.F., Sensabaugh, G.F., Perdreau-Remington, F., 2006. Complete genome sequence of USA300, an epidemic clone of community-acquired meticillin-resistant *Staphylococcus aureus*. Lancet 367 (9512), 731–739.

Drougka, E., Foka, A., Koutinas, C.K., Jelastopulu, E., Giormezis, N., Farmaki, O., Sarrou, S., Anastassiou, E.D., Petinaki, E., Spiliopoulou, I., 2016. Interspecies spread of *Staphylococcus aureus* clones among companion animals and human close contacts in a veterinary teaching hospital. A cross-sectional study in Greece. Prev. Vet. Med. 126 (4), 190–198.

Eady, E.A., Ross, J.I., Tipper, J.L., Walters, C.E., Cove, J.H., Noble, W.C., 1993. Distribution of genes encoding erythromycin ribosomal methylases and an erythromycin efflux pump in epidemiologically distinct groups of staphylococci. J. Antimicrob. Chemother. 31 (2), 211–217.

Feßler, A.T., Kadlec, K., Schwarz, S., 2011. Novel apramycin resistance gene *apmA* in bovine and porcine methicillin-resistant *Staphylococcus aureus* ST398 isolates. Antimicrob. Agents Chemother. 55 (1), 373–375.

Feßler, A., Scott, C., Kadlec, K., Ehricht, R., Monecke, S., Schwarz, S., 2010. Characterization of methicillin-resistant *Staphylococcus aureus* ST398 from cases of bovine mastitis. J. Antimicrob. Chemother. 65 (4), 619–625.

Feßler, A.T., Zhao, Q., Schoenfelder, S., Kadlec, K., Michael, G.B., Wang, Y., Ziebuhr, W., Shen, J., Schwarz, S., 2017. Complete sequence of a plasmid from a bovine methicillin-resistant *Staphylococcus aureus* harbouring a novel *ica*-like gene cluster in addition to antimicrobial and heavy metal resistance genes. Vet. Microbiol. 200, 95–100.

Fulham, K.S., Lemarie, S.L., Hosgood, G., Dick, H.L., 2011. In vitro susceptibility testing of meticillin-resistant and meticillin-susceptible staphylococci to mupirocin and novobiocin. Vet. Dermatol 22 (1), 88–94.

García-Álvarez, L., Holden, M.T., Lindsay, H., Webb, C.R., Brown, D.F., Curran, M.D., Walpole, E., Brooks, K., Pickard, D.J., Teale, C., Parkhill, J., Bentley, S.D., Edwards, G.F., Girvan, E.K., Kearns, A.M., Pichon, B., Hill, R.L., Larsen, A.R., Skov, R.L., Peacock, S.J., Maskell, D.J., Holmes, M.A., 2011. Meticillin-resistant *Staphylococcus aureus* with a novel *mecA* homologue in human and bovine populations in the UK and Denmark: a descriptive study. Lancet Infect. Dis. 11 (8), 595–603.

Gentry, D.R., McCloskey, L., Gwynn, M.N., Rittenhouse, S.F., Scangarella, N., Shawar, R., Holmes, D.J., 2008. Genetic characterization of Vga ABC proteins conferring reduced susceptibility of *Staphylococcus aureus* to pleuromutilins. Antimicrob. Agents Chemother. 52 (12), 4507–4509.

Gharsa, H., Ben Sallem, R., Ben Slama, K., Gómez-Sanz, E., Lozano, C., Jouini, A., Klibi, N., Zarazaga, M., Boudabous, A., Torres, C., 2012a. High diversity of genetic lineages and virulence genes in nasal *Staphylococcus aureus* isolates from donkeys destined to food consumption in Tunisia with predominance of the ruminant associated CC133 lineage. BMC Vet. Res. 8, 203.

Gharsa, H., Ben Slama, K., Lozano, C., Gómez-Sanz, E., Klibi, N., Ben Sallem, R., Gómez, P., Zarazaga, M., Boudabous, A., Torres, C., 2012b. Prevalence, antibiotic resistance, virulence traits and genetic lineages of *Staphylococcus aureus* in healthy sheep in Tunisia. Vet. Microbiol. 156 (3–4), 367–373.

Glaser, P., Martins-Simões, P., Villain, A., Barbier, M., Tristan, A., Bouchier, C., Ma, L., Bes, M., Laurent, F., Guillemot, D., Wirth, T., Vandenesch, F., 2016. Demography and intercontinental spread of the USA300 community-acquired methicillin-resistant *Staphylococcus aureus* lineage. mBio 7 (1), e02183–e02215.

Gómez-Sanz, E., Kadlec, K., Feßler, A.T., Zarazaga, M., Torres, C., Schwarz, S., 2013a. Novel *erm*(T)-carrying multiresistance plasmids from porcine and human isolates of methicillin-resistant *Staphylococcus aureus* ST398 that also harbor cadmium and copper resistance determinants. Antimicrob. Agents Chemother. 57 (7), 3275–3282.

Gómez-Sanz, E., Zarazaga, M., Kadlec, K., Schwarz, S., Torres, C., 2013b. Chromosomal integration of the novel plasmid pUR3912 from methicillin-susceptible *Staphylococcus aureus* ST398 of human origin. Clin. Microbiol. Infect. 19 (11), E519–E522.

Grøntvedt, C.A., Elstrøm, P., Stegger, M., Skov, R.L., Skytt Andersen, P., Larssen, K.W., Urdahl, A.M., Angen, Ø., Larsen, J., Åmdal, S., Løtvedt, S.M., Sunde, M., Bjørnholt, J.V., 2016. Methicillin-resistant *Staphylococcus aureus* CC398 in humans and pigs in Norway: a "One Health" perspective on introduction and transmission. Clin. Infect. Dis. 63 (11), 1431–1438.

Guran, H.S., Kahya, S., 2015. Species diversity and pheno- and genotypic antibiotic resistance patterns of staphylococci isolated from retail ground meats. J. Food Sci. 80 (6), M1291–M1298.

Haenni, M., Châtre, P., Dupieux, C., Métayer, V., Maillard, K., Bes, M., Madec, J.Y., Laurent, F., 2015. *mecC*-positive MRSA in horses. J. Antimicrob. Chemother. 70 (12), 3401–3402.

Hampele, I.C., D'Arcy, A., Dale, G.E., Kostrewa, D., Nielsen, J., Oefner, C., Page, M.G., Schönfeld, H.J., Stüber, D., Then, R.L., 1997. Structure and function of the dihydropteroate synthase from *Staphylococcus aureus*. J. Mol. Biol. 268 (1), 21–30.

Harrison, E.M., Paterson, G.K., Holden, M.T., Larsen, J., Stegger, M., Larsen, A.R., Petersen, A., Skov, R.L., Christensen, J.M., Bak Zeuthen, A., Heltberg, O., Harris, S.R., Zadoks, R.N., Parkhill, J., Peacock, S.J., Holmes, M.A., 2013. Whole genome sequencing identifies zoonotic transmission of MRSA isolates with the novel *mecA* homologue *mecC*. EMBO Mol. Med. 5 (4), 509–515.

Haroche, J., Allignet, J., Buchrieser, C., El Solh, N., 2000. Characterization of a variant of *vga*(A) conferring resistance to streptogramin A and related compounds. Antimicrob. Agents Chemother. 44 (9), 2271–2275.

Hauschild, T., Feßler, A.T., Billerbeck, C., Wendlandt, S., Kaspar, H., Mankertz, J., Schwarz, S., Kadlec, K., 2012a. Target gene mutations among methicillin-resistant *Staphylococcus aureus* and methicillin-susceptible *S. aureus* with elevated MICs of enrofloxacin obtained from diseased food-producing animals or food of animal origin. J. Antimicrob. Chemother. 67 (7), 1791–1809.

Hauschild, T., Feßler, A.T., Kadlec, K., Billerbeck, C., Schwarz, S., 2012b. Detection of the novel *vga*(E) gene in methicillin-resistant *Staphylococcus aureus* CC398 isolates from cattle and poultry. J. Antimicrob. Chemother. 67 (2), 503–504.

Hodgson, J.E., Curnock, S.P., Dyke, K.G., Morris, R., Sylvester, D.R., Gross, M.S., 1994. Molecular characterization of the gene encoding high-level mupirocin resistance in *Staphylococcus aureus* J2870. Antimicrob. Agents Chemother. 38 (5), 1205–1208.

Hisatsune, J., Hirakawa, H., Yamaguchi, T., Fudaba, Y., Oshima, K., Hattori, M., Kato, F., Kayama, S., Sugai, M., 2013. Emergence of *Staphylococcus aureus* carrying multiple drug resistance genes on a plasmid encoding exfoliative toxin B. Antimicrob. Agents Chemother. 57 (12), 6131–6140.

Horii, T., Suzuki, Y., Monji, A., Morita, M., Muramatsu, H., Kondo, Y., Doi, M., Takeshita, A., Kanno, T., Maekawa, M., 2003. Detection of mutations in quinolone resistance-determining regions in levofloxacin- and methicillin-resistant *Staphylococcus aureus*: effects of the mutations on fluoroquinolone MICs. Diagn. Microbiol. Infect. Dis. 46 (2), 139–145.

Horinouchi, S., Weisblum, B., 1982. Nucleotide sequence and functional map of pC194, a plasmid that specifies inducible chloramphenicol resistance. J. Bacteriol. 150 (2), 815–825.

Iihara, H., Suzuki, T., Kawamura, Y., Ohkusu, K., Inoue, Y., Zhang, W., Monir Shah, M., Katagiri, Y., Ohashi, Y., Ezaki, T., 2006. Emerging multiple mutations and high-level fluoroquinolone resistance in methicillin-resistant *Staphylococcus aureus* isolated from ocular infections. Diagn. Microbiol. Infect. Dis. 56 (3), 297–303.

International Working Group on the Classification of Staphylococcal Cassette Chromosome Elements (IWG-SCC), 2009. Classification of staphylococcal cassette chromosome *mec* (SCC*mec*): guidelines for reporting novel SCC*mec* elements. Antimicrob. Agents Chemother. 53 (12), 4961–4967.

Jamrozy, D.M., Coldham, N.G., Butaye, P., Fielder, M.D., 2014. Identification of a novel plasmid-associated spectinomycin adenyltransferase gene *spd* in methicillin-resistant *Staphylococcus aureus* ST398 isolated from animal and human sources. J. Antimicrob. Chemother. 69 (5), 1193–1196.

Jensen, S.O., Lyon, B.R., 2009. Genetics of antimicrobial resistance in *Staphylococcus aureus*. Future Microbiol. 4, 565–582.

Kaatz, G.W., Moudgal, V.V., Seo, S.M., 2002. Identification and characterization of a novel efflux-related multi-drug resistance phenotype in *Staphylococcus aureus*. J. Antimicrob. Chemother. 50 (6), 833–838.

Kaatz, G.W., Seo, S.M., 1997. Mechanisms of fluoroquinolone resistance in genetically related strains of *Staphylococcus aureus*. Antimicrob. Agents Chemother. 41 (12), 2733–2737.

Kaatz, G.W., Seo, S.M., Ruble, C.A., 1993. Efflux-mediated fluoroquinolone resistance in *Staphylococcus aureus*. Antimicrob. Agents Chemother. 37 (5), 1086–1094.

Kadlec, K., Ehricht, R., Monecke, S., Steinacker, U., Kaspar, H., Mankertz, J., Schwarz, S., 2009. Diversity of antimicrobial resistance pheno- and genotypes of methicillin-resistant *Staphylococcus aureus* ST398 from diseased swine. J. Antimicrob. Chemother. 64 (6), 1156–1164.

Kadlec, K., Feßler, A.T., Couto, N., Pomba, C.F., Schwarz, S., 2012a. Unusual small plasmids carrying the novel resistance genes *dfrK* or *apmA* isolated from methicillin-resistant or -susceptible staphylococci. J. Antimicrob. Chemother. 67 (10), 2342–2345.

Kadlec, K., Feßler, A.T., Hauschild, T., Schwarz, S., 2012b. Novel and uncommon antimicrobial resistance genes in livestock-associated methicillin-resistant *Staphylococcus aureus*. Clin. Microbiol. Infect. 18 (8), 745–755.

Kadlec, K., Pomba, C.F., Couto, N., Schwarz, S., 2010. Small plasmids carrying *vga*(A) or *vga*(C) genes mediate resistance to lincosamides, pleuromutilins and streptogramin A antibiotics in methicillin-resistant *Staphylococcus aureus* ST398 from swine. J. Antimicrob. Chemother. 65 (12), 2692–2693.

Kadlec, K., Schwarz, S., 2009a. Identification of a novel ABC transporter gene, *vga*(C), located on a multiresistance plasmid from a porcine methicillin-resistant *Staphylococcus aureus* ST398 strain. Antimicrob. Agents Chemother. 53 (8), 3589–3591.

Kadlec, K., Schwarz, S., 2009b. Identification of a novel trimethoprim resistance gene, *dfrK*, in a methicillin-resistant *Staphylococcus aureus* ST398 strain and its physical linkage to the tetracycline resistance gene *tet*(L). Antimicrob. Agents Chemother. 53 (2), 776–778.

Kadlec, K., Schwarz, S., 2010a. Identification of a plasmid-borne resistance gene cluster comprising the resistance genes *erm*(T), *dfrK*, and *tet*(L) in a porcine methicillin-resistant *Staphylococcus aureus* ST398 strain. Antimicrob. Agents Chemother. 54 (2), 915–918.

Kadlec, K., Schwarz, S., 2010b. Identification of the novel *dfrK*-carrying transposon Tn*559* in a porcine methicillin-susceptible *Staphylococcus aureus* ST398 strain. Antimicrob. Agents Chemother. 54 (8), 3475–3477.

Kehrenberg, C., Cuny, C., Strommenger, B., Schwarz, S., Witte, W., 2009. Methicillin-resistant and -susceptible *Staphylococcus aureus* strains of clonal lineages ST398 and ST9 from swine carry the multidrug resistance gene *cfr*. Antimicrob. Agents Chemother. 53 (2), 779–781.

Kehrenberg, C., Schwarz, S., 2004. *fexA*, a novel *Staphylococcus lentus* gene encoding resistance to florfenicol and chloramphenicol. Antimicrob. Agents Chemother. 48 (2), 615–618.

Kehrenberg, C., Schwarz, S., 2005. Florfenicol-chloramphenicol exporter gene *fexA* is part of the novel transposon Tn*558*. Antimicrob. Agents Chemother. 49 (2), 813–815.

Kehrenberg, C., Schwarz, S., 2006. Distribution of florfenicol resistance genes *fexA* and *cfr* among chloramphenicol-resistant *Staphylococcus* isolates. Antimicrob. Agents Chemother. 50 (4), 1156–1163.

Kehrenberg, C., Schwarz, S., Jacobsen, L., Hansen, L.H., Vester, B., 2005. A new mechanism for chloramphenicol, florfenicol and clindamycin resistance: methylation of 23S ribosomal RNA at A2503. Mol. Microbiol. 57 (4), 1064–1073.

Kennedy, A.D., Porcella, S.F., Martens, C., Whitney, A.R., Braughton, K.R., Chen, L., Craig, C.T., Tenover, F.C., Kreiswirth, B.N., Musser, J.M., DeLeo, F.R., 2010. Complete nucleotide sequence analysis of plasmids in strains of *Staphylococcus aureus* clone USA300 reveals a high level of identity among isolates with closely related core genome sequences. J. Clin. Microbiol. 48 (12), 4504–4511.

Khan, S.A., Novick, R.P., 1983. Complete nucleotide sequence of pT181, a tetracycline-resistance plasmid from *Staphylococcus aureus*. Plasmid 10 (3), 251–259.

Kondo, Y., Ito, T., Ma, X.X., Watanabe, S., Kreiswirth, B.N., Etienne, J., Hiramatsu, K., 2007. Combination of multiplex PCRs for staphylococcal cassette chromosome *mec* type assignment: rapid identification system for *mec*, *ccr*, and major differences in junkyard regions. Antimicrob. Agents Chemother. 51 (1), 264–274.

Landy, M., Larkum, N.W., Oswald, E.J., Streighoff, P., 1943. Increased synthesis of p-amino-benzoic acid associated with the development of sulfonamide resistance in *Staphylococcus aureus*. Science 97 (2516), 265–267.

Laurent, F., Chardon, H., Haenni, M., Bes, M., Reverdy, M.E., Madec, J.Y., Lagier, E., Vandenesch, F., Tristan, A., 2012. MRSA harboring *mecA* variant gene *mecC*, France. Emerg. Infect. Dis. 18 (9), 1465–1467.

Li, B., Wendlandt, S., Yao, J., Liu, Y., Zhang, Q., Shi, Z., Wei, J., Shao, D., Schwarz, S., Wang, S., Ma, Z., 2013. Detection and new genetic environment of the pleuromutilin-lincosamide-streptogramin A resistance gene *lsa*(E) in methicillin-resistant *Staphylococcus aureus* of swine origin. J. Antimicrob. Chemother. 68 (6), 1251–1255.

Li, D., Wu, C., Wang, Y., Fan, R., Schwarz, S., Zhang, S., 2015a. Identification of multiresistance gene *cfr* in methicillin-resistant *Staphylococcus aureus* from pigs: plasmid location and integration into a staphylococcal cassette chromosome *mec* complex. Antimicrob. Agents Chemother. 59 (6), 3641–3644.

Li, L., Feng, W., Zhang, Z., Xue, H., Zhao, X., 2015b. Macrolide-lincosamide-streptogramin resistance phenotypes and genotypes of coagulase-positive *Staphylococcus aureus* and coagulase-negative staphylococcal isolates from bovine mastitis. BMC Vet. Res. 11, 168.

Li, J., Feßler, A.T., Jiang, N., Fan, R., Wang, Y., Wu, C., Shen, J., Schwarz, S., 2016. Molecular basis of rifampicin resistance in multiresistant porcine livestock-associated MRSA. J. Antimicrob. Chemother. 71, 3313–3315.

Li, S., Skov, R.L., Han, X., Larsen, A.R., Larsen, J., Sørum, M., Wulf, M., Voss, A., Hiramatsu, K., Ito, T., 2011. Novel types of staphylococcal cassette chromosome *mec* elements identified in clonal complex 398 methicillin-resistant *Staphylococcus aureus* strains. Antimicrob. Agents Chemother. 55 (6), 3046–3050.

Locke, J.B., Hilgers, M., Shaw, K.J., 2009. Mutations in ribosomal protein L3 are associated with oxazolidinone resistance in staphylococci of clinical origin. Antimicrob. Agents Chemother. 53 (12), 5275–5278.

Loeffler, A., Baines, S.J., Toleman, M.S., Felmingham, D., Milsom, S.K., Edwards, E.A., Lloyd, D.H., 2008. In vitro activity of fusidic acid and mupirocin against coagulase-positive staphylococci from pets. J. Antimicrob. Chemother. 62 (6), 1301–1304.

Loeza-Lara, P.D., Soto-Huipe, M., Baizabal-Aguirre, V.M., Ochoa-Zarzosa, A., Valdez-Alarcón, J.J., Cano-Camacho, H., López-Meza, J.E., 2004. pBMSa1, a plasmid from a dairy cow isolate of *Staphylococcus aureus*, encodes a lincomycin resistance determinant and replicates by the rolling-circle mechanism. Plasmid 52 (1), 48–56.

Long, K.S., Poehlsgaard, J., Kehrenberg, C., Schwarz, S., Vester, B., 2006. The Cfr rRNA methyltransferase confers resistance to phenicols, lincosamides, oxazolidinones, pleuromutilins, and streptogramin A antibiotics. Antimicrob. Agents Chemother. 50 (7), 2500–2505.

Lozano, C., Aspiroz, C., Rezusta, A., Gómez-Sanz, E., Simon, C., Gómez, P., Ortega, C., Revillo, M.J., Zarazaga, M., Torres, C., 2012a. Identification of novel *vga*(A)-carrying plasmids and a Tn*5406*-like transposon in meticillin-resistant *Staphylococcus aureus* and *Staphylococcus epidermidis* of human and animal origin. Int. J. Antimicrob. Agents 40 (4), 306–612.

Lozano, C., Aspiroz, C., Sáenz, Y., Ruiz-García, M., Royo-García, G., Gómez-Sanz, E., Ruiz-Larrea, F., Zarazaga, M., Torres, C., 2012b. Genetic environment and location of the *lnu*(A) and *lnu*(B) genes in methicillin-resistant *Staphylococcus aureus* and other staphylococci of animal and human origin. J. Antimicrob. Chemother. 67 (12), 2804–2808.

Luna, V.A., Heiken, M., Judge, K., Ulep, C., Van Kirk, N., Luis, H., Bernardo, M., Leitao, J., Roberts, M.C., 2002. Distribution of *mef*(A) in gram-positive bacteria from healthy Portuguese children. Antimicrob. Agents Chemother. 46 (8), 2513–2517.

Lüthje, P., Schwarz, S., 2006. Antimicrobial resistance of coagulase-negative staphylococci from bovine subclinical mastitis with particular reference to macrolide-lincosamide resistance phenotypes and genotypes. J. Antimicrob. Chemother. 57 (5), 966–969.

Lüthje, P., Schwarz, S., 2007. Molecular basis of resistance to macrolides and lincosamides among staphylococci and streptococci from various animal sources collected in the resistance monitoring program BfT-GermVet. Int. J. Antimicrob. Agents 29 (5), 528–535.

Lyon, B.R., Skurray, R., 1987. Antimicrobial resistance of *Staphylococcus aureus*. Genet. Basis Microbiol. Rev. 51 (1), 88–134.

Manian, F.A., 2003. Asymptomatic nasal carriage of mupirocin-resistant, methicillin-resistant *Staphylococcus aureus* (MRSA) in a pet dog associated with MRSA infection in household contacts. Clin. Infect. Dis. 36 (2), e26–e28.

Matsuoka, M., Endou, K., Kobayashi, H., Inoue, M., Nakajima, Y., 1998. A plasmid that encodes three genes for resistance to macrolide antibiotics in *Staphylococcus aureus*. FEMS Microbiol. Lett. 167 (2), 221–227.

Matsuoka, M., Inoue, M., Endo, Y., Nakajima, Y., 2003. Characteristic expression of three genes, *msr*(A), *mph*(C) and *erm*(Y), that confer resistance to macrolide antibiotics on *Staphylococcus aureus*. FEMS Microbiol. Lett. 220 (82), 287–293.

McKenzie, T., Hoshino, T., Tanaka, T., Sueoka, N., 1987. Correction. A revision of the nucleotide sequence and functional map of pUB110. Plasmid 17 (1), 83–85.

Medhus, A., Slettemeås, J., Marstein, L., Larssen, K., Sunde, M., 2013. MRSA with the novel *mecC* gene variant isolated from a cat suffering from chronic conjunctivitis. J. Antimicrob. Chemother. 68 (4), 968–969.

Meka, V.G., Pillai, S.K., Sakoulas, G., Wennersten, C., Venkataraman, L., DeGirolami, P.C., Eliopoulos, G.M., Moellering Jr., R.C., Gold, H.S., 2004. Linezolid resistance in sequential *Staphylococcus aureus* isolates associated with a T2500A mutation in the 23S rRNA gene and loss of a single copy of rRNA. J. Infect. Dis. 190 (2), 311–317.

Mick, V., Domínguez, M.A., Tubau, F., Liñares, J., Pujol, M., Martín, R., 2010. Molecular characterization of resistance to rifampicin in an emerging hospital-associated methicillin-resistant *Staphylococcus aureus* clone ST228, Spain. BMC Microbiol. 10, 68.

Miller, K., Dunsmore, C.J., Fishwick, C.W., Chopra, I., 2008. Linezolid and tiamulin cross-resistance in *Staphylococcus aureus* mediated by point mutations in the peptidyl transferase center. Antimicrob. Agents Chemother. 52 (5), 1737–1742.

Monecke, S., Coombs, G., Shore, A.C., Coleman, D.C., Akpaka, P., Borg, M., Chow, H., Ip, M., Jatzwauk, L., Jonas, D., Kadlec, K., Kearns, A., Laurent, F., O'Brien, F.G., Pearson, J., Ruppelt, A., Schwarz, S., Scicluna, E., Slickers, P., Tan, H.L., Weber, S., Ehricht, R., 2011. A field guide to pandemic, epidemic and sporadic clones of methicillin-resistant *Staphylococcus aureus*. PLoS One 6 (4), e17936.

Monecke, S., Müller, E., Schwarz, S., Hotzel, H., Ehricht, R., 2012a. Rapid microarray-based identification of different *mecA* alleles in staphylococci. Antimicrob. Agents Chemother. 56 (11), 5547–5554.

Monecke, S., Ruppelt, A., Wendlandt, S., Schwarz, S., Slickers, P., Ehricht, R., Jäckel, S.C., 2012b. Genotyping of *Staphylococcus aureus* isolates from diseased poultry. Vet. Microbiol. 162 (2–4), 806–812.

Murphy, E., 1985. Nucleotide sequence of a spectinomycin adenyltransferase AAD(9) determinant from *Staphylococcus aureus* and its relationship to AAD(3″) (9). Mol. Genet. Genom. 200 (1), 33–39.

Murphy, E., Reinheimer, E., Huwyler, L., 1991. Mutational analysis of att*554*, the target of the site-specific transposon Tn*554*. Plasmid 26 (1), 20–29.

Nakaminami, H., Sato-Nakaminami, K., Noguchi, N., 2014. A novel GyrB mutation in meticillin-resistant *Staphylococcus aureus* (MRSA) confers a high level of resistance to third-generation quinolones. Int. J. Antimicrob. Agents 43 (5), 478–479.

Nawaz, M.S., Khan, S.A., Khan, A.A., Khambaty, F.M., Cerniglia, C.E., 2000. Comparative molecular analysis of erythromycin-resistance determinants in staphylococcal isolates of poultry and human origin. Mol. Cell. Probes 14 (5), 311–319.

Needham, C., Rahman, M., Dyke, K.G., Noble, W.C., 1994. An investigation of plasmids from *Staphylococcus aureus* that mediate resistance to mupirocin and tetracycline. Microbiology 140, 2577–2583.

Nurjadi, D., Olalekan, A.O., Layer, F., Shittu, A.O., Alabi, A., Ghebremedhin, B., Schaumburg, F., Hofmann-Eifler, J., Van Genderen, P.J., Caumes, E., Fleck, R., Mockenhaupt, F.P., Herrmann, M., Kern, W.V., Abdulla, S., Grobusch, M.P., Kremsner, P.G., Wolz, C., Zanger, P., 2014. Emergence of trimethoprim resistance gene *dfrG* in *Staphylococcus aureus* causing human infection and colonization in sub-Saharan Africa and its import to Europe. J. Antimicrob. Chemother. 69 (9), 2361–2368.

O'Neill, A.J., Chopra, I., 2006. Molecular basis of *fusB*-mediated resistance to fusidic acid in *Staphylococcus aureus*. Mol. Microbiol. 59 (2), 664–676.

O'Neill, A.J., Huovinen, T., Fishwick, C.W., Chopra, I., 2006. Molecular genetic and structural modeling studies of *Staphylococcus aureus* RNA polymerase and the fitness of rifampin resistance genotypes in relation to clinical prevalence. Antimicrob. Agents Chemother. 50 (1), 298–309.

O'Neill, A.J., McLaws, F., Kahlmeter, G., Henriksen, A.S., Chopra, I., 2007. Genetic basis of resistance to fusidic acid in staphylococci. Antimicrob. Agents Chemother. 51 (5), 1737–1740.

Overesch, G., Büttner, S., Rossano, A., Perreten, V., 2011. The increase of methicillin-resistant *Staphylococcus aureus* (MRSA) and the presence of an unusual sequence type ST49 in slaughter pigs in Switzerland. BMC Vet. Res. 7, 30.

Paterson, G.K., Larsen, A.R., Robb, A., Edwards, G.E., Pennycott, T.W., Foster, G., Mot, D., Hermans, K., Baert, K., Peacock, S.J., Parkhill, J., Zadoks, R.N., Holmes, M.A., 2012. The newly described *mecA* homologue, $mecA_{LGA251}$, is present in methicillin-resistant *Staphylococcus aureus* isolates from a diverse range of host species. J. Antimicrob. Chemother. 67 (12), 2809–2813.

Peacock, S.J., Paterson, G.K., 2015. Mechanisms of methicillin resistance in *Staphylococcus aureus*. Annu. Rev. Biochem. 84, 577–601.

Périchon, B., Courvalin, P., 2009. VanA-type vancomycin-resistant *Staphylococcus aureus*. Antimicrob. Agents Chemother. 53 (11), 4580–4587.

Projan, S.J., Kornblum, J., Moghazeh, S.L., Edelman, I., Gennaro, M.L., Novick, R.P., 1985. Comparative sequence and functional analysis of pT181 and pC221, cognate plasmid replicons from *Staphylococcus aureus*. Mol. Genet. Genom. 199 (3), 452–464.

Qin, X., Poon, B., Kwong, J., Niles, D., Schmidt, B.Z., Rajagopal, L., Gantt, S., 2011. Two paediatric cases of Skin and soft-tissue infections due to clindamycin-resistant *Staphylococcus aureus* carrying a plasmid-encoded *vga*(A) allelic variant for a putative efflux pump. Int. J. Antimicrob. Agents 38 (1), 81–83.

Ramirez, M.S., Tolmasky, M.E., 2010. Aminoglycoside modifying enzymes. Drug Resist. Updat. 13 (6), 151–171.

Robb, A., Pennycott, T., Duncan, G., Foster, G., 2012. *Staphylococcus aureus* carrying divergent *mecA* homologue ($mecA_{LGA251}$) isolated from a free-ranging wild bird. Vet. Microbiol. 162 (1), 300–301.

Roberts, M.C., 1996. Tetracycline resistance determinants: mechanisms of action, regulation of expression, genetic mobility, and distribution. FEMS Microbiol. Rev. 19 (1), 1–24.

Roberts, M.C., Sutcliffe, J., Courvalin, P., Jensen, L.B., Rood, J., Seppala, H., 1999. Nomenclature for macrolide and macrolide-lincosamide-streptogramin B resistance determinants. Antimicrob. Agents Chemother. 43 (12), 2823–2830.

Román, F., Roldán, C., Trincado, P., Ballesteros, C., Carazo, C., Vindel, A., 2013. Detection of linezolid-resistant *Staphylococcus aureus* with 23S rRNA and novel L4 riboprotein mutations in a cystic fibrosis patient in Spain. Antimicrob. Agents Chemother. 57 (5), 2428–2429.

Ross, J.I., Eady, E.A., Cove, J.H., Cunliffe, W.J., Baumberg, S., Wootton, J.C., 1990. Inducible erythromycin resistance in staphylococci is encoded by a member of the ATP-binding transport super-gene family. Mol. Microbiol. 4 (7), 1207–1214.

Rouch, D.A., Byrne, M.E., Kong, Y.C., Skurray, R.A., 1987. The *aacA-aphD* gentamicin and kanamycin resistance determinant of Tn*4001* from *Staphylococcus aureus*: expression and nucleotide sequence analysis. J. Gen. Microbiol. 133 (11), 3039–3052.

Rowland, S.J., Dyke, K.G., 1989. Characterization of the staphylococcal beta-lactamase transposon Tn*552*. EMBO J. 8 (9), 2761–2773.

Sarrou, S., Liakopoulos, A., Tsoumani, K., Sagri, E., Mathiopoulos, K.D., Tzouvelekis, L.S., Miriagou, V., Petinaki, E., 2016. Characterization of a novel *lsa*(E)- and *lnu*(B)-carrying structure located in the chromosome of a *Staphylococcus aureus* sequence type 398 strain. Antimicrob. Agents Chemother. 60 (2), 1164–1166.

Schmitz, F.J., Petridou, J., Milatovic, D., Verhoef, J., Fluit, A.C., Schwarz, S., 2002. In vitro activity of new ketolides against macrolide-susceptible and -resistant *Staphylococcus aureus* isolates with defined resistance gene status. J. Antimicrob. Chemother. 49 (3), 580–582.

Schnellmann, C., Gerber, V., Rossano, A., Jaquier, V., Panchaud, Y., Doherr, M.G., Thomann, A., Straub, R., Perreten, V., 2006. Presence of new *mecA* and *mph*(C) variants conferring antibiotic resistance in *Staphylococcus* spp. isolated from the skin of horses before and after clinic admission. J. Clin. Microbiol. 44 (12), 4444–4454.

Schwarz, S., Feßler, A.T., Hauschild, T., Kehrenberg, C., Kadlec, K., 2011. Plasmid-mediated resistance to protein biosynthesis inhibitors in staphylococci. Ann. N.Y. Acad. Sci. 1241, 82–103.

Schwarz, S., Kadlec, K., Strommenger, B., 2008. Methicillin-resistant *Staphylococcus aureus* and *Staphylococcus pseudintermedius* detected in the BfT-GermVet monitoring programme 2004-2006 in Germany. J. Antimicrob. Chemother. 61 (2), 282–285.

Schwarz, S., Kehrenberg, C., Doublet, B., Cloeckaert, A., 2004. Molecular basis of bacterial resistance to chloramphenicol and florfenicol. FEMS Microbiol. Rev. 28 (5), 519–542.

Schwarz, S., Loeffler, A., Kadlec, K., 2017. Bacterial resistance to antimicrobial agents and its impact on veterinary and human medicine. Vet. Dermatol. 28 (1), 82-e19.

Schwarz, S., Shen, J., Kadlec, K., Wang, Y., Michael, G.B., Feßler, A.T., Vester, B., 2016. Lincosamides, streptogramins, phenicols, and pleuromutilins: mode of action and mechanisms of resistance. Cold Spring Harb. Perspect. Med. 6 (11). pii:a027037.

Schwarz, S., Roberts, M.C., Werckenthin, C., Pang, Y., Lange, C., 1998. Tetracycline resistance in *Staphylococcus* spp. from domestic animals. Vet. Microbiol. 63 (2–4), 217–227.

Schwarz, S., Shen, J., Wendlandt, S., Feßler, A.T., Wang, Y., Kadlec, K., Wu, C.M., 2014. Plasmid-mediated antimicrobial resistance in staphylococci and other firmicutes. Microbiol. Spectr. 2 (6). http://dx.doi.org/10.1128/microbiolspec.PLAS-0020-2014.

Schwendener, S., Perreten, V., 2011. New transposon Tn*6133* in methicillin-resistant *Staphylococcus aureus* ST398 contains *vga*(E), a novel streptogramin A, pleuromutilin, and lincosamide resistance gene. Antimicrob. Agents Chemother. 55 (10), 4900–4904.

Shen, J., Wang, Y., Schwarz, S., 2013. Presence and dissemination of the multiresistance gene *cfr* in Gram-positive and Gram-negative bacteria. J. Antimicrob. Chemother. 68 (8), 1697–1706.

Shore, A.C., Brennan, O.M., Ehricht, R., Monecke, S., Schwarz, S., Slickers, P., Coleman, D.C., 2010. Identification and characterization of the multidrug resistance gene *cfr* in a Panton-Valentine leukocidin-positive sequence type 8 methicillin-resistant *Staphylococcus aureus* IVa (USA300) isolate. Antimicrob. Agents Chemother. 54 (12), 4978–4984.

Shore, A.C., Deasy, E.C., Slickers, P., Brennan, G., O'Connell, B., Monecke, S., Ehricht, R., Coleman, D.C., 2011. Detection of staphylococcal cassette chromosome *mec* type XI carrying highly divergent *mecA*, *mecI*, *mecR1*, *blaZ*, and *ccr* genes in human clinical isolates of clonal complex 130 methicillin-resistant *Staphylococcus aureus*. Antimicrob. Agents Chemother. 55 (8), 3765–3773.

Shore, A.C., Lazaris, A., Kinnevey, P.M., Brennan, O.M., Brennan, G.I., O'Connell, B., Feßler, A.T., Schwarz, S., Coleman, D.C., 2016. First report of *cfr*-carrying plasmids in the pandemic sequence type 22 methicillin-resistant *Staphylococcus aureus* Staphylococcal Cassette Chromosome *mec* Type IV clone. Antimicrob. Agents Chemother. 60 (5), 3007–3015.

Sieber, S., Gerber, V., Jandova, V., Rossano, A., Evison, J.M., Perreten, V., 2011. Evolution of multidrug-resistant *Staphylococcus aureus* infections in horses and colonized personnel in an equine clinic between 2005 and 2010. Microb. Drug Resist. 17 (3), 471–478.

Smith, M.C., Thomas, C.D., 2004. An accessory protein is required for relaxosome formation by small staphylococcal plasmids. J. Bacteriol. 186 (11), 3363–3373.

Smith, T.C., Gebreyes, W.A., Abley, M.J., Harper, A.L., Forshey, B.M., Male, M.J., Martin, H.W., Molla, B.Z., Sreevatsan, S., Thakur, S., Thiruvengadam, M., Davies, P.R., 2013. Methicillin-resistant *Staphylococcus aureus* in pigs and farm workers on conventional and antibiotic-free swine farms in the USA. PLoS One 8 (5), e63704.

Stegmann, R., Perreten, V., 2010. Antibiotic resistance profile of *Staphylococcus rostri*, a new species isolated from healthy pigs. Vet. Microbiol. 145 (1–2), 165–171.

Sugiyama, M., Yuasa, K., Bhuiyan, M.Z., Iwai, Y., Masumi, N., Ueda, K., 1996. IS*431mec*-mediated integration of a bleomycin-resistance gene into the chromosome of a methicillin-resistant *Staphylococcus aureus* strain isolated in Japan. Appl. Microbiol. Biotechnol. 46 (1), 61–66.

Takahashi, H., Kikuchi, T., Shoji, S., Fujimura, S., Lutfor, A.B., Tokue, Y., Nukiwa, T., Watanabe, A., 1998. Characterization of *gyrA*, *gyrB*, *grlA* and *grlB* mutations in fluoroquinolone-resistant clinical isolates of *Staphylococcus aureus*. J. Antimicrob. Chemother. 41 (1), 49–57.

Townsend, D.E., Bolton, S., Ashdown, N., Annear, D.I., Grubb, W.B., 1986. Conjugative staphylococcal plasmids carrying hitch-hiking transposons similar to Tn*554*: intra- and interspecies dissemination of erythromycin resistance. Aust. J. Exp. Biol. Med. Sci. 64 (4), 367–379.

Truong-Bolduc, Q.C., Dunman, P.M., Strahilevitz, J., Projan, S.J., Hooper, D.C., 2005. MgrA is a multiple regulator of two new efflux pumps in *Staphylococcus aureus*. J. Bacteriol. 187 (7), 2395–2405.

Turutoglu, H., Hasoksuz, M., Ozturk, D., Yildirim, M., Sagnak, S., 2009. Methicillin and aminoglycoside resistance in *Staphylococcus aureus* isolates from bovine mastitis and sequence analysis of their *mecA* genes. Vet. Res. Commun. 33 (8), 945–956.

van den Eede, A., Martens, A., Floré, K., Denis, O., Gasthuys, F., Haesebrouck, F., Van den Abeele, A., Hermans, K., 2013. MRSA carriage in the equine community: an investigation of horse-caretaker couples. Vet. Microbiol. 163 (3–4), 313–318.

Vandendriessche, S., Kadlec, K., Schwarz, S., Denis, O., 2011. Methicillin-susceptible *Staphylococcus aureus* ST398-t571 harbouring the macrolide-lincosamide-streptogramin B resistance gene *erm*(T) in Belgian hospitals. J. Antimicrob. Chemother. 66 (11), 24552459.

Villar, M., Marimón, J.M., García-Arenzana, J.M., de la Campa, A.G., Ferrándiz, M.J., Pérez-Trallero, E., 2011. Epidemiological and molecular aspects of rifampicin-resistant *Staphylococcus aureus* isolated from wounds, blood and respiratory samples. J. Antimicrob. Chemother. 66 (5), 997–1000.

Walters, M.S., Eggers, P., Albrecht, V., Travis, T., Lonsway, D., Hovan, G., Taylor, D., Rasheed, K., Limbago, B., Kallen, A., 2015. Vancomycin-resistant *Staphylococcus aureus* - Delaware, 2015. Morb. Mortal. Wkly. Rep. 64 (37), 1056.

Walther, B., Monecke, S., Ruscher, C., Friedrich, A.W., Ehricht, R., Slickers, P., Soba, A., Wleklinski, C.G., Wieler, L.H., Lübke-Becker, A., 2009. Comparative molecular analysis substantiates zoonotic potential of equine methicillin-resistant *Staphylococcus aureus*. J. Clin. Microbiol. 47 (3), 704–710.

Walther, B., Wieler, L.H., Vincze, S., Antão, E.-M., Brandenburg, A., Stamm, I., Kopp, P.A., Kohn, B., Semmler, T., Lübke-Becker, A., 2012. MRSA variant in companion animals. Emerg. Infect. Dis. 18 (12), 2017–2020.

Wang, Y., Zhang, W., Wang, J., Wu, C., Shen, Z., Fu, X., Yan, Y., Zhang, Q., Schwarz, S., Shen, J., 2012a. Distribution of the multidrug resistance gene *cfr* in *Staphylococcus* species isolates from swine farms in China. Antimicrob. Agents Chemother. 56 (3), 1485–1490.

Wang, X.M., Zhang, W.J., Schwarz, S., Yu, S.Y., Liu, H., Si, W., Zhang, R.M., Liu, S., 2012b. Methicillin-resistant *Staphylococcus aureus* ST9 from a case of bovine mastitis carries the genes *cfr* and *erm*(A) on a small plasmid. J. Antimicrob. Chemother. 67 (5), 1287–1289.

Weigel, L.M., Clewell, D.B., Gill, S.R., Clark, N.C., McDougal, L.K., Flannagan, S.E., Kolonay, J.F., Shetty, J., Killgore, G.E., Tenover, F.C., 2003. Genetic analysis of a high-level vancomycin-resistant isolate of *Staphylococcus aureus*. Science 302 (5650), 1569–1571.

Wendlandt, S., Feßler, A.T., Kadlec, K., van Duijkeren, E., Schwarz, S., 2014a. Identification of the novel spectinomycin resistance gene *spd* in a different plasmid background among methicillin-resistant *Staphylococcus aureus* CC398 and methicillin-susceptible *S. aureus* ST433. J. Antimicrob. Chemother. 69 (7), 2000–2003.

Wendlandt, S., Li, J., Ho, J., Porta, M.A., Feßler, A.T., Wang, Y., Kadlec, K., Monecke, S., Ehricht, R., Boost, M., Schwarz, S., 2014b. Enterococcal multiresistance gene cluster in methicillin-resistant *Staphylococcus aureus* from various origins and geographical locations. J. Antimicrob. Chemother. 69 (9), 2573–2575.

Wendlandt, S., Feßler, A.T., Monecke, S., Ehricht, R., Schwarz, S., Kadlec, K., 2013a. The diversity of antimicrobial resistance genes among staphylococci of animal origin. Int. J. Med. Microbiol. 303 (6–7), 338–349.

Wendlandt, S., Li, B., Lozano, C., Ma, Z., Torres, C., Schwarz, S., 2013b. Identification of the novel spectinomycin resistance gene *spw* in methicillin-resistant and methicillin-susceptible *Staphylococcus aureus* of human and animal origin. J. Antimicrob. Chemother. 68 (7), 1679–1680.

Wendlandt, S., Li, B., Ma, Z., Schwarz, S., 2013c. Complete sequence of the multi-resistance plasmid pV7037 from a porcine methicillin-resistant *Staphylococcus aureus*. Vet. Microbiol. 166 (3–4), 650–654.

Wendlandt, S., Kadlec, K., Feßler, A.T., Schwarz, S., 2015. Identification of ABC transporter genes conferring combined pleuromutilin-lincosamide-streptogramin A resistance in bovine methicillin-resistant *Staphylococcus aureus* and coagulase-negative staphylococci. Vet. Microbiol. 177 (3–4), 353–358.

Werckenthin, C., Cardoso, M., Martel, J.-L., Schwarz, S., 2001. Antimicrobial resistance in staphylococci from animals with particular reference to bovine *Staphylococcus aureus*, porcine *Staphylococcus hyicus*, and canine *Staphylococcus intermedius*. Vet. Res. 32, 341–362.

Wichelhaus, T.A., Böddinghaus, B., Besier, S., Schäfer, V., Brade, V., Ludwig, A., 2002. Biological cost of rifampin resistance from the perspective of *Staphylococcus aureus*. Antimicrob. Agents Chemother. 46 (11), 3381–3385.

Wichelhaus, T.A., Schäfer, V., Brade, V., Böddinghaus, B., 1999. Molecular characterization of *rpoB* mutations conferring cross-resistance to rifamycins on methicillin-resistant *Staphylococcus aureus*. Antimicrob. Agents Chemother. 43 (11), 2813–2816.

Wipf, J.R., Perreten, V., 2016. Methicillin-resistant *Staphylococcus aureus* isolated from dogs and cats in Switzerland. Schweiz. Arch. Tierheilkd 158 (6), 443–450.

Xiong, L., Kloss, P., Douthwaite, S., Andersen, N.M., Swaney, S., Shinabarger, D.L., Mankin, A.S., 2000. Oxazolidinone resistance mutations in 23S rRNA of *Escherichia coli* reveal the central region of domain V as the primary site of drug action. J. Bacteriol. 182 (19), 5325–5331.

Yu, F., Lu, C., Liu, Y., Sun, H., Shang, Y., Ding, Y., Li, D., Qin, Z., Parsons, C., Huang, X., Li, Y., Hu, L., Wang, L., 2014. Emergence of quinupristin/dalfopristin resistance among livestock-associated *Staphylococcus aureus* ST9 clinical isolates. Int. J. Antimicrob. Agents 44 (5), 416–419.

Zhou, W., Shan, W., Ma, X., Chang, W., Zhou, X., Lu, H., Dai, Y., 2012. Molecular characterization of rifampicin-resistant *Staphylococcus aureus* isolates in a Chinese teaching hospital from Anhui, China. BMC Microbiol. 12, 240.

CHAPTER

5

BIOFILM FORMATION OF *STAPHYLOCOCCUS AUREUS*

Daniel Vázquez-Sánchez[1], Pedro Rodríguez-López[2]

[1]*"Luiz de Queiroz" College of Agriculture (ESALQ), University of São Paulo (USP), Sao Paulo, Brazil;* [2]*Instituto de Investigaciones Marinas (IIM-CSIC), Vigo, Spain*

1. INTRODUCTION

Biofilms are the normal lifestyle of *Staphylococcus aureus*. Biofilm formation provides this bacterial pathogen several survival advantages in comparison to the planktonic counterparts, such as beneficial cell–cell interactions, increased protection against external stresses along with an increased dispersal capacity. Consequently, biofilms increase the persistence and proliferation of *S. aureus* both in biotic and abiotic habitats, being of high concern in the case of hospital and food facilities. Along this chapter, authors will show and explain the major processes known to date regarding biofilm development in *S. aureus*.

2. BIOFILMS—AN OVERVIEW

A biofilm is a sessile community of microorganisms formed by cells attached to a biotic or abiotic surface, embedded in a self-produced extracellular matrix and with altered phenotypic and genotypic characteristics that allow the adaptation to adverse environmental conditions. Biofilms are considered the prevailing microbial lifestyle in natural habitats due to their protector ability during the growth, allowing the survival under extreme environmental conditions of temperature (e.g., thermal waters and glaciers), acidity (e.g., sulfuric pools and geysers), and humidity (e.g., deserts and rainforests), among others (Dufour et al., 2012). Microorganisms also grow predominantly in form of biofilms on practically any kind of industrial surface, causing food and water contamination, metal surface corrosion, and the obstruction of equipments (Beech et al., 2005; Srey et al., 2013). Particularly in the food industry, biofilm formation may contribute to the persistence of spoilage and pathogenic microorganisms in food-processing environments, consequently increasing cross-contamination possibilities, which may involve a serious risk for the consumer health as well as subsequent economic losses due to recalls of contaminated food products. Concerning the medical ambit, the US National Institute of Health estimates that biofilms are involved in up to 75% of microbial infections in humans. However, most laboratory studies and many standard methods are still performed using only planktonic cells, though this state is considered a transitory phase in which aims to translocate microorganisms to other surfaces.

The development of biofilms is a dynamic process affected by the substratum (i.e., texture or roughness, hydrophobicity, surface chemistry, charge, and preconditioning film), the medium

Staphylococcus aureus. http://dx.doi.org/10.1016/B978-0-12-809671-0.00005-X

(i.e., nutrient levels, ionic strength, temperature, pH, flow rate, and presence of antimicrobial agents), and intrinsic properties of cells (i.e., cell surface hydrophobicity, extracellular appendages, extracellular polymeric substances (EPS), signaling molecules) (Donlan, 2002; Renner and Weibel, 2011). As shown in Fig. 5.1, the process of biofilm formation comprises (1) a reversible attachment of cells by weak interactions (i.e., Van der Waals forces) to a preconditioning film formed on the abiotic or biotic surface (Bos et al., 1999; Donlan, 2002); (2) an irreversible adsorption to the surface by hydrophilic/hydrophobic interactions, electrostatic forces, and Lewis acid–base interactions mediated by several attachment structures (e.g., flagella, fimbriae, lipopolysaccharides, and adhesive proteins) (Bos et al., 1999; Donlan, 2002); (3) the proliferation of adsorbed cells and production of a self-produced EPS matrix mainly composed by polysaccharides, proteins, and extracellular DNA (eDNA) (Branda et al., 2005; Flemming et al., 2007); (4) the formation of a mature biofilm, whose structure can be flat or mushroom-shaped and that contains water channels that effectively distribute nutrients and signaling molecules within the biofilm (Dufour et al., 2012; Hall-Stoodley et al., 2004); (5) the detachment of biofilm cells individually or in clumps as a response to external or internal factors; and, finally, (6) the spread and colonization of other niches (Srey et al., 2013).

3. *STAPHYLOCOCCUS AUREUS* BIOFILMS

Biofilms formed by *S. aureus* have been detected on diverse biotic and abiotic surfaces, including human tissues, indwelling medical devices (e.g., implanted catheters, artificial heart valves, bone, and joint prostheses), food products, and food-processing facilities (Simon and Sanjeev, 2007; Trampuz and Widmer, 2006; Vázquez-Sánchez et al., 2012). The development of biofilms increases the persistence of this bacterial pathogen in environments that could generate a risk for the human health. Therefore, a deep understanding of the processes involved in the biofilm formation of *S. aureus* is necessary before the selection of an effective control strategy against this pathogen.

3.1 INITIAL ADHERENCE

The strategy of *S. aureus* cells to attach initially varies between abiotic and biotic surfaces. Thus, the initial adhesion to abiotic surfaces mostly depends on the physicochemical characteristics both of the cell and the contact surface. Hydrophobic and electrostatic interactions participate generally in this initial attachment, but some specific bacterial surface molecules such as autolysins or teichoic acids have been also described to be involved in staphylococci (Gross et al., 2001; Heilmann et al., 1997). Probably, these determinants are able to modify the physicochemical properties of the bacterial surface to adapt to those present in the abiotic surface. Interestingly, high ionic strength conditions, such as the presence of seawater in fisheries or saline serum in hospitals, can also improve the adhesion of *S. aureus* to some abiotic surfaces. For example, high ionic strength conditions can attenuate the repulsive electrostatic interactions between highly negatively charged bacteria and negatively charged surfaces such as polystyrene (Vázquez-Sánchez et al., 2013).

In contrast, the attachment of *S. aureus* to biotic surfaces requires much more specific interactions, which are mediated by a great variety of cell wall-anchored proteins. Among the four families of cell wall-anchored proteins proposed by Foster et al. (2014), the microbial surface component

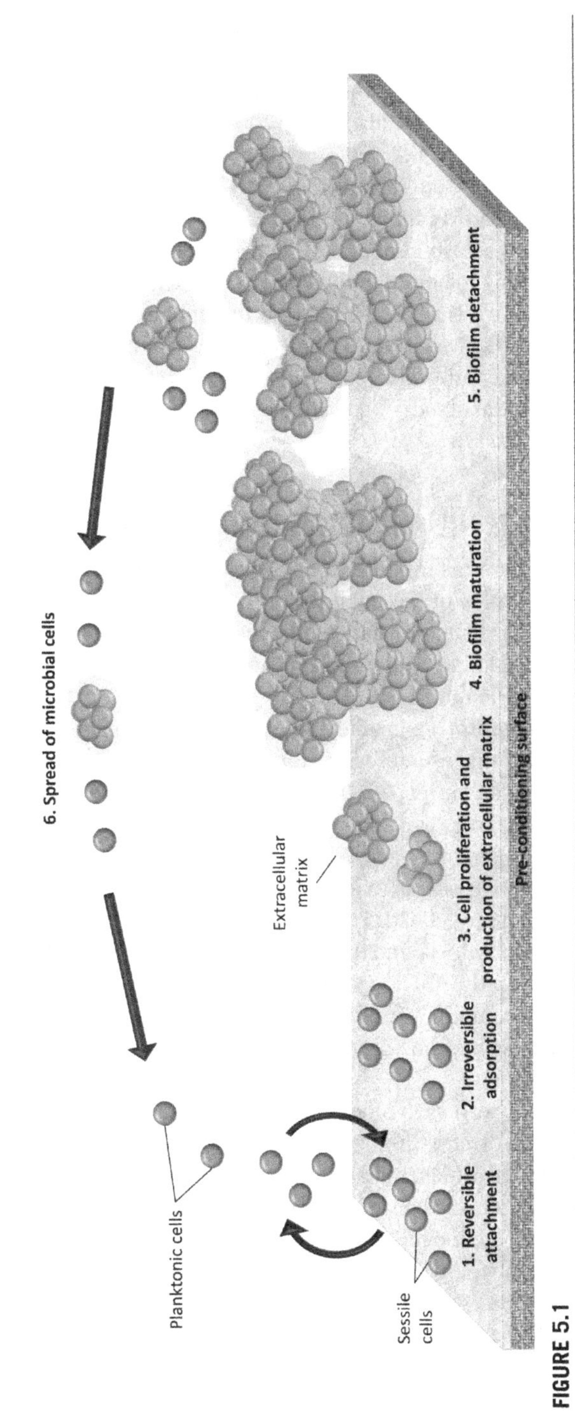

FIGURE 5.1
Processes involved in the development of a biofilm.

recognizing adhesive matrix molecules (MSCRAMMs), the near iron transporter (NEAT) motif family, and the G5–E repeat family are involved in the adhesion of *S. aureus* to biotic surfaces. For example, the collagen adhesin is a MSCRAMMs related with the adhesion to collagen-rich tissues (Zong et al., 2005). Other MSCRAMMs are also involved in the adhesion to the extracellular matrix, such as the fibronectin-binding proteins A (FnBPA) and B (FnBPB), and the bone sialoprotein-binding protein (Burke et al., 2011; Geoghegan et al., 2013; McCourt et al., 2014; Vazquez et al., 2011). Meanwhile, some MSCRAMMs intercede in the adhesion to immobilized fibrinogen (e.g., the clumping factor A or ClfA) (Ganesh et al., 2008; Geoghegan et al., 2010a,b), in the adhesion to desquamated epithelial cells, and in the nasal colonization (e.g., ClfB) (Ganesh et al., 2011; Mulcahy et al., 2012; O'Brien et al., 2002; Walsh et al., 2004; Wertheim et al., 2008; Xiang et al., 2012). The iron-regulated surface protein A (IsdA), a NEAT motif protein with the ability to capture heme from hemoglobin under iron-restricted conditions in the host, the *S. aureus* surface protein G (SasG), and a G5–E repeat protein, are also involved in the adhesion to desquamated epithelial cells (Clarke et al., 2006, 2009; Roche et al., 2003).

3.2 BIOFILM DEVELOPMENT: THE IMPORTANCE OF THE EXTRACELLULAR MATRIX

After initial adhesion, bacteria secrete a mixture of EPS that embed cells (Flemming and Wingender, 2010). This matrix has different roles in the integrity of the biofilm structure, not only as a surface-anchoring substance but also contributes to the stabilization and maintenance of mature structures, while offering protection to the residing cells from physical, chemical, and biological environmental aggressions (Branda et al., 2005; Flemming, 2011).

Because cells residing into a biofilm are unable to move physically, matrix plays an essential role to avoid desiccation and starvation acting as a reservoir for water and nutrients, respectively. However, to be immobilized does not mean to be entrapped. In fact, matrix is a high dynamic structure due to the presence of various hydrolytic enzymes capable of degrading compounds already present that can be further used as a nutrient source (Flemming and Wingender, 2010). These hydrolytic enzymes are also able to degrade matrix structural components to promote detachment phenomena and thus contributing to the dispersion of the cells (Archer et al., 2011; Foulston et al., 2014). Another important function of the staphylococcal biofilm matrix is to maintain cells into close proximity, facilitating the cell–cell communication due to an optimized diffusion of the signals produced by the cells (Flemming and Wingender, 2010). This cell–cell communication allows thus to establish local foci of communal cooperation between cells, making the biofilm to act as a multicellular organism (Branda et al., 2005; Flemming and Wingender, 2010).

Biofilm matrix in *S. aureus*, as well as in other Gram-positive species, is complex, heterogeneous, and strongly influenced by growth conditions and environmental stimuli (Foulston et al., 2014). Consequently, providing a complete biochemical profile of *S. aureus* matrix components still remains an important challenge for researchers due to the difficulty of extract and purifies all components (Flemming et al., 2007). Besides, ad hoc methods for each one of the fractions are normally required for efficient EPS characterization (Flemming and Wingender, 2010). Nevertheless, it is known that three different main fractions are present in the composition of the extracellular matrix produced by this pathogen: proteins, polysaccharides, and eDNA (Lister and Horswill, 2014). A general perspective of these three main components of the extracellular matrix of *S. aureus* is given next in this chapter.

3.2.1 Proteins

Several proteins have been identified to date in the biofilm matrix of *S. aureus*. The functionality of these molecules is diverse and play important roles, especially in the adhesion and accumulation of biomass onto surfaces (Lister and Horswill, 2014; Paharik and Horswill, 2016).

Many of these proteins are found to be anchored to the peptidoglycan walls, providing a bond between cells and abiotic surface. Among them, biofilm associated protein (Bap) is one of the major components. It was firstly identified in *S. aureus* biofilm-defective mutants and it facilitates the primary adhesion in biofilm formation and the host interaction in bovine mastitis (Di Martino, 2016; Lister and Horswill, 2014). Recently, a study carried out by Taglialegna et al. (2016) have demonstrated that native Bap proteins containing at least the N-terminus can aggregate forming amyloid-like fibers into the biofilm matrix in low pH and low Ca^{2+}, providing a scaffold between cells and, thus, improving biofilm stabilization. Modifications in the protein composition of the extracellular matrix in response to low pH were also observed by Foulston et al. (2014). These authors perceived that a high proportion of the proteins present in the extracellular matrix have a cytoplasmic origin and they are somehow nonspecifically incorporated and recycled into the matrix after cell lysis, although with different roles.

Another group of structural matrix proteins are the surface-wall associated proteins, which function as bacterial adhesins. In *S. aureus*, the MSCRAMMs represent the main group, comprising a wide diversity of proteins characterized by containing IgG-like domains linked tandemly (Foster et al., 2014). Among MSCRAMMs involved in the development of biofilms, the staphylococcal ClfA and ClfB and the FnBPA and FnBPB are the most prominent. Both ClfA and ClfB present in the cell wall of *S. aureus* mediate bacterial clumping to fibrinogen in solution, although only ClfB seems to promote biofilm formation in vitro in absence of Ca^{2+} when other matrix components such as poly-β(1,6)-N-acetyl-D-glucosamine glycans (PNAG) or eDNA become not functional (Abraham and Jefferson, 2012). Meanwhile, FnBPA and FnBPB have elastin and fibrinogen-linking domains that promote bacterial aggregation and accumulation (Mccourt et al., 2014; O'Neill et al., 2008; Paharik and Horswill, 2016). A different surface-wall associated protein, the SasG, has been also related with the cell accumulation and biofilm formation in *S. aureus*, but in a Zn^{2+}-dependent manner (Corrigan et al., 2007; Geoghegan et al., 2010a,b). Similarly, Schroeder et al. (2009) exposed the key role of the surface protein SasC in the cluster formation and intercellular adhesion (ica) of *S. aureus* cells during host infection.

The expansion of *S. aureus* biofilms is also controlled by secreted proteins, particularly the phenol-soluble modulins (PSMs). PSMs are involved in the formation of channels into mature biofilms, which allow the delivery of nutrients to deeper layers (Periasamy et al., 2012). The expression of these surfactant-like molecules is induced by the quorum-sensing system accessory gene regulator (*agr*), which is stimulated by auto-inducing peptides (Boles and Horswill, 2008; Lauderdale et al., 2010; Yarwood et al., 2004). Besides, Merino et al. (2009) observed that the multifactorial virulence factor SpA has also a significant role in the development of biofilm-associated infections when expressed at high levels, inducing bacterial aggregation and biofilm formation without being covalently anchored to the cell wall.

3.2.2 Polysaccharides

Polysaccharides, as well as proteins, constitute an essential part of the biofilm matrix in *S. aureus* and they are involved in various processes. The main component of this fraction is the polysaccharide intercellular adhesin (PIA), a PNAG with partially deacetylated residues. PNAGs are synthetized by the products of the chromosomal *ica* operon (also named *ica*ADBC) carried by most *S. aureus* strains (Cramton et al., 1999; Fitzpatrick et al., 2005; Maira-Litrán et al., 2002). Subunits IcaA and IcaD form a transmembrane protein N-acetyl-glucosamine transferase that produce N-acetyl-glucosamine oligomers, which are elongated and translocated to the cell surface by the membrane protein IcaC (Gerke et al., 1998), where the surface-attached protein IcaB finally deacetylates the polymers, introducing positive charges that enhance the adhesion of PIA to the bacterial surface (Vuong et al., 2004). The *ica* operon is fundamentally repressed by the *icaR* gene products (Jefferson et al., 2003), which are triggered by the stress-induced sigma factor B, the staphylococcus accessory regulator A, the LuxS/AI-2 quorum-sensing system and, indirectly, by the *rbf* gene (Cerca et al., 2008; Cue et al., 2009; Yu et al., 2012). The role of the *ica* operon in the biofilm formation of *S. aureus* is complex and likely to be both strain- and environment-dependent (O'Gara, 2007). In fact, the expression of the *ica* operon is influenced by different environmental factors such as anaerobic conditions, glucose, ethanol, osmolarity, temperature, and antimicrobial agents such as tetracycline (Cramton et al., 2001; Fitzpatrick et al., 2005).

PIAs play an important role in accumulation and cohesion of biofilms, and they are involved in vivo in a mechanism for immune evasion both in *S. aureus* and in coagulase negative staphylococci. However, PIAs seem not to be essential for biofilm formation, because some PIA-defective strains of *S. aureus* exhibit both in vivo and in vitro strong biofilm forming features despite its structural properties (Fitzpatrick et al., 2005; Lister and Horswill, 2014). Nevertheless, a study carried out by Izano et al. (2008) demonstrated that *S. aureus* biofilms growth in polystyrene is impaired by Dispersin B (a PNAG-degrading enzyme) if it is added to the medium, but does not affect the already formed biofilms. Therefore, they concluded that PIA is not a key structural component in *S. aureus* biofilms and may be only relevant in primary formation stages. Hence, the role of PIAs still remains unclear in *S. aureus* because different strains display various phenotypes regardless of the presence of this molecule.

Although not commonly associated with biofilm matrix, teichoic acids are nowadays also considered to be part of it (Archer et al., 2011; Payne and Boles, 2016). They are complex polysaccharides mainly anchored to the cell wall. Teichoic acids are formed by a chain of phosphate alternated by ribitol or glycerol, which are usually substituted by D-alanine and N-acetylglucosamine residues that confer a positive net charge to the cell surface (Gross et al., 2001). It has been postulated that teichoic acids play a key role in the initial adhesion and subsequent formation of the biofilm by electrochemical interactions between the cell wall and the surface (Götz, 2002; Gross et al., 2001). In fact, mutations in the *dlt* operon, responsible of the incorporation of D-alanine to the teichoic acids, modify the final structure of the molecule, producing a subsequent electrical misbalance in the cell wall that increase its negative charge and alter the biofilm formation capacity (Fitzpatrick et al., 2005). Biofilm forming ability can be restored in *dlt*-defective mutants by the addition of Mg^{2+} to the medium, which remarks the importance of the presence of a positive charge in the cell wall during biofilm formation (Götz, 2002).

3.2.3 Extracellular DNA

Likewise to teichoic acids, eDNA has been proposed to be an essential structural component of the biofilm matrix in *S. aureus* as well as in other staphylococci species (Izano et al., 2008). Its negative electric charge nature allows eDNA to act as a binder in cell–cell, cell–surface, and cell–host interactions (Lister and Horswill, 2014). It has been postulated that eDNA has a genomic origin due to the size of its functional fragments (Izano et al., 2008). The presence of eDNA outside the cell seems to be due to preprogrammed lysis of a subpopulation of cells into the biofilm, leading to a release of DNA into the matrix (Thomas and Hancock, 2009) after degradation of *S. aureus* cell wall by murein hydrolases (Lister and Horswill, 2014; Rice et al., 2007). This process is controlled by *cidA* and *lgrA* encoding holins and antiholins, respectively (Rice et al., 2007).

eDNA participates both in biofilm formation and maturation, as well as in maintenance of mature structures. This dual role of eDNA has been ascertained in a study performed by Izano et al. (2008), who observed that the addition of deoxyribonuclease I into an *S. aureus* culture inhibited biofilm formation, being also capable to disperse preformed biofilm on polystyrene plates. Besides, it has been demonstrated that eDNA can have protective roles because its depolymerization sensitizes *S. aureus* biofilms to antimicrobials, mainly due to a deep alteration in the matrix structure (Izano et al., 2008). It has been also reported that eDNA is able to interact with other molecules of biofilm matrices, acting as a cohesion molecule that form a scaffold between different components to further increase the stability of the final structure (Huseby et al., 2010; Schwartz et al., 2016).

Because eDNA is present in large quantities into the biofilm matrix, it represents a considerable genetic pool that can promote the horizontal gene transfer, especially in biofilms with several strains with different characteristics (Flemming et al., 2007; Schwartz et al., 2016). Consequently, it can promote a possible transference and subsequent spreading of antimicrobial resistances, representing a serious threat in clinically associated biofilms.

3.3 BIOFILM DISPERSION

Detachment of cells is the last step of the life cycle of a biofilm, allowing *S. aureus* to colonize new niches. Perturbations in the environmental conditions such as an increased fluid shear can generate an unintentional dispersion of biofilm cells (O'Toole et al., 2000). However, *S. aureus* normally use self-controlled processes, including both broad-spectrum and specific mechanisms, which facilitate the release of biofilm cells in the environment. Curiously, the efficacy of each of these mechanisms is highly dependent on the composition of the polymeric matrix (Chaignon et al., 2007; Kiedrowski et al., 2011).

A broad-spectrum dispersal mechanism of *S. aureus* is the production of D-amino acids during the late stationary phase. D-amino acids are incorporated into the peptidoglycan, perturbing the attachment of the major matrix protein TasA to the cell wall (Hochbaum et al., 2011; Kolodkin-Gal et al., 2010). As a result, the *ica* is decreased and clumps of biofilm cells are dispersed. The application of D-amino acids to polymeric surfaces has also showed a high efficacy to diminish *S. aureus* biofilm formation in vitro (Hochbaum et al., 2011; Sanchez et al., 2013, 2014), leading to interesting prevention strategies to control biofilm formation.

The stringent response also seems to play some role in dispersal of *S. aureus*, as observed by Fuente-Nunez et al. (2014) with the application of a potent antibiofilm peptide that causes the degradation of the ppGpp alarmone, which generates regulatory changes that inactive the metabolism of biofilm cells (Srivatsan and Wang, 2008). However, the effect of stringent response in *S. aureus* has been only demonstrated during infection (Geiger et al., 2010). Thus, further studies are still necessary to clarify the role of stringent response in the dispersal of *S. aureus* biofilms.

Biofilm cells can be also dispersed by the unspecific action of PSMs, leading to the dissemination of the infection (Periasamy et al., 2012). As aforementioned, PSMs also originate the characteristic channels of mature biofilms.

The secretion of proteases is a more specific dispersal mechanism of *S. aureus*, which cause the degradation of proteinaceous matrix components and, thus, the disruption of the biofilm. Ten different secreted proteases are produced by *S. aureus* (Shaw et al., 2004), including the V8 serine protease (SspA), the aureolysin (Aur), and the staphopains (SspB and ScpA). The SspA is able to degrade both FnBPs and Bap proteins (McGavin et al., 1997; O'Neill et al., 2008; Marti et al., 2010). Aur disturbs the functional properties of ClfB and Bap (Marti et al., 2010; Abraham and Jefferson, 2012). Meanwhile, the staphopains can disrupt the biofilm matrix (Mootz et al., 2013), although the target proteins which they affect are still unknown. In contrast, several proteins have been proposed as possible targets for degradation, such as Atl, Spa, and SasG (Lauderdale et al., 2009; Kolar et al., 2013), but they have not been associated with some specific protease yet. More proteomic analyses are therefore required to explain clearly the process of protease-mediated dispersal. Similarly to PSMs, the production of proteases is also positively regulated by the quorum-sensing system *agr* (Thoendel et al., 2011).

The secretion of thermostable nucleases by *S. aureus* seems to be also implicated in the biofilm dispersion occurred during the infection (Berends et al., 2010; Olson et al., 2013) and in the formation of channels within the biofilm (Kiedrowski et al., 2014). In fact, the expression of the nuclease Nuc is inhibited during biofilm formation to accumulate and preserve high molecular weight eDNA (Mann et al., 2009; Kiedrowski et al., 2011; Olson et al., 2013). However, further investigations are necessary to clarify the role of nucleases during biofilm dispersion.

The presence of diverse enzymes secreted by other microorganisms in the environment can also affect the biofilm growth of *S. aureus* in bacterial communities. For example, it was found that *Staphylococcus epidermidis* produces the serine protease Esp, which is able to cleave several matrix proteins of *S. aureus* such as Atl, Eap, and FnBPA, and also avoids the release of eDNA by disrupting murein hydrolases (Chen et al., 2013; Sugimoto et al., 2013). Meanwhile, *Actinobacillus actinomycetemcomitans* secretes the enzyme dispersin B, which hydrolyzes the glycosidic linkages of PIA in *S. epidermidis* and *S. aureus* biofilms (Kaplan et al., 2004). The potential of these dispersal enzymes as a strategy to control the biofilm formation of bacterial pathogens such as *S. aureus*, as well as the search of some other dispersal molecules, must be further implemented.

4. WHY BEING IN A BIOFILM? ADVANTAGES OF MULTICELLULAR STRUCTURES

4.1 CELL–CELL COMMUNICATION

The establishment of bacteria on surfaces generates a higher degree of stability in the cell growth and enables beneficial cell–cell interactions, such as quorum sensing and genetic exchange (Daniels et al., 2004; Davey and O'Toole, 2000; Elias and Banin, 2012; Hall-Stoodley et al., 2004; Watnick and Kolter, 2000). In the case of quorum sensing, low-molecular mass signaling molecules called auto-inducers (AI) regulate gene expression, metabolic cooperativity and competition, physical contact, and bacteriocin production, providing thus a mechanism for self-organization and regulation of biofilm cells (Daniels et al., 2004; Donlan, 2002; Elias and Banin, 2012; Parsek and Greenberg, 2005; van-Houdt and Michiels, 2010). The quorum sensing effects depend on the concentration of AI, which increases in a cell-density-dependent manner (Parsek and Greenberg, 2005). Meanwhile, the transmission of mobile genetic elements between nearby biofilm cells allows the acquisition of new antimicrobial resistance or virulence factors, and environmental survival capabilities (Madsen et al., 2012).

4.2 INCREASED RESISTANCE TO EXTERNAL STIMULUS

Bacterial cells in biofilms are protected against a wide range of environmental stresses, leading to a higher survival and persistence when compared with planktonic state, which is particularly problematic in clinical and food environments. Several physiological characteristics and mechanisms are responsible of this increase of resistance, such as the presence of the extracellular matrix, the expression of specific mechanisms of defense, and the changes in the physiology of biofilm cells. The presence of extracellular matrix acts as a barrier that slows down the infiltration, neutralizes, binds, and effectively diffuses to sublethal concentrations of antimicrobial agents (e.g., chlorine species, oxacillin, and vancomycin) before they can reach cell targets (Bridier et al., 2011; Singh et al., 2010). The extracellular matrix can also reduce other adverse external effects such as UV light, toxic metals, acidity, desiccation, salinity, and host defenses (Hall-Stoodley et al., 2004). Biofilm cells can also activate chromosomal β-lactamases and efflux pumps (e.g., QAC efflux system of *S. aureus*), induce mutations in antimicrobial target molecules or, indirectly, release eDNA that promotes the synthesis of biofilm matrix to counteract the effect of antimicrobials (Anderson and O'Toole, 2008; Bridier et al., 2011; Høiby et al., 2010; Kaplan et al., 2012). Changes in the expression of specific genes associated with sessile growth and the physiological differentiation between biofilm cells actively growing in the outer surface, those of oxygen- and nutrient-deprived inner interfaces and the dormant cells also reduce the susceptibility to antimicrobials (Bridier et al., 2011; Gilbert et al., 2002; Høiby et al., 2010). For example, persistent cells are able to produce particular cellular toxins that block cellular processes like translation, thus rendering protection against biocides that act only against active cells (Lewis, 2010).

4.3 INCREASED DISPERSAL CAPACITY

The dispersal of biofilm cells in clumps may provide a sufficient number of cells for an infective dose that is not typically found in bulk fluid, enabling an enhanced transmission and infection of bacterial pathogens (Hall-Stoodley and Stoodley, 2005). Detachment can be promoted by environmental changes such as nutrient starvation or by internal biofilm processes such as endogenous enzymatic degradation, or the release of EPS or surface-binding proteins (Srey et al., 2013).

5. MULTISPECIES *STAPHYLOCOCCUS AUREUS*—CARRYING BIOFILMS

Most investigations regarding different aspects of *S. aureus* biofilms have been carried out using monospecies cultures. Nevertheless, it has been extensively demonstrated that the majority of the staphylococcal species, including *S. aureus*, are able to associate with other microorganisms such as bacteria and yeasts and form polymicrobial structures (Elias and Banin, 2012; Götz, 2002). In these multispecies communities, the existing relationships between *S. aureus* and the accompanying microbiota clearly influence the formation kinetics, morphology, and physicochemical characteristics of the final structure if compared to monospecies biofilms (Elias and Banin, 2012; Harriott and Noverr, 2009; Kucera et al., 2014).

It has been detected that the distribution of *S. aureus* in a multispecies biofilm seems to be affected by the actual species composition of the community. Harriott and Noverr (2009) observed that *S. aureus* tends to be located on the top layers of a dual-species biofilm with *Candida albicans*. Similar results were obtained by Kucera et al. (2014), who showed that *S. aureus* have a physical preference to be located in the upper zones of the biofilm, when forming part of a polymicrobial structure with *Pseudomonas aeruginosa*, *Enterococcus faecalis*, and *Bacillus subtilis*. However, the organization of the different microorganisms inside the biofilm can be also influenced by different environmental factors (Gutiérrez et al., 2012).

One of the main concerns of *S. aureus* in multispecies biofilms is the increased resistance to antimicrobials, being of a special relevance in clinical environments (Elias and Banin, 2012; Hoffman et al., 2006; Kart et al., 2014). A particular case of antimicrobial resistance in *S. aureus* occurred when it is associated with *P. aeruginosa* in patients with cystic fibrosis (Elias and Banin, 2012). *P. aeruginosa* generates several toxic products that negatively affect the growth of *S. aureus* (Biswas et al., 2009), mainly the 4-hydroxy-2-heptylquinoline-N-oxide, which specifically suppresses the *S. aureus* respiration (Hoffman et al., 2006). This adverse situation leads to the selection of *S. aureus* small-colony variants within the biofilm, which have high antimicrobial resistance and persistence. Thus, polymicrobial structures such as these are commonly associated with colonization of indwelling medical devices (e.g., stents, implants, catheters, etc.), which can cause severe infections with high morbidity and mortality rates (Kart et al., 2014). Moreover, Harriott and Noverr (2009) demonstrated that *S. aureus* associated with *C. albicans* is not only able to form biofilm in conditions in which otherwise it would not be able to adhere, but also to show a complete lack of response to vancomycin due to its embedding into the polymeric matrix secreted by *C. albicans*. Despite these observations, the mechanisms of this increasing antimicrobial resistance in *S. aureus* are still not completely understood and they require further investigation.

On the other hand, it has been observed that in a biofilm coculture with *C. albicans* and *P. aeruginosa*, *S. aureus* appears to be more susceptible to disinfectants such as cetrimide and chlorhexidine, among others (Kart et al., 2014).

6. BIOFILM PRODUCING *STAPHYLOCOCCUS AUREUS* IN THE FOOD INDUSTRY

As *S. aureus* is a major component of the human microbiome, a high degree of handling can enhance the spread of *S. aureus* to food and food-contact surfaces (DeVita et al., 2008; Sattar et al., 2001; Simon and Sanjeev, 2007; Sospedra et al., 2012). Once there, formation of biofilms increases the resistance of *S. aureus* to food-processing stresses, such as antimicrobial agents, relatively high temperatures, high salt contents, etc (Bridier et al., 2011; van-Houdt and Michiels, 2010; Vázquez-Sánchez et al., 2013). Consequently, the persistence of *S. aureus* increases as well as the cross-contamination possibilities, which may compromise the safety of food products.

S. aureus is able to form biofilms on diverse work surfaces widely used in industrial facilities such as polystyrene, polypropylene, stainless steel and glass, and also in food products (DeVita et al., 2008; Sattar et al., 2001; Simon and Sanjeev, 2007; Vázquez-Sánchez et al., 2013, 2014). Alterations in both substratum properties and environmental factors such as osmolarity, nutrient content, and temperature, may lead to staphylococcal biofilm development (Pagedar et al., 2010; Rode et al., 2007; Xu et al., 2010; Vázquez-Sánchez et al., 2013). Therefore, a proper maintenance of food-processing equipment, a rigorous monitoring of environmental conditions, and the application of effective sanitizing methods are indispensable to control the proliferation of *S. aureus* biofilms.

7. CONCLUSION

Biofilm formation has a key role in the survival and dispersion of *S. aureus* in environments that could generate a risk for the human health. Thus, to know and understand completely all the underlying processes involved in the development of biofilms, in the cell-to-cell communication, in the resistance to external stresses, and in the dispersion of cells from mature structures, among others, are essential to find effective control strategies against *S. aureus*.

REFERENCES

Abraham, N.M., Jefferson, K.K., 2012. *Staphylococcus aureus* clumping factor B mediates biofilm formation in the absence of calcium. Microbiology 158, 1504–1512.

Archer, N.K., Mazaitis, M.J., Costerton, J.W., Leid, J.G., Powers, M.E., Shirtliff, M.E., 2011. *Staphylococcus aureus* biofilms: properties, regulation, and roles in human disease. Virulence 2, 445–459.

Anderson, G.G., O'Toole, G.A., 2008. Innate and induced resistance mechanisms of bacterial biofilms. Curr. Top. Microbiol. Immunol. 322, 85–105.

Beech, I.B., Sunner, J.A., Hiraoka, K., 2005. Microbe-surface interactions in biofouling and biocorrosion processes. Int. Microbiol. 8, 157–168.

Berends, E.T., Horswill, A.R., Haste, N.M., Monestier, M., Nizet, V., Kockritz-Blickwede, M., 2010. Nuclease expression by *Staphylococcus aureus* facilitates escape from neutrophil extracellular traps. J. Innate Immun. 2, 576–586.

Biswas, L., Biswas, R., Schlag, M., Bertram, R., Götz, F., 2009. Small-colony variant selection as a survival strategy for *Staphylococcus aureus* in the presence of *Pseudomonas aeruginosa*. Appl. Environ. Microbiol. 75, 6910–6912.

Boles, B.R., Horswill, A.R., 2008. Agr-mediated dispersal of *Staphylococcus aureus* biofilms. PLoS Pathog. 4, e1000052.

Bos, R., Mei, H.C., Busscher, H.J., 1999. Physico-chemistry of initial microbial adhesive interactions-its mechanisms and methods for study. FEMS Microbiol. Rev. 23, 179–230.

Branda, S.S., Vik, S., Friedman, L., Kolter, R., 2005. Biofilms: the matrix revisited. Trends Microbiol. 13, 20–26.

Bridier, A., Briandet, R., Thomas, V., Dubois-Brissonnet, F., 2011. Resistance of bacterial biofilms to disinfectants: a review. Biofouling 27, 1017–1032.

Burke, F.M., Di Poto, A., Speziale, P., Foster, T.J., 2011. The A domain of fibronectin-binding protein B of *Staphylococcus aureus* contains a novel fibronectin binding site. FEBS J. 278, 2359–2371.

Cerca, N., Brooks, J.L., Jefferson, K.K., 2008. Regulation of the intercellular adhesin locus regulator (*icaR*) by SarA, sigmaB, and IcaR in *Staphylococcus aureus*. J. Bacteriol. 190, 6530–6533.

Chaignon, P., Sadovskaya, I., Ragunah, C., Ramasubbu, N., Kaplan, J.B., Jabbouri, S., 2007. Susceptibility of staphylococcal biofilms to enzymatic treatments depends on their chemical composition. Appl. Microbiol. Biotechnol. 75, 125–132.

Chen, C., Krishnan, V., Macon, K., Manne, K., Narayana, S.V.L., Schneewind, O., 2013. Secreted proteases control autolysin-mediated biofilm growth of *Staphylococcus aureus*. J. Biol. Chem. 288, 29440–29452.

Clarke, S.R., Brummell, K.J., Horsburgh, M.J., Mcdowell, P.W., Mohamad, S.A., Stapleton, M.R., Acevedo, J., Read, R.C., Day, N.P., Peacock, S.J., Mond, J.J., Kokai-Kun, J.F., Foster, S.J., 2006. Identification of *in vivo*-expressed antigens of *Staphylococcus aureus* and their use in vaccinations for protection against nasal carriage. J. Infect. Dis. 193, 1098–1108.

Clarke, S.R., Andre, G., Walsh, E.J., Dufrêne, Y.F., Foster, T.J., Foster, S.J., 2009. Iron-regulated surface determinant protein A mediates adhesion of *Staphylococcus aureus* to human corneocyte envelope proteins. Infect. Immun. 77, 2408–2416.

Corrigan, R.M., Rigby, D., Handley, P., Foster, T.J., 2007. The role of *Staphylococcus aureus* surface protein SasG in adherence and biofilm formation. Microbiology 153, 2435–2446.
Cramton, S., Ulrich, M., Götz, F., Döring, G., 2001. Anaerobic conditions induce expression of polysaccharide intercellular adhesin in *Staphylococcus aureus* and *Staphylococcus epidermidis*. Infect. Immun. 69, 4079–4085.
Cramton, S.E., Gerke, C., Schnell, N.F., Nichols, W.W., Götz, F., 1999. The intercellular adhesion (*ica*) locus is present in *Staphylococcus aureus* and is required for biofilm formation. Infect. Immun. 67, 5427–5433.
Cue, D., Lei, M.G., Luong, T.T., Kuechenmeister, L., Dunman, P.M., O'donnell, S., Rowe, S., O'gara, J.P., Lee, C.Y., 2009. *Rbf* promotes biofilm formation by *Staphylococcus aureus* via repression of *icaR*, a negative regulator of *icaADBC*. J. Bacteriol. 191, 6363–6373.
Daniels, R., Vanderleyden, J., Michiels, J., 2004. Quorum sensing and swarming migration in bacteria. FEMS Microbiol. Rev. 28, 261–289.
Davey, M.E., O'Toole, G.A., 2000. Microbial biofilms: from ecology to molecular genetics. Microbiol. Mol. Biol. Rev. 64, 847–867.
DeVita, M.D., Wadhera, R.K., Theis, M.L., Ingham, S.C., 2008. Assessing the potential of *Streptococcus pyogenes* and *Staphylococcus aureus* transfer to foods and customers via a survey of hands, hand-contact surfaces and food-contact surfaces at foodservice facilities. J. Foodserv. 18, 76–79.
Di Martino, P., 2016. Bap: a new type of functional amyloid. Trends Microbiol. 24 (9), 682–684.
Donlan, R.M., 2002. Biofilms: microbial life on surfaces. Emerg. Infect. Dis. 8, 881–890.
Dufour, D., Leung, V., Lévesque, C.M., 2012. Bacterial biofilm: structure, function, and antimicrobial resistance. Endod. Top. 22, 2–16.
Elias, S., Banin, E., 2012. Multi-species biofilms: living with friendly neighbors. FEMS Microbiol. Rev. 36, 990–1004.
Fitzpatrick, F., Humphreys, H., O'gara, J.P., 2005. The genetics of staphylococcal biofilm formation—will a greater understanding of pathogenesis lead to better management of device-related infection. Clin. Microbiol. Infect. 11, 967–973.
Flemming, H., Wingender, J., 2010. The biofilm matrix. Nat. Rev. Microbiol. 8, 623–633.
Flemming, H.-C., 2011. The perfect slime. Colloids Surf. B Biointerfaces 86, 251–259.
Flemming, H.-C., Neu, T.R., Wozniak, D.J., 2007. The EPS matrix: the "house of biofilm cells". J. Bacteriol. 189, 7945–7947.
Foster, T.J., Geoghegan, J.A., Ganesh, V.K., Höök, M., 2014. Adhesion, invasion and evasion: the many functions of the surface proteins of *Staphylococcus aureus*. Nat. Rev. Microbiol. 12, 49–62.
Foulston, L., Elsholz, A.K.W., Defrancesco, A.S., Losick, R., 2014. The extracellular matrix of *Staphylococcus aureus* biofilms comprises cytoplasmic proteins that associate with the cell surface in response to decreasing pH. mBio 5, 1–9.
Fuente-Nunez, C., Reffuveille, F., Haney, E.F., Straus, S.K., Hancock, R.E., 2014. Broad-spectrum anti-biofilm peptide that targets a cellular stress response. PLoS Pathog. 10, e1004152.
Ganesh, V.K., Rivera, J.J., Smeds, E., YA-Ping, K., Bowden, M.G., Wann, E.R., Gurusiddappa, S., Fitzgerald, J.R., Höök, M., 2008. A structural model of the *Staphylococcus aureus* ClfA-fibrinogen interaction opens new avenues for the design of anti-staphylococcal therapeutics. PLoS Pathog. 4, e1000226.
Ganesh, V.K., Barbu, E.M., Deivanayagam, C.C., Le, B., Anderson, A.S., Matsuka, Y.V., Lin, S.L., Foster, T.J., Narayana, S.V., Höök, M., 2011. Structural and biochemical characterization of *Staphylococcus aureus* clumping factor B/ligand interactions. J. Biol. Chem. 286, 25963–25972.
Geiger, T., Goerke, C., Fritz, M., Schafer, T., Ohlsen, K., Liebeke, M., Lalk, M., Wolz, C., 2010. Role of the (p) ppGpp synthase RSH, a RelA/SpoT homolog, instringent response and virulence of *Staphylococcus aureus*. Infect. Immun. 78, 1873–1883.
Geoghegan, J.A., Corrigan, R.M., Gruszka, D.T., Speziale, P., Gara, J.P.O., Potts, J.R., Foster, T.J., 2010a. Role of surface protein SasG in biofilm formation by *Staphylococcus aureus*. J. Bacteriol. 192, 5663–5673.

Geoghegan, J.A., Ganesh, V.K., Smeds, E., Liang, X., Hook, M., Foster, T.J., 2010b. Molecular characterization of the interaction of staphylococcal microbial surface components recognizing adhesive matrix molecules (MSCRAMM) ClfA and Fbl with fibrinogen. J. Biol. Chem. 285, 6208–6216.

Geoghegan, J.A., Monk, I.R., O'gara, J.P., Foster, T.J., 2013. Subdomains N2N3 of fibronectin binding protein A mediate *Staphylococcus aureus* biofilm formation and adherence to fibrinogen using distinct mechanisms. J. Bacteriol. 195, 2675–2683.

Gerke, C., Kraft, A., Sübmuth, R., Schweitzer, O., Götz, F., 1998. Characterization of the N-acetylglucosaminyltransferase activity involved in the biosynthesis of the *Staphylococcus epidermidis* polysaccharide intercellular adhesin. J. Biol. Chem. 273, 18586–18593.

Gilbert, P., Allison, D., Mcbain, A., 2002. Biofilms *in vitro* and *in vivo*: do singular mechanisms imply cross-resistance. J. Appl. Microbiol. Sympo. Suppl. 92, 98–110.

Götz, F., 2002. Staphylococcus and biofilms. Mol. Microbiol. 43, 1367–1378.

Gross, M., Cramton, S.E., Gotz, F., Peschel, A., 2001. Key role of teichoic acid net charge in *Staphylococcus aureus* colonization of artificial surfaces. Infect. Immun. 69, 3423–3426.

Gutiérrez, D., Delgado, S., Vázquez-Sánchez, D., Martínez, B., Cabo, M.L., Rodríguez, A., Herrera, J.J., García, P., 2012. Incidence of *Staphylococcus aureus* and analysis of associated bacterial communities on food industry surfaces. Appl. Environ. Microbiol. 78, 8547–8554.

Hall-Stoodley, L., Costerton, J.W., Stoodley, P., 2004. Bacterial biofilms: from the natural environment to infectious diseases. Nat. Rev. 2, 95–108.

Hall-Stoodley, L., Stoodley, P., 2005. Biofilm formation and dispersal and the transmission of human pathogens. Trends Microbiol. 13, 7–10.

Harriott, M.M., Noverr, M.C., 2009. *Candida albicans* and *Staphylococcus aureus* form polymicrobial biofilms: effects on antimicrobial resistance. Antimicrob. Agents Chemother. 53, 3914–3922.

Heilmann, C., Hussain, M., Peters, G., Götz, F., 1997. Evidence for autolysin-mediated primary attachment of *Staphylococcus epidermidis* to a polystyrene surface. Mol. Microbiol. 24, 1013–1024.

Hochbaum, A.I., Kolodkin-Gal, I., Foulston, L., Kolter, R., Aizenberg, J., Losick, R., 2011. Inhibitory effects of D-amino acids on *Staphylococcus aureus* biofilm development. J. Bacteriol. 193, 5616–5622.

Hoffman, L.R., Deziel, E., D'argenio, D.A., Lepine, F., Emerson, J., Mcnamara, S., Gibson, R.L., Ramsey, B.W., Miller, S.I., 2006. Selection for *Staphylococcus aureus* small-colony variants due to growth in the presence of *Pseudomonas aeruginosa*. Proc. Natl. Acad. Sci. 103, 19890–19895.

Høiby, N., Bjarnsholt, T., Givskov, M., Molin, S., Ciofu, O., 2010. Antibiotic resistance of bacterial biofilms. Int. J. Antimicrob. Agents 35, 322–332.

Huseby, M.J., Kruse, A.C., Digre, J., Kohler, P.L., Vocke, J.A., Mann, E.E., Bayles, K.W., Bohach, G.A., Schlievert, P.M., Ohlendorf, D.H., Earhart, C.A., 2010. Beta toxin catalyzes formation of nucleoprotein matrix in staphylococcal biofilms. Proc. Natl. Acad. Sci. 107, 14407–14412.

Izano, E.A., Amarante, M.A., Kher, W.B., Kaplan, J.B., 2008. Differential roles of poly-N-acetylglucosamine surface polysaccharide and extracellular DNA in *Staphylococcus aureus* and *Staphylococcus epidermidis* biofilms. Appl. Environ. Microbiol. 74, 470–476.

Jefferson, K.K., Cramton, S.E., Götz, F., Pier, G.B., 2003. Identification of a 5-nucleotide sequence that controls expression of the *ica* locus in *Staphylococcus aureus* and characterization of the DNA-binding properties of IcaR. Mol. Microbiol. 48, 889–899.

Kaplan, J.B., Izano, E.A., Gopal, P., Karwacki, M.T., Kim, S., Bose, J.L., Bayles, K.W., 2012. Low levels of β-lactam antibiotics induce extracellular DNA release and biofilm formation in *Staphylococcus aureus*. mBio 3, e00198–e00212.

Kaplan, J.B., Velliyagounder, K., Ragunath, C., Rohde, H., Mack, D., Knobloch, J.K., Ramasubbu, N., 2004. Genes involved in the synthesis and degradation of matrix polysaccharide in *Actinobacillus actinomycetemcomitans* and *Actinobacillus pleuropneumoniae* biofilms. J. Bacteriol. 186, 8213–8220.

Kart, D., Tavernier, S., Van Acker, H., Nelis, H.J., Coenye, T., 2014. Activity of disinfectants against multispecies biofilms formed by *Staphylococcus aureus*, *Candida albicans* and *Pseudomonas aeruginosa*. Biofouling 30, 377–383.

Kiedrowski, M.R., Crosby, H.A., Hernandez, F.J., Malone, C.L., Mcnamara 2ND, J.O., Horswill, A. R, 2014. *Staphylococcus aureus* Nuc2 is a functional, surface-attached extracellular nuclease. PLoS One 9, e95574.

Kiedrowski, M.R., Kavanaugh, J.S., Malone, C.L., Mootz, J.M., Voyich, J.M., Smeltzer, M.S., Bayles, K.W., Horswill, A.R., 2011. Nuclease modulates biofilm formation in community-associated methicillin-resistant *Staphylococcus aureus*. PLoS One 6, e26714.

Kolar, S.L., Ibarra, J.A., Rivera, F.E., Mootz, J.M., Davenport, J.E., Stevens, S.M., Horswill, A.R., Shaw, L.N., 2013. Extracellular proteases are key mediators of *Staphylococcus aureus* virulence via the global modulation of virulence-determinant stability. Microbiologyopen 2, 18–34.

Kolodkin-Gal, I., Romero, D., Cao, S., Clardy, J., Kolter, R., Losick, R., 2010. D-amino acids trigger biofilm disassembly. Science 328, 627–629.

Kucera, J., Sojka, M., Pavlik, V., Szuszkiewicz, K., Velebny, V., Klein, P., 2014. Multispecies biofilm in an artificial wound bed-A novel model for *in vitro* assessment of solid antimicrobial dressings. J. Microbiol. Methods 103, 18–24.

Lauderdale, K.J., Boles, B.R., Cheung, A.L., Horswill, A.R., 2009. Interconnections between *SigmaB*, *agr*, and proteolytic activity in *Staphylococcus aureus* biofilm maturation. Infect. Immun. 77, 1623–1635.

Lauderdale, K.J., Malone, C.L., Boles, B.R., Morcuende, J., Horswill, A.R., 2010. Biofilm dispersal of community-associated methicillin-resistant *Staphylococcus aureus* on orthopedic implant material. J. Orthop. Res. 28, 55–61.

Lewis, K., 2010. Persister cells. Annu. Rev. Microbiol. 64, 357–372.

Lister, J.L., Horswill, A.R., 2014. *Staphylococcus aureus* biofilms: recent developments in biofilm dispersal. Front. Cell. Infect. Microbiol. 4, 178.

Madsen, J.S., Burmølle, M., Hansen, L.H., Sørensen, S.J., 2012. The interconnection between biofilm formation and horizontal gene transfer. FEMS Immunol. Med. Microbiol. 65, 183–195.

Maira-Litrán, T., Kropec, A., Abeygunawardana, C., Joyce, J., Mark III, G., Goldmann, D.A., Pier, G.B., 2002. Immunochemical properties of the staphylococcal poly-N-acetylglucosamine surface polysaccharide. Infect. Immun. 70, 4433–4440.

Mann, E.E., Rice, K.C., Boles, B.R., Endres, J.L., Ranjit, D., Chandramohan, L., Tsang, L.H., Smeltzer, M.S., Horswill, A.R., Bayles, K.W., 2009. Modulation of eDNA release and degradation affects *Staphylococcus aureus* biofilm maturation. PLoS One 4, e5822.

Marti, M., Trotonda, M.P., Tormo-Mas, M.A., Vergara-Irigaray, M., Cheung, A.L., Lasa, I., Penadés, J.R., 2010. Extracellular proteases inhibit protein-dependent biofilm formation in *Staphylococcus aureus*. Microbes Infect. 12, 55–64.

McCourt, J., O'halloran, D.P., Mccarthy, H., O'gara, J.P., Geoghegan, J.A., 2014. Fibronectin-binding proteins are required for biofilm formation by community-associated methicillin-resistant *Staphylococcus aureus* strain LAC. FEMS Microbiol. Lett. 353, 157–164.

Mcgavin, M.J., Zahradka, C., Rice, K.C., Scott, J.E., 1997. Modification of the *Staphylococcus aureus* fibronectin binding phenotype by V8 protease. Infect. Immun. 65, 2621–2628.

Merino, N., Toledo-Arana, A., Valle, J., Solano, C., Lopez, J.A., Foster, T.J., José, R., Vergara-Irigaray, M., Calvo, E., Penade, R., 2009. Protein A-mediated multicellular behavior in *Staphylococcus aureus*. J. Bacteriol. 191, 832–843.

Mootz, J.M., Malone, C.L., Shaw, L., Horswill, A.R., 2013. Staphopains modulate *Staphylococcus aureus* biofilm integrity. Infect. Immun. 81, 3227–3238.

Mulcahy, M.E., Geoghegan, J.A., Monk, I.R., O'keeffe, K.M., Walsh, E.J., Foster, T.J., Mcloughlin, R.M., 2012. Nasal colonization by *Staphylococcus aureus* depends upon clumping factor B binding to the squamous epithelial cell envelope protein loricrin. PLoS Pathog. 8, e1003092.

Olson, M.E., Nygaard, T.K., Ackermann, L., Watkins, R.L., Zurek, O.W., Pallister, K.B., Griffith, S., Kiedrowski, M.R., Flack, C.E., Kavanaugh, J.S., Kreiswirth, B.N., Horswill, A.R., Voyich, J.M., 2013. *Staphylococcus aureus* nuclease is an SaeRS-dependent virulence factor. Infect. Immun. 81, 1316–1324.

O'Brien, L.M., Walsh, E.J., Massey, R.C., Peacock, S.J., Foster, T.J., 2002. *Staphylococcus aureus* clumping factor B (ClfB) promotes adherence to human type I cytokeratin 10: implications for nasal colonization. Cell Microbiol. 4, 759–770.

O'Gara, J.P., 2007. Ica and beyond: biofilm mechanisms and regulation in *Staphylococcus epidermidis* and *Staphylococcus aureus*. FEMS Microbiol. Lett. 270, 179–188.

O'Neill, E., Pozzi, C., Houston, P., Humphreys, H., Robinson, D.A., Loughman, A., Foster, T.J., O'gara, J.P., 2008. A novel *Staphylococcus aureus* biofilm phenotype mediated by the fibronectin-binding proteins, FnBPA and FnBPB. J. Bacteriol. 190, 3835–3850.

O'Toole, G., Kaplan, H.B., Kolter, R., 2000. Biofilm formation as microbial development. Annu. Rev. Microbiol. 54, 49–79.

Pagedar, A., Singh, J., Batish, V.K., 2010. Surface hydrophobicity, nutritional contents affect *Staphylococcus aureus* biofilms and temperature influences its survival in preformed biofilms. J. Basic Microbiol. 50, S98–S106.

Paharik, A.E., Horswill, A.R., 2016. The staphylococcal biofilm: adhesins, regulation, and host response. Microbiol. Spectr. 4, 1–27.

Parsek, M.R., Greenberg, E.P., 2005. Sociomicrobiology: the connections between quorum sensing and biofilms. Trends Microbiol. 13, 27–33.

Payne, D.E., Boles, B.R., 2016. Emerging interactions between matrix components during biofilm development. Curr. Genet. 62, 137–141.

Periasamy, S., Chatterjee, S.S., Cheung, G.Y., Otto, M., 2012. Phenol-soluble modulins in staphylococci: what are they originally for? Commun. Integr. Biol. 5, 275–277.

Renner, L.D., Weibel, D.B., 2011. Physicochemical regulation of biofilm formation. MRS Bulletin 36, 347–355.

Rice, K.C., Mann, E.E., Endres, J.L., Weiss, E.C., Cassat, J.E., Smeltzer, M.S., Bayles, K.W., 2007. The *cidA* murein hydrolase regulator contributes to DNA release and biofilm development in *Staphylococcus aureus*. Proc. Natl. Acad. Sci. 104, 8113–8118.

Roche, F.M., Meehan, M., Foster, T.J., 2003. The *Staphylococcus aureus* surface protein SasG and its homologues promote bacterial adherence to human desquamated nasal epithelial cells. Microbiology 149, 2759–2767.

Rode, T.M., Langsrud, S., Holck, A., Møretrø, T., 2007. Different patterns of biofilm formation in *Staphylococcus aureus* under food-related stress conditions. Int. J. Food Microbiol. 116, 372–383.

Sanchez, C.J.J.R., Prieto, E.M., Krueger, C.A., Zienkiewicz, K.J., Romano, D.R., Ward, C.L., Akers, K.S., Guelcher, S.A., Wenke, J.C., 2013. Effects of local delivery of D-amino acids from biofilm-dispersive scaffolds on infection in contaminated rat segmental defects. Biomaterials 34, 7533–7543.

Sanchez, C.J.J.R., Akers, K.S., Romano, D.R., Woodbury, R.L., Hardy, S.K., Murray, C.K., Wenke, J.C., 2014. D-amino acids enhance the activity of antimicrobials against biofilms of clinical wound isolates of *Staphylococcus aureus* and *Pseudomonas aeruginosa*. Antimicrob. Agents Chemother. 58 (8), 4353–4361.

Sattar, S.A., Springthorpe, S., Mani, S., Gallant, M., Nair, R.C., Scott, E., Kain, J., 2001. Transfer of bacteria from fabrics to hands and other fabrics: development and application of a quantitative method using *Staphylococcus aureus* as a model. J. Appl. Microbiol. 90, 962–970.

Schroeder, K., Jularic, M., Horsburgh, S.M., Hirschhausen, N., Neumann, C., Bertling, A., Schulte, A., Foster, S., Kehrel, B.E., Peters, G., Heilmann, C., 2009. Molecular characterization of a novel *Staphylococcus aureus* surface protein (SasC) involved in cell aggregation and biofilm accumulation. PLoS One 4, e7567.

Schwartz, K., Ganesan, M., Payne, D.E., Solomon, M.J., Boles, B.R., 2016. Extracellular DNA facilitates the formation of functional amyloids in *Staphylococcus aureus* biofilms. Mol. Microbiol. 99, 123–134.

Shaw, L., Golenka, E., Potempa, J., Foster, S.J., 2004. The role and regulation of the extracellular proteases of *Staphylococcus aureus*. Microbiology 150, 217–228.

Simon, S.S., Sanjeev, S., 2007. Prevalence of enterotoxigenic *Staphylococcus aureus* in fishery products and fish processing factory workers. Food Control 18, 1565–1568.

Singh, R., Ray, P., Das, A., Sharma, M., 2010. Penetration of antibiotics through *Staphylococcus aureus* and *Staphylococcus epidermidis* biofilms. J. Antimicrob. Chemother. 65, 1955–1958.

Sospedra, I., Mañes, J., Soriano, J.M., 2012. Report of toxic shock syndrome toxin 1 (TSST-1) from *Staphylococcus aureus* isolated in food handlers and surfaces from food service establishments. Ecotoxicol. Environ. Saf. 80, 288–290.

Srey, S., Jahid, I.K., Ha, S., 2013. Biofilm formation in food industries: a food safety concern. Food Control 31, 572–585.

Srivatsan, A., Wang, J.D., 2008. Control of bacterial transcription, translation and replication by (p)ppGpp. Curr. Opin. Microbiol. 11, 100–105.

Sugimoto, S., Iwamoto, T., Takada, K., Okuda, K., Tajima, A., Iwase, T., Mizunoe, Y., 2013. *Staphylococcus epidermidis* Esp degrades specific proteins associated with *Staphylococcus aureus* biofilm formation. J. Bacteriol. 195, 1645–1655.

Taglialegna, A., Navarro, S., Ventura, S., Garnett, J.A., Matthews, S., Penades, J.R., Lasa, I., Valle, J., 2016. Staphylococcal Bap proteins build amyloid scaffold biofilm matrices in response to environmental signals. PLoS Pathog. 12, e1005711.

Thoendel, M., Kavanaugh, J.S., Flack, C.E., Horswill, A.R., 2011. Peptide signaling in the staphylococci. Chem. Rev. 111, 117–151.

Thomas, V.C., Hancock, L.E., 2009. Suicide and fratricide in bacterial biofilms. Int. J. Artif. Organs 32 (9), 537–544.

Trampuz, A., Widmer, A.F., 2006. Infections associated with orthopedic implants. Curr. Opin. Infect. Dis. 19, 349–356.

van-Houdt, R., Michiels, C.W., 2010. Biofilm formation and the food industry, a focus on the bacterial outer surface. J. Appl. Microbiol. 109, 1117–1131.

Vazquez, V., Liang, X., Horndahl, J.K., Ganesh, V.K., Smeds, E., Foster, T.J., Hook, M., 2011. Fibrinogen is a ligand for the *Staphylococcus aureus* microbial surface components recognizing adhesive matrix molecules (MSCRAMM) bone sialoprotein-binding protein (Bbp). J. Biol. Chem. 286, 29797–29805.

Vázquez-Sánchez, D., Cabo, M.L., SAÁ-Ibusquiza, P., Rodríguez-Herrera, J.J., 2012. Incidence and characterization of *Staphylococcus aureus* in fishery products marketed in Galicia (Northwest Spain). Int. J. Food Microbiol. 157, 286–296.

Vázquez-Sánchez, D., Habimana, O., Holck, A., 2013. Impact of food-related environmental factors on the adherence and biofilm formation of natural *Staphylococcus aureus* isolates. Curr. Microbiol. 66, 110–121.

Vázquez-Sánchez, D., Cabo, M.L., Ibusquiza, P.S., Rodríguez-Herrera, J.J., 2014. Biofilm-forming ability and resistance to industrial disinfectants of *Staphylococcus aureus* isolated from fishery products. Food Control 39, 8–16.

Vuong, C., Kocianova, S., Voyich, J.M., Yao, Y., Fischer, E.R., Deleo, F.R., Otto, M., 2004. A crucial role for exopolysaccharide modification in bacterial biofilm formation, immune evasion, and virulence. J. Biol. Chem. 279, 54881–54886.

Walsh, E.J., O'brien, L.M., Liang, X., Hook, M., Foster, T.J., 2004. Clumping factor B, a fibrinogen-binding MSCRAMM (microbial surface components recognizing adhesive matrix molecules) adhesin of *Staphylococcus aureus*, also binds to the tail region of type I cytokeratin 10. J. Biol. Chem. 279, 50691–50699.

Watnick, P., Kolter, R., 2000. Biofilm, city of microbes. J. Bacteriol. 182, 2675–2679.

Wertheim, H.F., Walsh, E., Choudhurry, R., Melles, D.C., Boelens, H.A., Miajlovic, H., Verbrugh, H.A., Foster, T., Van Belkum, A., 2008. Key role for clumping factor B in *Staphylococcus aureus* nasal colonization of humans. PLoS Med. 5, e17.

Xiang, H., FENG, Y., Wang, J., Liu, B., Chen, Y., Liu, L., Deng, X., Yang, M., 2012. Crystal structures reveal the multiligand binding mechanism of *Staphylococcus aureus* ClfB. PLoS Pathog. 8, e1002751.

Xu, H., Zou, Y., Lee, H.Y., Ahn, J., 2010. Effect of NaCl on the biofilm formation by foodborne pathogens. J. Food Sci. 75 (9), M580–M585.

Yarwood, J.M., Bartels, D.J., Volper, E.M., Greenberg, E.P., 2004. Quorum sensing in *Staphylococcus aureus* biofilms. J. Bacteriol. 186, 1838–1850.

Yu, D., Zhao, L., Xue, T., Sun, B., 2012. *Staphylococcus aureus* autoinducer-2 quorum sensing decreases biofilm formation in an *icaR*-dependent manner. BMC Microbiol. 12, 288.

Zong, Y., Xu, Y., Liang, X., Keene, D.R., Höök, A., Gurusiddappa, S., Höök, M., Narayana, S.V., 2005. A 'Collagen Hug' model for *Staphylococcus aureus* CNA binding to collagen. EMBO J. 24, 4224–4236.

CHAPTER

METHODS FOR THE IDENTIFICATION, CHARACTERIZATION, AND TRACKING THE SPREAD OF *STAPHYLOCOCCUS AUREUS*

6

Maria de Lourdes Ribeiro de Souza da Cunha

Botucatu Institute of Biosciences, UNESP – University Estadual Paulista, Botucatu, Brazil

1. INTRODUCTION

The rapid and efficient identification of *Staphylococcus aureus* is important for routine testing in microbiology laboratories. The development of reliable, simple, and inexpensive techniques is essential for identifying outbreaks and new strains of *S. aureus*. In recent years, different commercial systems for the identification of *S. aureus* have been developed as an alternative to classical identification methods. Automated systems and commercial kits based on miniature biochemical tests are widely used today both, in routine and research laboratories. However, these diagnostic systems have some disadvantages such as their high cost. In addition, they are unable to reliably distinguish between different *Staphylococcus* species due to the variable expression of phenotypic characteristics because they are based on colorimetric results and subjectivity in their interpretation can lead to ambiguity (Couto et al., 2001).

Molecular techniques have been the preferred method for identifying microorganisms because of their higher specificity and sensitivity. Staphylococcal strains not identified to the species level or erroneously identified by conventional phenotypic tests can be correctly identified by genotypic techniques such as polymerase chain reaction (PCR), multiplex PCR, real-time PCR, DNA–DNA hybridization, and sequence-based identification systems. New sequencing technologies, called next-generation sequencing (NGS), are a powerful alternative for structural and functional genomics studies and strain characterization. These whole-genome technologies and also microarrays have demonstrated the existence of an enormous variety of *S. aureus* strains. Techniques of protein profile analysis have been introduced for the identification of microorganisms and detection of toxins. One of the methods most commonly used today for the analysis of biomolecules is matrix-assisted laser desorption/ionization (MALDI) mass spectrometry, followed by detection in a time-of-flight (TOF) mass analyzer (Croxatto et al., 2012).

Staphylococcus aureus. http://dx.doi.org/10.1016/B978-0-12-809671-0.00006-1

The development of robust typing methods, such as *spa* typing, multilocus sequence typing (MLST), multiple-locus variable number of tandem repeat analysis (MLVA), and pulsed-field gel electrophoresis (PFGE), has permitted to trace the clonal relationship of different *S. aureus* lineages. This has led to surprising discoveries about how *S. aureus* spread and may permit better predictions and control of staphylococcal food poisoning (SFP) outbreaks in the future.

For the detection of staphylococcal enterotoxins (SEs) in food, various techniques, such as immunological (e.g., enzyme-linked immunosorbent assays (ELISA)) and molecular-based methodologies, have been integrated in the diagnosis strategy. The PCR approach is known to provide information on the presence or absence of genes encoding SEs, but not their expression. On the other hand, MALDI-TOF tools may overcome specific technical limitations of existing ELISA methods for detecting and quantifying SEs, its throughput and cost per analysis compare unfavorably with ELISA. Further limitations are lack of methods allowing for the detection of SE types other than the classical ones (staphylococcal enterotoxin A (SEA)–staphylococcal enterotoxin E (SEE)) in routine diagnostics.

2. IDENTIFICATION AND CHARACTERIZATION OF *STAPHYLOCOCCUS AUREUS* BY PHENOTYPIC METHODS

2.1 CULTURE-BASED METHODS AND BIOCHEMICAL TESTS

Bacterial identification in the clinical microbiology laboratory is traditionally performed by isolation of the microorganism and phenotypic analysis of its characteristics. Staphylococci are isolated from foods using selective and differential media to select for members of the genus *Staphylococcus*, such as Baird-Parker (BP) agar and Mannitol Salt agar.

BP agar was developed by Baird-Parker and contains lithium chloride, potassium tellurite, and egg yolk (Baird-Parker, 1962; Baird-Parker and Davenport, 1965). The presence of the salt and potassium tellurite selects for the genus *Staphylococcus*, which reduces tellurite forming black colonies. The medium is differential by the formation of a dual halo resulting from the production of lipase and lecithinase by coagulase-positive staphylococcal strains, which break down lipids and lecithin in the egg yolk. There is an ISO method available where BP agar is used for the enumeration of staphylococci in food (ISO 6888-1: 1999: Microbiology of food and animal feeding stuffs—Horizontal method for the enumeration of coagulase-positive *staphylococci* (*Staphylococcus aureus* and other species)—Part 1: Technique using Baird-Parker agar medium).

Another widely used medium for the isolation of *S. aureus* is Mannitol Salt Agar, which inhibits the growth of other bacteria due to the presence of 7.5% NaCl. *S. aureus* ferments mannitol and forms yellow colonies on the medium. In addition, blood agar is often used for the isolation of *Staphylococcus* spp. from clinical materials. *S. aureus* produces yellow, circular, and medium-sized colonies surrounded by a clear halo that indicates hemolytic activity on blood agar.

Typical colonies obtained on the media described above are submitted to Gram staining to determine their purity and to observe their morphology and specific color. After confirmation of these characteristics, the isolates may be submitted to catalase and coagulase tests.

The tube coagulase test for the detection of free coagulase continues to be the gold standard for the identification of *S. aureus* and other coagulase-positive staphylococci. The test should always be analyzed after 4 h of incubation since some strains will produce fibrinolysin when incubated for

24 h, causing dissolution of the clot during the period of incubation, and may be erroneously identified as coagulase-negative staphylococci (CoNS) if the test is read only after 24 h. Although the cell wall of most *S. aureus* strains contains a bound coagulase, called "clumping factor," that can react directly with the fibrinogen in plasma mounted on a slide and produce rapid agglutination, the slide coagulase test is not recommended for the differentiation from CoNS because some of these species, such as *Staphylococcus lugdunensis* and *Staphylococcus schleiferi* subsp. *schleiferi*, can also produce the clumping factor and may be erroneously identified as coagulase-positive staphylococci.

Although the tube coagulase test is sufficient to identify *S. aureus* in human clinical samples, because it is the only species present in humans that is able to clot plasma, the exclusive use of this test is not sufficient for the identification of *S. aureus* isolated from foods of animal origin where other coagulase-positive staphylococci such as *Staphylococcus hyicus* may be present, too. To overcome this issue and to distinguish *S. aureus* from other species, testing of biochemical properties of strains such as the fermentation of sugars (trehalose, maltose, mannitol, and xylose) and the production of acetoin are suitable (Table 6.1).

Further differentiation of the genus *Staphylococcus* from the genus *Micrococcus* is based on the oxidation and fermentation of glucose, resistance to bacitracin (0.04 U) indicated by the absence of an inhibition halo or formation of a halo of up to 9 mm, and susceptibility to furazolidone (100 µg) characterized by an inhibition halo ranging in diameter from 15 to 35 mm (Baker, 1984).

For the enumeration of coagulase-positive staphylococci including *S. aureus* in food are currently three different horizontal ISO methods available: (1) ISO 6888-1:1999 using Baird-Parker agar (see above), (2) ISO 6888-2:1999: Microbiology of food and animal feeding stuffs—Horizontal method for the enumeration of coagulase-positive staphylococci (*Staphylococcus aureus* and other species)—Part 2: Technique using rabbit plasma fibrinogen agar medium, and (3) ISO 6888-3:2003: Microbiology of food and animal feeding stuffs—Horizontal method for the enumeration of coagulase-positive staphylococci (*Staphylococcus aureus* and other species)—Part 3: Detection and MPN (Most Probable Number) technique for low numbers; all are widely used in routine diagnostics and for food control.

Table 6.1 Selected Biochemical Properties of Coagulase-Positive Staphylococci

Species	Trehalose	Maltose	Mannitol	Xylose	Acetoin
Staphylococcus intermedius	+	V	V	–	–
Staphylococcus hyicus	–	–	–	–	–
Staphylococcus aureus	+	+	+	–	+
Staphylococcus schleiferi subsp. *coagulans*	–	–	V	–	+
Staphylococcus delphini	–	+	+	NA	–
Staphylococcus lutrae	+	+	V	+	–
Staphylococcus agnetis	V	–	–	–	–

NA, *not analyzed;* V, *variable.*

2.2 AUTOMATED SYSTEMS

Automated systems and commercial kits based on miniature biochemical tests are widely used today in both routine and research laboratories. The commercial API Staph system kit (bioMérieux) is used for the identification of *Micrococcus* spp. and *Staphylococcus* spp. The kit consists of 19 miniature and dehydrated tests that are resuspended during inoculation with the microbial suspension. The biochemical reactions are read after incubation for 18–24h at 37°C, generating a seven-digit profile number for each microorganism that can then be identified in the database of the manufacturer. However, false-positive classification of strains as *S. aureus* occurs, too (Renneberg et al., 1995; Cunha et al., 2004); most likely due to the fact that the API Staph system does not test on the coagulase reaction (Koneman et al., 1997).

Automation in clinical microbiology is at an early stage of development compared to the level of automation that has been achieved in other clinical laboratories such as hematology and biochemistry laboratories. In the last 20 years, a variety of automated systems for the identification of microorganisms and antimicrobial susceptibility testing of the identified organisms have been developed based on the automated interpretation of the results of biochemical tests or using microdilution trays after overnight incubation and photometric determination of growth (McDonald et al., 2001). Technological advances that permit rapid bacterial identification and antimicrobial susceptibility testing of microorganisms are now seen as having clinical and financial benefits.

Automated identification systems continue to be unable to provide a fully reliable differentiation between different *Staphylococcus* species due to the variable expression of phenotypic characteristics by these microorganisms. The Microscan (Dade/Microscan) panels contain 27 miniature biochemical tests for conventional identification; 18 of them are used to identify staphylococci. The lyophilized panels are seeded and incubated for 24–48h. The panels are read visually or automatically by the Microscan Walk/Away system that generates a six-digit code number, which is used to identify the microorganism in the database. One study showed only 53.6% of correct identifications among a total of 233 staphylococcal isolates (Grant et al., 1994).

Another automated identification system is the VITEK1 system (bioMérieux), which originated in the 1970s as an automated system for the identification and susceptibility testing of microorganisms. The VITEK1 card consists of 30 microcavities (28 tests and 2 controls) that contain substrates for the identification of *Staphylococcus* spp. A suspension of the microorganism is prepared in saline and the card is filled with the bacterial suspension using a transfer tube and sealed with the VITEK filling/sealing module. The card is then inserted into the incubation/reading module of the VITEKsystem and analyzed optically after 10–13h. Evaluation of the updated database of the Gram-positive identification (GPI) card used in the automated VITEK1 system for 500 clinical isolates showed overall agreement of 89% between the GPI card and conventional methods (Bannerman et al., 1993). In another study of the same system, the GPI card correctly identified only 67% of 185 isolates (Koneman et al., 1997). Another more recent study on the identification of *Staphylococcus* spp. using the VITEK1 system reported a sensitivity of 84.2% and specificity of 100.0%. Despite the correct identification of all *S. aureus* isolates, the VITEK1 system failed to identify CoNS species (Ferreira et al., 2012).

Several studies demonstrated the inability of the GPI card to reliably identify less common staphylococcal species and that the database of this system needed to be improved. The system thus evolved into the VITEK2 system, which automatically performs all of the steps necessary for bacterial identification and antimicrobial susceptibility testing after a primary inoculum has been prepared and standardized. This system allows kinetic analysis of the inoculated culture by reading each test every 15 min. The optical system of the device combines a multichannel fluorimeter and photometer readings to record data such as turbidity, fluorescence, and colorimetric signals (Stager and Davis, 1992). The VITEK2 system consists of four identification cards, called reagent cards. Each card contains 64 wells with individual substrates that measure different metabolic activities, such as acidity, alkalinity, hydrolysis, and growth in the presence of inhibitors. To guarantee bacterial growth, the technology used in the VITEK2 system maintains oxygenation levels through a film covering both sides of the card. At the same time, this film prevents contamination of the cards.

Studies using the VITEK2 identification system have reported more promising results. The first evaluation of the accuracy of the VITEK2 system using the GPI card for rapid identification of clinically significant isolates of staphylococci, streptococci, and enterococci showed 98% agreement (Funke et al., 1998). Recently, Monteiro et al. (2016) showed that the GPI card used to identify Gram-positive cocci resulted in a rate of correct identification of 92.6% (199/215) of the microorganisms studied. The system correctly identified all *S. aureus* isolates (17/17) and incorrectly identified 15 of the 186 (91.9%) isolates belonging to the CoNS group. Despite this lower performance, the kappa values indicated reliability of the results obtained with the VITEK2 system. Specificity tests showed a performance higher than the 90% required for commercial devices, demonstrating that this system is suitable for the identification of microorganisms isolated in routine microbiology laboratories.

2.3 MATRIX-ASSISTED LASER DESORPTION/IONIZATION TIME-OF-FLIGHT

In recent years, techniques of protein profile analysis have been introduced for the identification of microorganisms. Mass spectrometry is a powerful analytical technique used to identify unknown compounds, to quantify known compounds, and to elucidate the structure and chemical properties of molecules. Basically, this technique consists of the ionization of atoms or molecules of a sample (breakage of atoms or molecules so that they carry more or less electrons than the original), separation of these atoms or molecules according to their mass-to-charge ratio (*m/z*), and their identification and quantification. One of the methods most commonly used today for the analysis of biomolecules is MALDI mass spectrometry, followed by detection in a TOF mass analyzer (Croxatto et al., 2012). Although the acquisition costs of the equipment are high, MALDI-TOF mass spectrometry (MS) offers low operation costs and short sample processing times.

MALDI-TOF MS has emerged as a promising technology for the identification of bacteria. Recent studies have shown high rates of agreement between the results produced by MALDI-TOF MS and reference identification techniques such as PCR and sequencing (Carbonnelle et al., 2007; Dubois et al., 2010). *Staphylococcus* spp. are a diverse group of microorganisms and the correct discrimination and identification of species represent a challenge because many of them are genetically very similar (Kim et al., 2008). In this technique, the sample is mixed with a matrix on a conductive metal plate. After crystallization of the matrix and sample, the metal plate is introduced into the mass spectrometer where it is bombarded with brief laser pulses. The desorbed and ionized molecules are accelerated through an electric field and enter a metal tube submitted to vacuum (flight tube through which the molecules pass) until they reach the detector. Small ions (*m/z*) travel faster through the flight tube than larger ions. Thus, the time of arrival at the detector (TOF) varies depending on the molecule, generating different peaks and mass spectra, according to their *m/z* ratio. The result is transferred to a graph, giving several peaks, and a specific graph is obtained for each bacterial or fungal species. A computerized database interprets and provides the result in a rapid manner. Comparison of the spectrum of the sample with the database permits to identify the bacterial species (Croxatto et al., 2012).

Two MALDI-TOF systems are commercially available, the MALDI Biotyper (Bruker Daltonics/BD) and VITEK MS (bioMérieux). The mass spectral data are compared to the database that contains reference spectra (Biotyper, Bruker Daltonics/BD) or "super spectra" (Myla/SARAMIS, bioMérieux) for the identification of species on these platforms, respectively.

The main advantage of the MALDI-TOF technology over other techniques for the identification of *S. aureus* is its rapidity, with the final identification result being obtained within less than 30 min (Croxatto et al., 2012). The main limiting factor of the use of MALDI-TOF MS is the high cost of the equipment; however, the low cost of analysis and the easy processing of a large number of samples within a short period of time compensate for the initial acquisition costs of the equipment.

The use of MALDI-TOF MS is more economical in terms of time and labor, is easier, requires less training, and the results are more easily interpreted when compared to traditional methods. Everything indicates that this technology will replace traditional assays as the identification system in routine laboratory testing. However, this does not mean that traditional assays will be extinct, also in case of identification of *S. aureus*. Depending on the equipment and database, the identification of some microorganisms may represent a challenge for MALDI-TOF mass spectrometry. If this were the case after verification by the laboratory or if it were reported in the literature as a problem, traditional tests may have to be continued along with MALDI-TOF MS to help distinguish different microorganisms that can potentially be misidentified by this system.

2.4 ANTIMICROBIAL SUSCEPTIBILITY TESTING

Antimicrobial susceptibility testing of *S. aureus* is an important task in any microbiological laboratory as this testing allows for the detection of possible drug resistances of a given isolate and assures susceptibility to drugs of choice for treatment of *S. aureus* infections. The most widely used testing methods include broth microdilution or rapid automated instrument methods that use commercially marketed materials and devices; also disk diffusion methods are widely used. Each method has strengths and weaknesses. Nevertheless, all current available testing methods provide accurate detection of common antimicrobial resistance mechanisms (Jorgensen and Ferraro, 2009). The reference methods recommended by the Clinical Laboratory Standards Institute (CLSI) for detecting resistance in *S. aureus* include the determination of minimal inhibitory concentrations (MICs) by the agar or broth dilution method and by the disk diffusion method (CLSI, 2015).

In general, all susceptibility testing methods require breakpoints, also known as interpretive criteria, so that the results of the tests can be interpreted as susceptible, intermediate, or resistant and reported as such to a broad range of clinicians. The term "breakpoint" has been used in a variety of ways. The first and most obvious one refers to the MIC for any given antibacterial that distinguishes wild-type populations of bacteria from those with acquired or selected resistance mechanisms ("wild-type breakpoints," also called, "epidemiological cut-off values"). In this context, the wild-type strain is defined as a strain of a bacterium that does not harbor any acquired or selected resistance to the particular antibacterial being examined or to antibacterials with the same mechanism/site of action (Turnidge and Paterson, 2007). The European Committee on Antimicrobial Susceptibility Testing publishes epidemiological cut-off values also for *S. aureus*. The second is so-called "clinical breakpoints," which refer to those concentrations (MICs) that separate strains where there is a high likelihood of treatment success from those bacteria where treatment is more likely to fail (Turnidge and Paterson, 2007); CLSI usually sets the MIC standards for *S. aureus*.

In the following, an overview of those methods most often applied for antimicrobial susceptibility testing of *S. aureus* strains is given.

One of the first methods developed was the broth dilution susceptibility test, which still serves as a reference method according to CLSI. Serial dilutions of the antimicrobial agent are prepared in broth and a standardized bacterial suspension is then added. Macrodilution testing uses broth volumes of about 1.0 mL in standard test tubes. The intensive labor involved in preparing the multitest tubes has been a major impediment to their routine use. The microdilution broth method is an adaptation of this test that uses small volumes in 96-well microtiter plates. The advantage of the microdilution method is the use of small volumes of reagents that can be tested against a panel of antibiotics.

In the agar dilution procedure, which is the second reference method according to CLSI, a standardized suspension of bacteria is inoculated onto a series of agar plates, each containing a different concentration of antibiotic, encompassing the therapeutic range of the drug.

Disk diffusion testing described by Bauer et al. (1966) is a qualitative test that provides the results in defined categories (sensitive, intermediate, and resistant). Commercially prepared disks, each of which is preimpregnated with a standard concentration of a particular antibiotic, is then evenly dispensed and lightly pressed onto the agar surface. A growth medium, usually Mueller–Hinton agar, is first evenly seeded throughout the plate with the isolate of interest that has been diluted at a standard concentration ($\sim 1\text{–}2 \times 10^8$ CFUs/mL). After an overnight incubation, the bacterial growth around each disk is observed. If the test isolate is susceptible to an antibiotic, a clear area of "no growth" will be observed around that particular disk.

Another method based on the diffusion of antibiotics in agar is the E-test. This procedure is a quantitative method using transparent, inert plastic strips measuring 60 mm in length by 5.5 mm in width, in which a predefined concentration gradient of the antimicrobial to be investigated is incorporated. The concentration of the various drugs in the E-test ranges from 0.016 to 256 μg/mL or from 0.002 to 32 μg/mL. The technical parameters of the E-test are Mueller–Hinton culture medium and an inoculum with a final concentration corresponding to a turbidity of 0.5 of the McFarland scale. The MIC is read from the intersection of the ellipse zone of inhibition of bacterial growth with the value printed on the strip.

There are also multiple automated systems for the testing of antimicrobial susceptibility of *S. aureus* available. The largest market is shared by two products: Vitek (bioMérieux) and Microscan (Dade International). The Vitek system uses computerized analysis of growth on cards for the calculation of MICs. In some cases, this calculation depends on bacterial identification. Microscan products are based on the traditional MIC methodology and include a range of panels and microtiter plates where reagents for bacterial identification may also be included. Automatic equipment can be extremely useful, permitting to reduce the time of these tests. However, in some situations this decrease in the time of reading of the tests can interfere with the adequate detection of specific types of resistance, compromising the result.

3. IDENTIFICATION AND CHARACTERIZATION OF *STAPHYLOCOCCUS AUREUS* BY GENOTYPIC METHODS

Over the last few years, different commercial systems for the identification of *S. aureus* have been developed as an alternative to the classical (phenotypic) identification as described before. Automated systems and commercial kits based on miniature biochemical tests are widely used today in both routine and research laboratories. However, these diagnostic systems have some disadvantages such as their high cost and the variable expression of phenotypic characteristics, because they are based on colorimetric results and subjectivity in their interpretation can lead to ambiguity (Couto et al., 2001).

Molecular techniques have been the preferred method for identifying microorganisms because of their higher specificity and sensitivity. In recent years, these techniques have evolved from previously difficult and laborious analyses to easier, less expensive, and rapid tests. In some cases, the test results are available within as little as 1 h. Staphylococcal strains not identified to the species level or erroneously identified by conventional phenotypic tests can be correctly identified by genotypic techniques such as PCR, multiplex PCR, real-time PCR, DNA–DNA hybridization, and sequence-based identification systems.

3.1 POLYMERASE CHAIN REACTION/REAL-TIME POLYMERASE CHAIN REACTION-BASED METHODS

The introduction of PCR in the diagnostic laboratory has been an important advancement. PCR-based analyses permit laboratory personnel to detect microorganisms that cannot be cultured, as well as fastidious or slow-growing organisms. However, the early methods used to detect PCR products were laborious and time-consuming and were therefore reserved for more important pathogens. The development of automated and semiautomated PCR platforms and of easier and faster detection methods has increased the application of these technologies to a wider range of pathogens.

The targets commonly used for PCR identification of *S. aureus* are the genes that encode ribosomal subunits (rRNA). These genes are particularly useful for the taxonomic classification of microorganisms because they contain both highly conserved and variable regions.

Multiplex PCR is an alternative method for the detection of multiple pathogens in a single reaction. In this technique, multiple sets of primers are used in a single PCR reaction, each of them having a specific pathogen as target. The amplification of an internal control is important to ensure that the amplification reaction was not inhibited. Rocchetti (2014) developed a multiplex PCR protocol designed to detect a broad spectrum of *Staphylococcus* spp. directly in blood cultures. The author used an extraction technique that was found to be efficient in eliminating interfering agents present in blood. The protocol includes a primer for the detection of the 16S rRNA gene specific for the genus *Staphylococcus*, investigation of the *coa* gene that distinguishes coagulase-positive staphylococci and CoNS, and the SAU primer specific for *S. aureus*, which was able to differentiate *S. aureus* from other coagulase-positive staphylococci. The multiplex PCR was found to be sensitive, specific, and fast, and showed good agreement with the phenotypic results. Thus, multiplex PCR is an effective method that is not more expensive than the automated identification method based on biochemical tests (VITEK 2) and can be used to help with the diagnosis of staphylococci, permitting to obtain a faster result.

The amplification of nucleic acids in real time, notably real-time PCR, contributes significantly to the analysis of a wide range of important microorganisms. Many microorganisms that used to be identified by conventional PCR can now be detected using real-time PCR approaches. This type of amplicon detection represents a significant advancement in nucleic acid chemistry because it is faster, allows for quantification and, in case that fluorescent reporter probes are used, is even more specific and/or sensitive than conventional methods. In addition, the risk of amplicon contamination in the laboratory is significantly reduced because there is no need to open the reaction tube for postamplification manipulation of the amplicon. Different real-time nucleic acid amplification platforms are available. Real-time PCR has been used for the identification of fastidious pathogens that have traditionally been investigated by conventional PCR. In addition to these applications, because the technique is now more widely available and more user friendly, real-time PCR is being applied to the investigation of commonly found bacteria including *S. aureus* identification. Studies have shown that real-time PCR is a reliable and rapid method for the differentiation of *S. aureus* from other coagulase-positive staphylococci and for the identification of CoNS species (Skow et al., 2005).

3.2 MICROARRAY-BASED METHODS

The DNA microarray technology permits to simultaneously evaluate the presence of thousands of genes in a given organism. The DNA arrays, also known as DNA chips, are an orderly arrangement of DNA segments on a solid surface. The microarrays consist of slides on which probes (DNA samples) were immobilized at precisely defined quantities and positions (spots) to permit hybridization with a pool of mRNAs extracted from biological samples (targets) that were previously labeled with fluorescent markers (fluorophores). Because the messenger RNA (mRNA) molecules are highly unstable when manipulated, most laboratory protocols recommend reverse transcription to convert them into the corresponding complementary DNA (cDNA) during labeling (Jaluria et al., 2007).

After hybridization, each slide is washed to remove excess target sequences that did not bind to the probes. The slide is then exposed to laser light that excites the fluorophores incorporated into the "targets," which thus emit fluorescence. In general, the higher the expression of a given gene, the higher will be the number of fluorophore-labeled targets and consequently the intensity of fluorescence emitted by the target-probe complex after hybridization (Jaluria et al., 2007). The microarray technology thus provides an indirect measure of the level of gene expression by quantifying the abundance of transcribed mRNAs.

Microarrays can provide detailed and relevant information for the study of *S. aureus* by permitting the simultaneous detection of the presence or absence of a large number of virulence-associated genes in a single experiment. Spence et al. (2008) developed a microarray (called VirEp) to study the virulence and epidemiology of *S. aureus*. The array comprised 84 clinically relevant gene targets for cost-effective characterization and molecular typing and was able to simultaneously detect 13 *S. aureus* isolates. Nucleic acids from a collection of 64 *S. aureus* isolates were hybridized to the VirEp microarray and all isolates were correctly identified as *S. aureus*. Analysis of the microarray results revealed that 36/84 gene targets were present in all 64 isolates, while 12 gene targets were absent in all isolates analyzed. The conserved genes included identification genes (11%), genes encoding adhesins (25%), proteases (22%) and toxins (16.5%), antimicrobial resistance genes (16.5%), and molecular typing genes (9%).

Glass slides with different surface modifications that permit the covalent binding of the probe is the conventional microarray format widely accepted in research laboratories. However, this format requires skilled handling and is not very well accepted in routine laboratories. Alternative formats have therefore been developed.

Microarrays prepared in tubes or 96-well plates represent an easy-to-use format, which is often applied in commercial products for the routine diagnostic market. One example is the *S. aureus* Genotyping Kit 2.0 from Alere Technologies GmbH (2016). This kit permits the detection of *S. aureus* resistance and virulence genes and the determination of similarity of *S. aureus* isolates with known reference strains. The set of gene targets comprises a variety of species markers, virulence-associated genes including genes encoding Leucocidin Panton-Valentine (PVL) and exotoxins, antibiotic resistance genes, genes encoding microbial surface components recognizing adhesive matrix molecules, as well as other markers. DNA from an *S. aureus* colony is amplified and labeled, and the resulting amplification products are hybridized to 336 oligonucleotide probes of the microarray. The overall pattern is analyzed automatically for the presence or absence of specific genes and is compared to a database of profiles that permits the assignment to clonal complexes.

3.3 SEQUENCE-BASED METHODS INCLUDING NEXT-GENERATION SEQUENCING TECHNIQUES

DNA sequencing, formerly a resource used exclusively in research laboratories, has become common in molecular pathology and microbiology laboratories. The first popular method for DNA sequencing was the chain termination method published by Sanger et al. (1977). The first automated DNA sequencer, the ABI 370, was launched in 1986, and the first capillary electrophoresis system, the ABI 3700, was introduced in 1998. This technology, which uses traditional Sanger sequencing, consists of the incorporation of dideoxynucleotides into the growing DNA strand. DNA sequencing of an amplified product is now a common method of postamplification analysis in the identification of microorganisms (Koneman et al., 1997).

The MicroSeq Microbial Identification System is a commercially available identification system that uses primers targeting the 16S rRNA gene. After PCR amplification, the amplified product is purified and sequenced. The sequence obtained is compared to a database (Basic Local Alignment Search Tool (BLAST)), which is maintained and updated by the server (Koneman et al., 1997). Several other genetic targets, like the *16S rRNA* gene, which contain highly conserved and variable regions have been used for the identification of microorganisms. These alternative targets for PCR identification of microorganisms and sequencing include RNA polymerase (*rpoB*), heat shock protein (*hsp*), and elongation factor (*tuf*) (Martineau et al., 2001; Drancourt et al., 2004; Kim et al., 2004).

New sequencing technologies, called NGS, started to be commercialized in 2005 and are rapidly evolving. The NGS platforms can generate data for millions of base pairs in a single run (Mardis, 2008). Despite considerable differences, all NGS platforms are based on the massive parallel sequencing of DNA fragments. Whereas an electrophoresis sequencer processes a maximum of 96 fragments per run, NGS systems can read up to billions of fragments at the same time (Kircher and Kelso, 2010).

The NGS platforms are a powerful alternative for structural and functional genomics studies. The systems are divided into first-generation (chemical degradation and Sanger chain termination), second-generation (454-Roche, Illumina Genome Analyzer-HiSeq/MiSeq, and SOLID), third-generation (Ion Torrent and PacBio RS), and fourth-generation sequencers (Nanopore) (Mayo et al., 2014).

NGS has emerged as a powerful tool to profile complex microbial communities. This new technology greatly reduces the time and cost of DNA sequencing, permitting a small laboratory to sequence the whole genome of the selected bacterium. Different strains of many bacterial species of interest have been sequenced.

In recent years, NGS has been used for the identification and characterization of infectious agents isolated from clinical samples, foods, and the environment. The approaches used are whole-genome sequencing of the infectious agent and sequencing of gene regions for the typing of microorganisms and antimicrobial resistance genes. The whole-genome and partial sequence data are analyzed against databases available at the Center for Genomic Epidemiology, BLAST, and PubMLST (Mayo et al., 2014).

A new testing methodology based on NGS technologies could accelerate the diagnosis of foodborne bacterial outbreaks, allowing public health officials to identify the microbial culprits in less than a day. The methodology could also identify coinfections with secondary microbes, determine the specific variant of the pathogen, and help alert health officials to the presence of new or unusual pathogens.

Researchers from the Georgia Institute of Technology and the U.S. Centers for Disease Control and Prevention recently compared the new methodology against traditional culture-based methods with samples from two severe outbreaks of *Salmonella*, a common foodborne pathogen (Huang et al., 2017). The metagenomics approach, which relies on DNA sequencing and bioinformatics analysis of the resulting sequencing data, not only correctly identified the bacterial culprit, but also found a possible coinfection with *S. aureus*. This study shows the potential to use stools from healthy and sick people directly to determine who is involved in an outbreak, which will revolutionize the way we detect and monitor for foodborne disease in the future. Metagenomics identifies microbes by sequencing the entire DNA present in a sample and comparing the genomic data to a database of known microbes. In addition to identifying the bacteria present in the samples, the methodology can also measure the relative abundance of each microbial species and their virulence potential.

4. GENOTYPING METHODS FOR MOLECULAR EPIDEMIOLOGY OF *STAPHYLOCOCCUS AUREUS*

Molecular typing techniques have emerged in an attempt to develop more specific and sensitive methods that are able to establish genetic relationships between *S. aureus* isolates. These techniques may represent an additional discriminatory tool, especially in the case of outbreaks of hospital infections and food poisoning (SFP). In this respect, the understanding of the relationship between different isolates is fundamental for the elucidation of these outbreaks.

Molecular typing systems can be used for the investigation of SFP outbreaks, confirmation, and characterization of transmission patterns of one or more *S. aureus* clones, hypothesis testing on the origin and vectors of transmission of these clones, and monitoring of their reservoirs. Molecular analysis is also useful for epidemiological surveys and for the evaluation of control measures by documenting the occurrence of certain *S. aureus* strains over time and the circulation of clones.

The application and interpretation of microbial typing tools in epidemiological studies require to understand their limitations. In addition to reliability, a technique is considered valid when its capacity to discriminate between strains is satisfactory and a biological basis for the grouping of strains with apparently distinct types is possible (Foxman and Riley, 2001).

In case of *S. aureus*, several molecular typing techniques exist that are based on the differentiation of genotype profiles analyzed in total DNA samples or in samples digested with enzymes. These so-called restriction enzymes cut the DNA into different fragments, generating different restriction profiles for analysis. However, the profiles are clearly complex and the isolates need to be separated by electrophoresis.

According to van Belkum et al. (2007), typing of *S. aureus* can be performed at different levels, depending on the situation locally, in hospitals or other healthcare facilities for small investigations; regionally or nationally, in a reference laboratory for public health and surveillance studies, or in international collaborations to define or monitor major clones. At each of these levels, different methods can be applied. Methods for the molecular typing of *S. aureus*, particularly *spa* typing, MLST, MLVA, PFGE, and staphylococcal cassette chromosome *mec* (SCC*mec*) typing have several clinical-epidemiological applications.

4.1 *SPA* TYPING

Single locus DNA sequencing of the repeat region of the protein A gene (*spa*) can be used for reliable, accurate, and discriminatory typing of *S. aureus*. This typing method is called *spa* typing, which uses sequences of the polymorphic X region, or short sequence repeats, of the protein A gene (*spa*) of *S. aureus*. The *spa* gene comprises ~2150 bp and contains a number of functionally distinct regions: an IgG Fc-binding region, called X region (a repetitive and highly polymorphic region, but stable enough to permit the discrimination of unrelated isolates), and a C-terminal sequence required for attachment to the cell wall (Frenay et al., 1996). For typing, part of the gene encoding protein A (staphylococcal surface protein known to contain polymorphic regions), containing 24 nucleotide repeats, is amplified and sequenced. Sequencing reveals the primary structure of the repeat units, which facilitates identification by an r-code. On the basis of the series of r-codes identified, a *spa* type (t-code) can be defined and a homology score between types can be calculated based on the relationships between t-codes. The sequences can be deposited in a database such as the RIDOM SpaServer (http://www.spaserver.ridom.de) (Harmsen et al., 2003). At time of writing of this chapter more than 16,600 different *spa*-types have been listed in the RIDOM database.

4.2 MULTILOCUS SEQUENCE TYPING

This molecular typing method characterizes isolates based on the sequence of 450-bp internal fragments of seven housekeeping genes of the microorganism in which different sequences correspond to different alleles of each gene and to a certain type of sequence, called sequence type (ST). Because many alleles exist for each of the seven genes, it is unlikely that the isolates have identical allelic profiles by chance and isolates with the same allelic profile can be considered members of the same clone (Enright et al., 2002). On the other hand, a clonal complex is defined as a group of STs in which every ST shares at least five of seven identical alleles with at least one other ST in the group (Day et al., 2001). One of the main advantages of MLST is the easy comparison of sequence date between laboratories; one other is the possibility to compare the results obtained in different studies through the Internet. The MLST results are entered into a digital database at http://www.mlst.net, permitting comparisons between *S. aureus* sequences described in different parts of the world. When writing this chapter, 3112 different S. *aureus* STs are included in the MLST database. If new alleles and STs are found, they are stored in the database after verification. Researchers perform epidemiological and phylogenetical studies by comparing STs of different clonal complexes. A huge set of data is produced during the sequencing and identification process, and bioinformatics techniques are used to arrange, manage, analyze, and merge these biological data.

4.3 MULTIPLE-LOCUS VARIABLE NUMBER OF TANDEM REPEAT ANALYSIS

MLVA, an abbreviation for multiple-locus variable number of tandem repeats analysis, is another method that can be used to type a large number of bacterial pathogens, including *S. aureus*. Microbiologists usually perform MLVA after PFGE—see below—to obtain further details about the bacterial type possibly been involved in a foodborne outbreak. This method is based on the analysis of polymorphic repeated sequences, called multiple-locus variable number tandem repeats (VNTR). The number of tandem repeats is a genetic marker that is characteristic for a particular strain and the number of tandem repeats is similar for related strains. Analysis of multiple variable loci increases the discriminatory power of the method.

Sobral et al. (2012) showed that MLVA is equally well adapted for studying *S. aureus* epidemiology, regardless of the origin of the strain (animal or human). The four steps of the MLVA procedure (DNA extraction, amplification of the 16 VNTRs in two multiplex PCRs, fragment analysis by capillary electrophoresis, and MLVA code assignment) were standardized to be usable and understandable by nonexpert users. The resulting data can be queried against freely accessible nternet MLVA databases such as http://mlva.u-psud.fr.

4.4 PULSED-FIELD GEL ELECTROPHORESIS

PFGE is the most commonly used method for bacterial molecular typing and consists of the direct digestion of genomic DNA with a group of enzymes, called restriction endonucleases, followed by separation of the fragments by gel electrophoresis in an alternating electrical field. These enzymes cleave DNA at specific sequences, referred to as restriction sites. PFGE produces chromosomal restriction patterns of 5–20 fragments that range in size from 10 to 800 kb. The type of restriction endonuclease used is generally defined by the bacterium analyzed. In the case of *S. aureus*, the most commonly used enzyme is *SmaI* (van Belkum et al., 1998). Softwares (Bionumerics, Applied Maths) are then applied to determine whether the strains are distinguishable or indistinguishable (Fig. 6.1).

However, not all *S. aureus* isolates can be successfully PFGE typed by *SmaI*; e.g., livestock-associated methicillin-resistant *S. aureus* (LA-MRSA) of MLST type ST398 are nontypeable by *SmaI*-PFGE. To overcome this problem, Argudin et al. (2009) established a PFGE protocol using Cfr91, a *SmaI* neoschizomer.

Careful standardization permits to define the levels of inter- and intralaboratory reproducibility, which led to the creation and maintenance of international databases such as the PulseNet network (Swaminathan et al., 2001). Started in the United States in 1996, and now including PulseNet Europe and other international databases, this network has become one of the major global typing systems available to date (van Belkum et al., 2007).

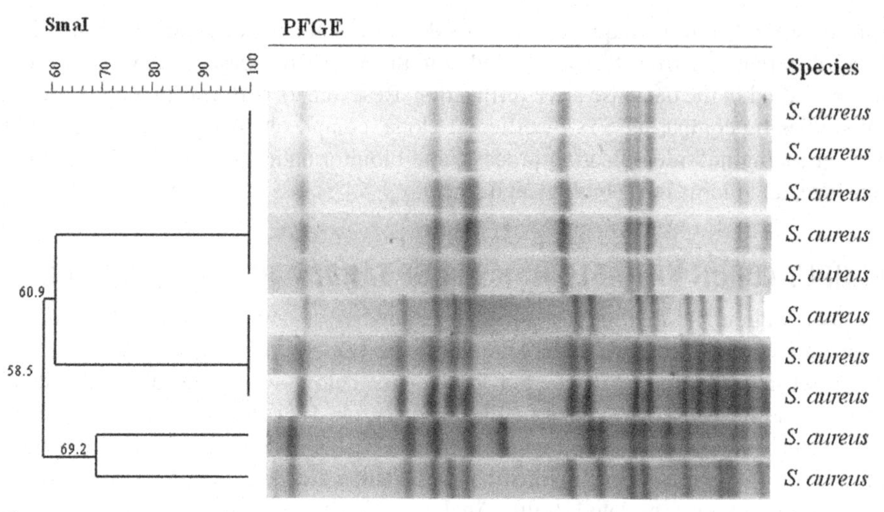

FIGURE 6.1

Dendrogram obtained for *Staphylococcus aureus* isolates generated by Dice/unweighted pair group method using arithmetic averages (UPGMA) analysis (Bionumerics; Applied Maths) of pulsed-field gel electrophoresis (PFGE) profiles (*SmaI*) (similarity ≥80%).

Although PFGE is an adequate technique to study outbreaks, it is not sufficient for long-term or global epidemiological studies. In addition to the limitations of the interpretation criteria related to the short period of time and restricted geographic areas, PFGE has problems of reproducibility. Different results may be obtained when the technique is performed in different laboratories, even when standardized conditions are used (van Belkum et al., 1998). Moreover, PFGE is a very time-consuming and laborious techniques that requires highly experienced technicians.

4.5 STAPHYLOCOCCAL CASSETTE CHROMOSOME *MEC* TYPING

Another typing technique, to be applied for MRSA isolates, only, is based on supplemental penicillin binding protein (PBP 2a) encoding gene *mec*A, which is carried by a mobile genetic element, the (SCC*mec*). So far, SCC*mec* types I, II, III, IV, V, VI, VII, VIII, IX, X, and XI have been described. Types are defined by the combination of the type of *ccr* gene complex and the class of the *mec* gene complex. Subtypes are defined by J region polymorphisms in the same combination of *mec* and *ccr* complexes (IWG-SCC, 2009).

SCC*mec* typing is a useful epidemiological tool because a predominance of different types in the hospital environment or community. Among the main SCC*mec* types (I, II, III, IV, and V), only types I, II, and III are found in healthcare-acquired MRSA, while types IV and V are observed in community-acquired *S. aureus* (CA-MRSA) and also among LA-MRSA.

5. METHODS FOR DETECTION OF STAPHYLOCOCCAL ENTEROTOXINS

SEs are the main cause of food poisoning and are associated with a type of gastroenteritis that manifests clinically as vomiting with or without diarrhea as a result of the ingestion of one or more enterotoxins present in food contaminated with these bacteria (Dinges et al., 2000).

Diagnosis of SFP outbreaks is generally confirmed either by the recovery of at least 10^5 *S. aureus* g−1 from food remnants or by the detection of SEs in food remnants. In some cases, confirmation of SFP is difficult because *S. aureus* is heat sensitive, whereas SEs are not. Thus, in heat-treated food matrices, *S. aureus* may be eliminated without inactivating SEs. In such cases, it is not possible to characterize a food poisoning outbreak by enumerating coagulase-positive staphylococci (CPS) in food remnants or detecting SE genes in isolated strains, i.e., it is crucial to confirm an outbreak by detection of SEs (Bergdoll, 1990; Hennekinne et al., 2012).

Three types of methods are mainly used to detect bacterial toxins in food: bioassays, molecular biology, and/or immunological techniques; more recently, mass spectrometry-based methods are also developed for SE detection.

First, bioassays were used to detect SEs in foods, which consist of observing the emetic response of monkeys after the intragastric administration of foods. As symptoms of SFP appear only at considerably higher numbers of SEs than those involved in human food poisoning (Asao et al., 2003; Ostyn et al., 2010), this technique is not appropriate for characterizing SFP outbreaks. Therefore, only approaches applying immunological and molecular biology techniques are described in the following sections. In addition, a brief overview of other SE detection methods, i.e., mass spectrometry-based techniques, is provided.

5.1 IMMUNOLOGICAL METHODS

Initially, specific precipitation reactions in agarose were developed for the detection of SEs, including single one-dimensional diffusion, double diffusion, micro slide technique, and capillary tube diffusion (Bergdoll, 1990). Among these methods, the optimal sensitivity plate (OSP) method, a modification of the Ouchterlony plate test, has been widely used for this purpose (Bergdoll, 1990; Robbins et al., 1974). The sensitivity of the OSP method is 0.5 μg/mL, which is sufficient to detect toxins in staphylococcal culture supernatants (Bergdoll, 1990). The use of supernatants concentrated by the cellophane-over-agar method (Robbins et al., 1974) or by dialysis sac culture (Donnelly et al., 1967) can increase the sensitivity to 0.1 μg/mL. However, a significant number of toxigenic strains produce concentrations ≤1 ng/mL, i.e., below the detection levels of diffusion methods. Thus, more sensitive detection methods have been developed. The reverse passive hemagglutination test, which uses adsorption of specific antibodies to red blood cells, was the first truly sensitive method that can detect enterotoxin levels of 1 ng/g in foods. The disadvantages of this method are nonspecific hemagglutinations and the difficulty in adsorbing certain antibodies to the surface of red blood cells (Silverman et al., 1968).The system has been improved by replacing erythrocytes with latex particles (reverse passive latex agglutination—RPLA), which increased the sensitivity to 0.25 ng enterotoxins/g. However, it continues to be difficult to distinguish nonspecific reactions in certain foods or culture supernatants (Cunha et al., 2006, 2007).

A variety of ELISA-based methods have been described for the detection of SEs (Poli et al., 2002), including the automated VIDAS Staph Enterotoxin II—SET2 screening system (bioMérieux), Tecra kit (3M), and RIDASCREEN SET Total test (R-Biopharm AG, Darmstadt, Germany). However, it is widely recognized that the use of immunological methods to detect contaminants in food matrices is a difficult task, mainly because of the lack of specificity and sensitivity of the assay. Many drawbacks impair the development and use of these techniques for detecting SEs (Hennekinne et al., 2012). First, highly purified toxins are needed to raise specific antibodies to develop an enzyme-immunoassay; purified toxins are difficult and expensive to obtain. Moreover, and until very recently, only antibodies against SEA to SEE, staphylococcal enterotoxin G (SEG), staphylococcal enterotoxin H (SEH), and staphylococcal enterotoxin-like toxin type Q (SElQ) were available (Schlievert and Case, 2007). The ELISA test will not detect the other SEs, which could partly explain why some outbreaks remained uncharacterized without a known etiological agent. Another drawback is the low specificity of some commercial kits, where false positives may occur depending on food components (Wieneke, 1991) as it is well known that some proteins, such as protein A, can interfere with binding to the Fc fragment (and, to a lesser extent, Fab fragments) in immunoglobulin G from several animal species, such as mouse or rabbit, but not rat or goat. Other interferences are associated with endogenous enzymes, such as alkaline phosphatase or lactoperoxidase.

Whatever the detection method used and owing to the low amount of SEs present in food, it is crucial to concentrate the extract before performing detection assays. For this purpose, various methodologies have been tested (Hennekinne et al., 2012). Among them, only extraction followed by dialysis concentration has been approved by the European Union for extracting SEs from food (Anon., 2007).

Very recently, attempts were made to establish the first international (ISO/CEN) standard method for the detection of SEs, SEA, staphylococcal enterotoxin B (SEB), staphylococcal enterotoxin C1, C2 and C3 (SECs), staphylococcal enterotoxin D (SED), and SEE in food stuff. The Standard consists of two main steps (1) extraction followed by a concentration based on dialysis principle and (2) an immunoenzymatic detection using commercially available detection kits; it is to be assumed that this Standard is available for official controls and routine check of foods shortly (Anon., 2017).

5.2 MOLECULAR BIOLOGICAL TECHNIQUES

Whereas immunological methods can yield false-positive results due to cross-reactivity between antigens and the occurrence of nonspecific reactions these disadvantages are not observed in the direct detection of the genes responsible for toxin production. According to Schmitz et al. (1998), for practical consideration, toxin gene-positive staphylococcal strains should be considered potential toxin producers because in vivo toxin production cannot be ruled out.

Currently, the PCR is one of the most common methods used because it permits the identification of enterotoxin genes with high sensitivity and specificity. PCR detects genes of interest regardless of their expression. In this respect, the genes responsible for the production of enterotoxins may be present but not active. Molecular techniques such as reverse transcriptase-PCR (RT-PCR) permit to detect mRNA, regardless of whether or not a gene is expressed. In this technique, mRNA is converted to cDNA by the reverse transcriptase enzyme. The cDNA is then amplified by PCR using specific primers to confirm expression, which leaves no doubt as to the toxigenic potential of the microorganism. In the study of Calsolari et al. (2011) using RT-PCR for the detection of SEA, SEB, SEC, SED, and toxic shock syndrome toxin 1 in *S. aureus*, expression of mRNA encoding the staphylococcal toxins was observed in 43 (39.8%) of the 108 strains that tested positive by PCR. RT-PCR was also used for the detection of mRNA encoding enterotoxins SEE, SEG, SEH, and staphylococcal enterotoxin I in 90 *S. aureus* isolates (Vasconcelos et al., 2011) and expression of mRNA was observed in 42 (50.6%) of the 83 isolates that had a positive PCR result.

Promising techniques that clearly confirm enterotoxin production are those permitting detection of the respective proteins. Sodium dodecyl sulfate polyacrylamide gel electrophoresis (SDS-PAGE), described by Laemmli (1970), is used to determine the mobility of proteins in a gel submitted to an electrical current. The samples are heated in the presence of SDS and reducing agents that denature proteins by destroying disulfide bridges and separate polypeptide chains, which can be identified based on their molecular weight. Another effective method for the detection of proteins is Western blotting, which uses gel electrophoresis to separate denatured proteins. The proteins are then transferred to a nitrocellulose membrane where they are stained with antibodies used as probes. This technique permits to examine the amount of protein in a given sample and to compare levels between different groups.

In a study investigating *S. aureus* enterotoxins in food, protein profiles of eight enterotoxigenic *S. aureus* strains were analyzed by SDS-PAGE. All enterotoxigenic isolates had one or two bands at 26–29 kDa, proving the production of enterotoxins by *S. aureus* strains isolated from raw milk, yoghurt, and chicken (El-Jakee et al., 2013).

5.3 OTHER STAPHYLOCOCCAL ENTEROTOXIN DETECTION METHODS

Considering the drawbacks with currently available detection methods and the lack of available antibodies against the newly described SEs, other strategies based on physicochemical techniques have been developed very recently. One of the methods developed uses MS to identify a protein toxin in a model food matrix. MS has emerged as a very promising and suitable technique for analyzing protein and peptide mixtures hence, and is one of the most sensitive techniques currently available because it provides specific, rapid, and reliable analytical quantification of the amount of enterotoxins (Hennekinne et al., 2012). The development of two soft ionization methods, such as electrospray ionization and MALDI, and the use of appropriate mass analyzers such as TOF) have revolutionized the analysis of biomolecules. Given the wide range of methodologies available, a single MS technique cannot be used for all proteins. The MS method thus requires the development of a series of techniques, individually suited for each particular case. Moreover, as any food matrix may contain many proteins, lipids, and other molecular species that interfere with the detection of the targeted toxin and may distort quantification, the critical step when applying MS-based methods remains the sample preparation as shown before in case of immune-enzymatic SE detection methods.

Taken together, key in detecting SEs (e.g., in case of SFP outbreak investigation) might be the combined application of classical microbiological techniques for enumerating CPS strains with immunological techniques, molecular biology, and MS-based methods (Hennekinne et al., 2012).

6. CONCLUSION

Until recently, microbiological methods dedicated to *S. aureus* identification were based on the use of selective growth media, which are time-consuming and preclude same-day diagnosis. The development of new methods for detecting *S. aureus* has been and will continue to be essential for our understanding of this organism and its relationship with food. Molecular assays have proven rapid, affordable, and successful in terms of sensitivity and specificity. The establishment of strain relatedness using genotyping methods (PFGE, MLST, and MLVA) allows documenting the spread of strains and potential SFP outbreaks. However, there is still a big need in developing detection methods for SEs in food.

Many tools and techniques are available for the study of the *S. aureus* genome.

Technologies that exploit whole genomes, microarrays, and proteomics have permitted major advances in our understanding of staphylococcal biology and will revolutionize the way we detect and monitor for foodborne disease in the future.

REFERENCES

Alere Technologies GmbH Homepage, 2016. Alere Technologies GmbH. Jena, Germany http://alere-technologies.com/en/products/lab-solutions.html/.

Anon., 2007. Commission regulation No. 1441/2007 of 5 December 2007. Off. J. Eur. Union L322, 12–29.

Anon., 2017. ISO TC 34/SC 9. Microbiology of Food Chain — Horizontal Method for the Immunoenzymatic Detection Of Staphylococcal Enterotoxins in Foodstuffs (Date: 2017-01-13).

Argudín, M.A., Rodicio, M.R., Guerra, B., January 2010. The emerging methicillin-resistant *Staphylococcus aureus* ST398 clone can easily be typed using the Cfr9I SmaI-neoschizomer. Lett. Appl. Microbiol. 50 (1), 127–130.

Asao, T., Kumeda, Y., Kawai, T., Shibata, T., Oda, H., Haruki, K., Nakazawa, H., Kozaki, S., 2003. An extensive outbreak of staphylococcal food poisoning due to low-fat milk in Japan: estimation of enterotoxin A in the incriminated milk and powdered skim milk. Epidemiol. Infect. 130, 33–40.

Baird-Parker, A.C., 1962. An improved diagnostic and selective medium for isolating coagulase-positive staphylococci. J. Appl. Microbiol. 25, 12–19.

Baird-Parker, A.C., Davenport, E., 1965. The effect of recovery medium on the isolation of *Staphylococcus aureus* after heat treatment and after the storage of frozen or dried cells. J. Appl. Microbiol. 28, 390–402.

Baker, J.S., 1984. Comparison of various methods for differentiation of staphylococci and micrococci. J. Clin. Microbiol. 9, 875–879.

Bannerman, T.L., Kleeman, K.T., Kloos, W.E., 1993. Evaluation of the Vitek Systems gram-positive identification card for species identification of coagulase-negative staphylococci. J. Clin. Microbiol. 31, 1322–1325.

Bergdoll, M.S., 1990. Analytical methods for *Staphylococcus aureus*. Int. J. Food Microbiol. 10 (2), 91–99.

Bauer, A.W., Kirby, W.M., Sherris, J.C., Turck, M., 1966. Antibiotic susceptibility testing by a standardized single disk method. Am. J. Clin. Pathol. 45, 493–496.

Calsolari, R.A.O., Pereira, V.C.P., AraújoJúnior, J.P., Cunha, M.L.R.S., 2011. Determination of toxigenic capacity by reverse transcription polymerase chain reaction in coagulase-negative staphylococci and *Staphylococcus aureus* isolated from newborns in Brazil. Microbiol. Immunol. 55, 394–407.

Carbonnelle, E., Beretti, J.L., Cottyn, S., Quesne, G., Berche, P., Nassif, X., et al., 2007. Rapid identification of Staphylococci isolated in clinical microbiology laboratories by matrix-assisted laser desorption ionization-time of flight mass spectrometry. J. Clin. Microbiol. 45, 2156–2161.

Clinical, Laboratory Standards Institute (CLSI), 2015. Performance Standards for Antimicrobial Susceptibility Testing: 25st Informational Supplement (M100–S25). (Wayne, PA).

Couto, I., Pereira, S., Miragaia, M., Sanches, I.S., de Lencastre, H., 2001. Identification of clinical staphylococcal isolates from humans by internal transcribed spacer PCR. J. Clin. Microbiol. 39 (9), 3099–3103.

Croxatto, A., Prod'hom, G., Greub, G., 2012. Applications of MALDI-TOF mass spectrometry in clinical diagnostic microbiology. FEMS Microbiol. Rev. 36 (2), 380–407.

Cunha, M.L.R.S., Sinzato, Y.K., Silveira, L.V.A., 2004. Comparison of methods for the identification of coagulase-negative staphylococci. Mem. Do Inst. Oswaldo Cruz. 99 (8), 855–860.

Cunha, M.L.R.S., Peresi, E., Calsolari, R.A., AraújoJúnior, J.P., 2006. Detection of enterotoxins genes in coagulase-negative staphylococci isolated from foods. Braz. J. Microbiol. 37 (1), 70–74.

Cunha, M.L.R.S., Calsolari, R.A.O., Araújo Jr., J.P., 2007. Detection of enterotoxin and toxic shock syndrome toxin 1 genes in *Staphylococcus*, with emphasis on coagulase-negative staphylococci. Microbiol. Immunol. 51, 381–390.

Day, N.P., Moore, C.E., Enright, M.C., Berendt, A.R., Smith, J.M., Murphy, M.F., Peacock, S.J., Spratt, B.G., Feil, E.J., 2001. A link between virulence and ecological abundance in natural populations of *Staphylococcus aureus*. Science 292, 114–116 Retraction, 295:971, 2002.

Dinges, M.M., Orwin, P.M., Schlievert, P.M., 2000. Exotoxins of *Staphylococcus aureus*. Clin. Microbiol. Rev. 13, 16–34.

Donnelly, C.B., Leslie, J.E., Black, L.A., Lewis, K.H., 1967. Serological identification of enterotoxigenic staphylococci from cheese. Appl. Microbiol. 15, 1382–1387.
Drancourt, M., Roux, V., Fournier, P.E., Raoult, D., 2004. rpoB gene sequence-based identification of aerobic gram-positive cocci of the genera *Streptococcus*, *Enterococcus*, *Gemella*, *Abiotrophia*, and *Granulicatella*. J. Clin. Microbiol. 42, 497–504.
Dubois, D., Leyssene, D., Chacornac, J.P., Kostrzewa, M., Schmit, P.O., Talon, R., et al., 2010. Identification of a variety of *Staphylococcus* species by matrix-assisted laser desorption ionization-time of flight mass spectrometry. J. Clin. Microbiol. 10 (3), 941–945.
El-Jakee, J., Marouf, S.A., Ata, N.S., Abdel-Rahman, E.H., Abd El-Moez, S.I., Samy, A.A., El-Sayed, W.E., 2013. Rapid method for detection of *Staphylococcus aureus* enterotoxins in food. Glob. Vet. 11, 335–341.
Enright, M.C., Robinson, D.A., Randle, G., Feil, E.J., Grundmann, H., Spratt, B.G., 2002. The evolutionary history of methicillin-resistant *Staphylococcus aureus* (MRSA). Proc. Natl. Acad. Sci. U.S.A 99 (11), 7687–7692.
Ferreira, A.M., Bonesso, M.F., Mondelli, A.L., Cunha, M.L.R.S., 2012. Identification of *Staphylococcus saprophyticus* isolated from patients with urinary tract infection using a simple set of biochemical tests correlating with 16S 23S interspace region molecular weight patterns. J. Microbiol. Methods 91 (3), 406–411.
Foxman, B., Riley, L., 2001. Molecular epidemiology: focus on infection. Am. J. Epidemiol. 153, 1135–1141.
Frenay, H.M., Bunschoten, A.E., Schouls, L.M., Van Leeuwen, W.J., Vandenbroucke-Grauls, C.M.J.E., Verhoef, J., et al., 1996. Molecular typing of methicillin resistant *Staphylococcus aureus* on the basis of protein A gene polymorphisms. Eur. J. Clin. Microbiol. Infect. Dis. 15, 768–770.
Funke, G., Monnet, D., de Bernardis, C., von Graevenitz, A., Freney, J., 1998. Evaluation of the VITEK2 system for rapid identification of medically relevant gram-negative rods. J. Clin. Microbiol. 36 (7), 1948–1952.
Grant, C.E., Sewell, D.L., Pfaller, M., Bumgardner, R.V.S., Williams, J.A., 1994. Evaluation of two commercial systems for identification of coagulase-negative staphylococci to species level. Diagn. Microbiol. Infect. Dis. 18, 1–5.
Harmsen, D., Claus, H., Witte, W., Rothganger, J., Claus, H., Turnwald, D., et al., 2003. Typing of methicillin resistant *Staphylococcus aureus* in a university hospital setting by using novel software for spa repeat determination and database management. J. Clin. Microbiol. 41, 5442–5448.
Hennekinne, J.A., De Buyser, M.L., Dragacci, S., 2012. *Staphylococcus aureus* and its food poisoning toxins: characterization and outbreak investigation. FEMS Microbiol. Rev. 36, 815–836.
Huang, A.D., Luo, C., Pena-Gonzalez, A., Weigand, M.R., Tarr, C.L., Konstantinidis, K.T., 2017. Metagenomics of two severe foodborne outbreaks provides diagnostic signatures and signs of coinfection not attainable by traditional methods. Appl. Environ. Microbiol. 83. http://dx.doi.org/10.1128/AEM.02577-16.
(IWG-SCC) International Working Group on the Classification of Staphylococcal Cassette Chromosome Elements, 2009. Classification of staphylococcal cassette chromosome *mec* (SCC*mec*): guidelines for reporting novel SCC*mec* elements. Antimicrob. Agents Chemother. 53 (12), 4961–4967.
Jaluria, P., Konstantopoulos, K., Betenbaugh, M., Shiloach, J., 2007. A perspective on microarrays: current applications, pitfalls, and potential uses. Microb. Cell Factories 13 (6), 4.
Jorgensen, J.H., Ferraro, M.J., 2009. Antimicrobial susceptibility testing: a review of general principles and contemporary practices. Clin. Infect. Dis. 49, 1749–1755.
Kim, B.J., Hong, S.K., Lee, K.H., Yun, Y.J., Kim, E.C., Park, Y.G., et al., 2004. Differential identification of *Mycobacterium tuberculosis* complex and nontuberculosis mycobacteria by duplex PCR assay using the RNA polymerase gene (rpoB). J. Clin. Microbiol. 42, 1308–1312.
Kim, M., Heo, S.R., Choi, S.H., Kwon, H., Park, J.S., Seong, M.W., et al., 2008. Comparison of the MicroScan, Vitek 2, and crystal GP with 16S rRNA sequencing and MicroSeq 500 v2.0 analysis for coagulase-negative staphylococci. BMC Microbiol. 8, 233.
Kircher, M., Kelso, J., 2010. High-throughput DNA sequencing–concepts and limitations. Bioessays 32, 524–536.
Koneman, E.W., Allen, S.D., Janda, W.M., Schreckenberger, P.C., 1997. Color Atlas and Textbook of Diagnostic Microbiology. Lippincott, Philadelphia.

Laemmli, U.K., 1970. Cleavage of structural proteins during the assembly of the head of bacteriophage T4. Nature 227 (5259), 680–685.

Mayo, B., Rachid, T.C.C., Alegría, A., Leite, A.M.O., Peixoto, R.S., Delgado, S., 2014. Impact of next generation sequencing techniques in food microbiology. Curr. Genom. 15 (4), 293–309.

Mardis, E.R., 2008. Next-generation DNA sequencing methods. Annu. Rev. Genom. Hum. Genet. 9, 387–402.

Martineau, F., Picard, F.J., Danbing, K., Paradis, S., Roy, P.H., Ouellette, M., Bergeron, M.G., 2001. Development of a PCR assay for identification of staphylococci at genus and species levels. J. Clin. Microbiol. 39, 2541–2547.

McDonald, L.C., Weinstein, M.P., Fune, J., Mirrett, S., Reimer, L.G., Reller, L.B., 2001. Controlled comparison of Bact/ALERT FAN aerobic medium and BACTEC fungal blood culture medium for detection of fungemia. J. Clin. Microbiol. 39 (2), 622–624.

Monteiro, A.C.M., Fortaleza, C.M.C.B., Ferreira, A.M., Cavalcante, R.S., Mondelli, A.L., Bagagli, E., et al., 2016. Comparison of methods for the identification of microorganisms isolated from blood cultures. Ann. Clin. Microbiol. Antimicrob. 15, 45.

Ostyn, A., De Buyser, M.L., Guillier, F., Groult, J., Felix, B., Salah, S., Delmas, G., Hennekinne, J.A., 2010. First evidence of a food-poisoning due to staphylococcal enterotoxin type E in France. Euro Surveill. 15, 19528.

Poli, M.A., Rivera, V.R., Neal, D., 2002. Sensitive and specific colorimetric ELISAs for *Staphylococcus aureus* enterotoxins A and B in urine and buffer. Toxicon 40 (12), 1723–1726.

Renneberg, J., Rieneck, K., Gutschik, E., 1995. Evaluation of Staph ID 32 system and Staph-Zym system for identification of coagulase-negative Staphylococci. J. Clin. Microbiol. 33 (5), 1150–1153.

Robbins, R., Gould, S., Bergdoll, M.S., 1974. Detecting the enterotoxigenicity of *Staphylococcus aureus* strains. Appl. Microbiol. 28, 946–950.

Rocchetti, T.T., 2014. Detecção do operonica da produção de biofilme, gene mecA de resistência à oxacilina e identificação de espécies de Staphylococcusspp. diretamente dos frascos de hemoculturas pela técnica de PCR multiplex. Dissertaçãoa presentada ao programa de Pós-Graduação em DoençasTropicais da Faculdade de Medicina de Botucatu. Universidade Estadual Paulista (FMB/UNESP), Brasil.

Sanger, F., Nicklen, S., Coulson, A.R., 1977. DNA sequencing with chain-terminating inhibitors. Proc. Natl. Acad. Sci. U.S. 74 (12), 5463–5467.

Silverman, S.J., Knott, A.R., Howard, M., 1968. Rapid, sensitive assay for staphylococcal enterotoxin and a comparison of serological methods. Appl. Microbiol. 16, 1019–1023.

Schlievert, P.M., Case, L.C., 2007. Molecular analysis of staphylococcal superantigens. Methods Mol. Biol. 391, 113–126.

Schmitz, F.J., Steiert, M., Hofmann, B., Verhoef, J., Hadding, U., Heinz, H.P., et al., 1998. Development of multiplex-PCR for direct of genes for enterotoxin B and C, and toxic shock syndrome toxin-1 in *Staphylococcus aureus* isolates. J. Med. Microbiol. 47, 335–340.

Skow, A., Mangold, K.A., Tajuddin, M., Huntington, A., Fritz, B., Thomson Jr., R.B., Kaul, K.L., 2005. Species-level identification of staphylococcal isolates by real-time PCR and melt curve analysis. J. Clin. Microbiol. 43 (6), 2876–2880.

Sobral, D., Schwarz, S., Bergonier, D., Brisabois, A., Feßler, A.T., et al., 2012. High throughput multiple locus variable number of tandem repeat analysis (MLVA) of *Staphylococcus aureus* from human, animal and food sources. PLoS One 7 (5), e33967.

Spence, R.P., Wright, V., Ala-Aldeen, D.A.A., Turner, D.P., Wooldridge, K.G., James, R., 2008. Validation of virulence and epidemiology DNA microarray for identification and characterization of *Staphylococcus aureus* isolates. J. Clin. Microbiol. 46 (5), 1620–1627.

Stager, C.E., Davis, J.R., 1992. Automated systems for identification of microorganisms. Clin. Microbiol. Rev. 5 (3), 302–327.

Swaminathan, B., Barret, T.J., Hunter, S.B., Tauxe, R.V., 2001. CDC Pulsenet Task Force. PulseNet: the molecular subtyping network for foodborne bacterial disease surveillance, United States. Emerg. Infect. Dis. 7, 382–389.

Turnidge, J., Paterson, D.L., 2007. Setting and revising antibacterial susceptibility breakpoints. Clin. Microbiol. Rev. 20, 391–408.

van Belkum, A., Tassios, P.T., Dijkshoorn, L., Haeggman, S., Cookson, B., Fry, N.K., et al., 2007. For the European society of clinical microbiology and infectious diseases (ESCMID) study group on epidemiological markers (ESGEM). Guidelines for the validation and application of typing methods for use in bacterial epidemiology. Clin. Microbiol. Infect. 13 (Suppl. 3), 1–46.

van Belkum, A., Scherer, S., Van Alphen, L., Verbrugh, H.A., 1998. Short sequence DNA repeats in prokaryotic genomes. Microbiol. Mol. Rev. 62, 275–293.

Vasconcelos, N.G., Pereira, V.C., AraújoJúnior, J.P., Cunha, M.L.R.S., 2011. Molecular detection of enterotoxins E, G, H and I in *Staphylococcus aureus* and coagulase-negative staphylococci isolated from clinical samples of newborns in Brazil. J. Appl. Microbiol. 111, 749–762.

Wieneke, A.A., 1991. Comparison of four kits for the detection of staphylococcal enterotoxin in foods from outbreaks of food poisoning. Int. J. Food Microbiol. 14, 305–312.

PART

STAPHYLOCOCCUS AUREUS FROM FARM TO FORK—FOOD SAFETY ASPECTS

CHAPTER

7

STAPHYLOCOCCUS AUREUS AS A LEADING CAUSE OF FOODBORNE OUTBREAKS WORLDWIDE

Jacques-Antoine Hennekinne

ANSES, French Agency for Food, Environmental and Occupational Health & Safety, Maisons-Alfort Cedex, France

1. INTRODUCTION

Staphylococcal food poisoning (SFP) is one of the most common foodborne diseases, resulting from ingestion of staphylococcal enterotoxins produced in food by enterotoxigenic strains of staphylococci, mainly coagulase positive staphylococci (CPS), with *Staphylococcus aureus* being the most prominent and important one, and only occasionally coagulase negative staphylococci (CNS). This chapter presents food-poisoning events due to staphylococci, mainly *S. aureus*. The first part is dedicated to data related to history and characteristics of staphylococci and *S. aureus* as well as their behavior in food. In the second part, examples of SFP events and a comparison between the European Union (EU) and United States of America reporting systems are presented.

2. HISTORY

SFP is one of the most common foodborne diseases in the world following the ingestion of staphylococcal enterotoxins (SEs), which are produced by enterotoxigenic strains of CPS, mainly *S. aureus* (Jablonski and Bohach, 1997) and very occasionally by other staphylococci species such as *Staphylococcus intermedius* (Genigeorgis, 1989; Khambaty et al., 1994). Whatever the reporting system, an outbreak of foodborne disease is defined as the occurrence of two or more cases of a similar illness resulting from ingestion of a common food.

When outbreaks occurred during large social events, chaotic situations resulted requiring the rapid implementation of medical care for a high number of cases (Bonnetain et al., 2003; Do Carmo et al., 2004).

The first description of foodborne disease involving staphylococci was investigated in Michigan (United States) in 1884 by Vaughan & Sternberg. This food-poisoning event was due to consumption

Staphylococcus aureus. http://dx.doi.org/10.1016/B978-0-12-809671-0.00007-3

of a cheese contaminated by staphylococci. The authors stated: "It seems not improbable that the poisonous principle is a ptomaine developed in the cheese as a result of the vital activity of the above mentioned Micrococcus or some other microorganisms which had preceded it, and had perhaps been killed by its own poisonous products".

Ten years later, Denys (1894) concluded that the illness of a family who had consumed meat from a cow that had died of vitullary fever was because of the presence of pyogenic staphylococci.

Proof of the involvement of staphylococci in food poisoning was first brought by Barber in 1914. He demonstrated with certainty that staphylococci were able to cause poisoning by consuming unrefrigerated milk from a cow suffering from mastitis, an inflammation of the udder that was caused by staphylococci. However, correlation between staphylococci-containing food and symptomatology was not recognized until other examples of food poisoning occurred later in the 20th century. It was Baerthlein, when reporting on a huge outbreak involving 2000 soldiers of the German army during World War I, who established in 1922 the possible role of bacteria. "I am going to report the case of an extended demonstration of poisoned sausages (approximately 2,000 cases) held in the spring 1918 during the military campaign of Verdun, which would probably have catastrophic military consequences. Early in June 1918, sudden and massive demonstrations that have the appearance of an acute and in some cases severe gastroenteritis, similar to cholera, affected the troops around Verdun; entire companies were disabled except just a few people, and within two days about 2000 men had been affected. The symptoms were so severe that some troops (more than 200) had to be transferred to field hospitals. The suspicion of food poisoning has been mentioned because, according to reports of the sick, the disease occurred 2 or 3 hours (some of the symptoms appeared after 6 to 8 hours) after eating a dish of sausages. Only troops who did not eat the meal were spared, such as soldiers who had returned to headquarters to receive orders, soldiers who for other reasons had not eaten sausages, and soldiers who were on leave and/or following a different diet. However, it was surprising that among the troops that were not present at the front, such as butchers, who ate the same sausage two days earlier, we did not observe any cases of disease" (Baerthlein, 1922).

In 1930, Dack et al. found that a sponge cake was responsible for the intoxication of 11 individuals; the authors highlighted that the disease was probably linked to a toxin called "enterotoxin" produced by yellow hemolytic *Staphylococcus*. Broth culture filtrates of this strain were administrated intravenously to a rabbit and orally to three human volunteers. The rabbit died, after first developing water diarrhea, and the three volunteers developed nausea, chilliness, and vomiting after 3 h. In the same year, Jordan (1930) showed that various strains of staphylococci exhibited cultural properties of generating a substance, which was purified from broth and, when taken orally, produced gastrointestinal disturbance.

A few years later, in 1934, Jordan & Burrows observed nine outbreaks related to the presence of staphylococci in food remnants, whereas Dolman explained that "the food poisoning substance is probably produced by only a few strains of staphylococci, and that it is a special metabolite whose formation and excretion are favoured in the laboratory by such environmental conditions as a semi-fluid medium and atmosphere containing a high percentage of carbon dioxide, conditions which promote, respectively, abundant growth and increased cellular permeability with partial buffering."

One of the first well-documented staphylococcal food poisoning outbreaks (SFPOs) was described by Denison in 1936. This outbreak occurred among high school students after they had eaten tainted cream puffs. He depicted the typical symptoms of 122 cases as follows: "Within 2-4 hours after eating there was first noticed a feeling of nausea. Severe abdominal cramps developed and were quickly followed by vomiting which was severe and continued at 5–20 minute intervals for 1–8 hours [...] A diarrhea of 1-7 liquid stools usually began with the vomiting and continued for several hours after its onset [...] During the acute stage the temperature was normal or subnormal, the pulse noticeably increased, there were cold sweats, prostration was severe and the patients were very definitely in a state of shock. Headache was mild and of a short duration. Muscular cramping [...] was present in the majority. Dehydration was marked in some. While the acute symptoms usually lasted only 1–8 hours, complete recovery [...] was delayed for 1–2 days."

SFP symptomatology has been extensively studied especially by the US Army: in a naturally occurring outbreak among the US Army personnel, involving 400 of 600 men, DeLay (1944) reported that about 25% of cases were classified as severe or shock cases. Numerous SFPOs have been described since the end of World War II. For example, Brink and Van Meter (1960) from the Institute for Cooperative Research of the University of Pennsylvania wrote a long report on an outbreak of SE food poisoning, which happened in 1960: "On a Saturday afternoon in the middle of summer, an epidemic of staphylococcal enterotoxin food poisoning occurred at a picnic held two miles from Gabriel, a small Midwestern town. (The name of the town and other names in this report are fictitious, in accordance with commitments to Task Surprise respondents.) About 1700 persons attended the picnic, which is an annual affair sponsored by the Johnson Co., of Croydon, some 60 miles away. Early in the morning, approximately seven hours before the picnic began, an unventilated, unrefrigerated truck containing a large supply of ham sandwiches was parked at the picnic grounds. The truck was exposed to the heat of direct sunlight, while the average ambient temperature for the day was close to 100 degrees Fahrenheit. In this environment, the staphylococcal organisms which elaborate the toxin multiplied rapidly. During the epidemic that followed, approximately 1100 persons became ill."

The main point highlighted by these reports is that any food that provides a convenient medium for CPS growth may be involved in an SFPO. The foods most frequently involved differ widely from one country to another, probably because of differing food habits (Le Loir et al., 2003). For instance, in the United Kingdom or the United States, meat or meat-based products are the food vehicles mostly involved in SFP (Genigeorgis, 1989), although poultry, salads, and cream-filled bakery items are other good examples of foods that have been involved in SFPOs (Minor and Marth, 1972). In France, various food types have been associated with SFPOs, but as the consumption of unpasteurized milk cheeses is much more common than in Anglo-Saxon countries, milk-based products are more frequently involved than in other countries (De Buyser et al., 2001).

To conclude this introductive section, as SFP is a short-term disease and usually results in full recovery, doctors do not take it very seriously, especially when the outbreak affects only a few people. Although such outbreaks should be reported to the sanitary authorities, this situation leads to underreporting (De Buyser et al., 2001). However, many researchers consider that SFP is one of the most common foodborne diseases worldwide (Balaban and Rasooly, 2000).

3. CHARACTERISTICS AND BEHAVIOR OF *STAPHYLOCOCCUS AUREUS* IN THE FOOD ENVIRONMENT

Staphylococcus is a spherical, nonsporulating, nonmotile coccoid bacterium that, when observed under the microscope, occurs in pairs, short chains, or grapelike clusters. These facultative aero–anaerobic bacteria are Gram- and catalase-positive. Staphylococci are ubiquitous in the environment and can be found in the air, dust, sewage, water, environmental surfaces, humans, and animals.

To date, more than 50 species and subspecies of staphylococci have been described according to their potential to produce coagulase. Their classification thus distinguishes between coagulase-producing strains, designated as CPS, and noncoagulase-producing strains, called CNS. Among CNS, some species are known to play an important role in the fermentation of meat and milk-based products and are therefore considered as food grade. The enterotoxigenic potential of CNS has always been a subject of controversy. Several investigations failed to detect enterotoxin production or enterotoxin-like genes in CNS (Becker et al., 2001; Rosec et al., 1997). However, some studies found that certain CNS strains were able to produce enterotoxins, which could lead to food poisoning (Vernozy-Rozand et al., 1996; Zell et al., 2008). More recently, another study demonstrated that, among 129 CNS strains isolated from fermented foodstuffs, only one carried SE genes (Even et al., 2010).

Among the seven described species belonging to the CPS group, *S. aureus* subsp. *aureus* is the main causative agent described in SFP incidents and therefore in the focus of this chapter. Among other CPS, Becker et al. (2001) highlighted the enterotoxinogenic potential of *S. intermedius*. The enterotoxinogenic potential (particularly for staphylococcal enterotoxin C (SEC)) of this species has been shown in strains isolated from dogs (Hirooka et al., 1988). The presence in the environment of strains producing toxins raises a possible health hazard, especially when carried by animals such as dogs that come in close contact with humans. *S. intermedius* was involved in one outbreak caused by blended margarine and butter involving over 265 cases in October 1991 in the United States (Khambaty et al., 1994; Bennett, 1996).

3.1 *STAPHYLOCOCCUS AUREUS* BIOTYPES AND RESERVOIRS

S. aureus belongs to the normal flora found on the skin and mucous membranes of mammals and birds. This bacterium can be disseminated in the environment of its hosts and survives for long periods in these areas. Several biotypes isolated from different hosts (human, poultry, cattle, sheep/goat) have been described within *S. aureus* species demonstrating the close adaptation of the bacterium to its host. They were identified according to four biochemical tests (staphylokinase, ß-hemolysin production, coagulation of bovine plasma, and growth type on crystal violet agar) following the simplified biotyping scheme described by Devriese (1984). However, many strains cannot be assigned to these host-specific biotypes and belong to non-host-specific biotypes, i.e., those associated with several hosts. Later, a poultry-like biotype associated with meat products and meat workers was tentatively designated as a "slaughterhouse" biotype by Isigidi et al. (1990). Indeed, introduction of an additional biochemical test, protein A production, and phage typing allowed researchers to differentiate the poultry biotype from this new biotype. However, as the protein A test is no longer commercially available, and as phage typing cannot be routinely used, these two biotypes cannot be easily distinguished. Several pitfalls were encountered when applying the biotyping method: discordant results because of the variety of test parameters, lack of standardized reagents, problematic interpretation for "hemolysin," "bovine plasma coagulation," and "crystal violet" tests when applied to some strains and, as previously mentioned, lack of commercially available tests to distinguish between the described biotypes. Despite these drawbacks, *S. aureus* biotyping has been useful in tracing or estimating the origin of this organism in various food products (Devriese et al., 1985; De Buyser et al., 1987; Rosec et al., 1997), in the food industry (Isigidi et al., 1990), and also for epidemiological investigations of food-poisoning outbreaks (Hennekinne et al., 2003; Kerouanton et al., 2007). Alves et al. (2009) performed pulsed field gel electrophoresis (PFGE) typing of *S. aureus* strains isolated from small (n=88) and large ruminants (n=65). The authors carried out a molecular analysis and confirmed that ovine and caprine strains, which could not be distinguished from one another, were nonetheless different from bovine strains. To suggest the source of contamination (animal or human origins), molecular-based methods have been used by various authors to study food-poisoning outbreaks (Chiou et al., 2000; Shimizu et al., 2000; Wei and Chiou, 2002; Kerouanton et al., 2007; Ostyn et al., 2010). Among these methods, PFGE and the *Staphylococcus* Protein A gene (*spa*) typing have been used alone or in association providing additional information to highlight the origin of the *S. aureus* contamination. For further information on methods applied in the course of SFPO events and to track the spread of *S. aureus*, please refer to Chapter 6 of this book.

3.2 MEANS OF CONTAMINATION

In the last few decades, numerous examples of SFPO events occurring around the globe were described in the literature—see summarizing Table 7.1).

The prerequisite of each of these SFPO was that food or one of its ingredients got contaminated with an enterotoxigenic *S. aureus* (or other *Staphylococcus* spp.) strain. Most SFPOs arise because of poor hygiene practices during processing (Asao et al., 2003), cooking, or distribution of the food product (Pereira et al., 1996). Moreover, after contamination, inadequate cooling of foods can induce *Staphylococcus* growth and/or stimulate toxin production, resulting in food poisoning (Barber, 1914; Anon., 1997).

Ideally, five conditions need to be fulfilled to induce SFPOs: (1) a source containing enterotoxin-producing strains, e.g., (raw) food materials, healthy, or infected carriers, (2) transfer of bacteria from source to food, e.g., unclean food preparation tools due to poor hygiene practices, (3) food composition with favorable physicochemical characteristics for *S. aureus* (or other *Staphylococcus* spp.) growth and toxinogenesis, (4) favorable temperature and sufficient time for bacterial growth and toxin production, and (5) ingestion of food containing sufficient amounts of enterotoxin to provoke symptoms. A good example of an outbreak where all of the five conditions required to induce an SFPO were fulfilled occurred in a takeaway restaurant in June 2015 in Cyprus. Two people complained for nausea, vomiting, abdominal pain, and diarrhea 2–3 h after consumption of sandwiches. As a result of an epidemiological case-control study, mixed roast pork sandwiches were suspected as incriminated food vehicle. In consequence, roast pork–based sandwiches and ingredients (roast pork, roast chicken, cheese, mayonnaise)—all with favorable physicochemical characteristics for *S. aureus* growth and toxin production—were submitted to microbiological analysis including CPS count. In addition, food samples were analyzed on the presence of SEs. Microbiological investigations revealed a high contamination level of roast pork–based sandwiches and roast pork with CPS (>10^8 cfu/g); SEs were also detected in both samples. SE-type identification and quantification succeeded with the roast pork sample and revealed a staphylococcal enterotoxin A (SEA) contamination at a concentration of 2.15 ng/g. In parallel to the investigation of food samples, workers of the takeaway restaurant were examined by taking swabs from their nose and hands. *S. aureus* isolated from these workers and samples of roasted pork sandwiches were positive for genes encoding SE types A and H and displayed identical PFGE profiles. These findings allow for the conclusion that healthy carriers were the source of enterotoxigenic *S. aureus*, which were transferred to the food because of poor hygiene practice. For further details, see Kourtis and Payani (2016).

Table 7.1 Examples of Staphylococcal Food Poisoning Outbreaks From the Literature, Years 1968–2014

Year	Location	Incriminated Food	Number of Cases	References
1968	School children, Texas	Chicken salad	1300	Anon. (1968)
1971	UK army	Sausages rolls, ham sandwiches	100	Morris et al. (1972)
1975	Flight from Japan to Denmark	Ham	197	Eisenberg et al. (1975)
1976	Flight from Rio to NYC	Chocolate eclairs	80	Anon. (1976)
1980	Canada	Cheese curd	62	Todd et al. (1981)
1982	North Carolina and Pennsylvania	Ham and cheese sandwich; stuffed chicken	121	Anon. (1983a)
1983	Caribbean cruise ship	Dessert cream pastry	215	Anon. (1983b)
1984	Scotland	Sheep's milk cheese	27	Bone et al. (1989)
1985	France, United Kingdom, Italy, Luxembourg	Dried lasagna	50	Woolaway et al. (1986)
1985	School children, Kentucky	2% chocolate milk	>1000	Evenson et al. (1988)
1986	Country Club, New Mexico	Turkey, poultry, gravy	67	Anon. (1986)
1989	Various US states	Canned mushrooms	102	Anon. (1989)
1990	Thailand	Eclairs	485	Thaikruea et al. (1995)
1992	Elementary school, Texas	Chicken salad	1364	Anon. (1992)
1997	Retirement party, Florida	Precooked ham	18	Anon. (1997)
1998	Minas Gerais, Brazil	Chicken, roasted beef, rice and beans	4000	Do Carmo et al. (2004)
2000	Osaka, Japan	Low-fat milk	13,420	Asao et al. (2003)
2006	Ile de France area, France	Coco nut pearls (Chinese dessert)	17	Hennekinne et al. (2009)
2007	Scouts' camp, Belgium	Hamburger	15	Fitz-James et al. (2008)
2007	Elementary school, Austria	Milk, cacao milk, vanilla milk	166	Schmid et al. (2009)
2008	Weeding dinner, Ile de France area, France	Caribbean meals	47	Authors personal communication
2008	French district	Pasta salad	100	Authors personal communication
2009	Nagoya university festival, Japan	crepes	75	Kitamoto et al. (2009)
2009	Various districts, France	Raw milk cheese	23	Ostyn et al. (2010)
2011	Catered dinner party, Turino, Italy	Seafood salad	26	Gallina et al. (2013)
2012	Sporting event, Australia	Fried rice, chicken stir-fry	22	Pillsbury et al. (2013)
2012	Work lunch party, military service, United States	"perlo" (chicken, sausage, and rice dish)	13	CDC (2012)
2013	Christening party, Freiburg, Germany	Ice cream	13	Fetsch et al. (2014)
2014	Boarding school, Switzerland	Tomme (soft cheese made from raw cow milk)	13	Johler et al. (2015)
2014	International equine sports event, Luxembourg	Pasta salad	31	Mossong et al. (2015)
2014	Workshop, Zimbabwe	Stewed chicken	53	Gumbo et al. (2015)

3.3 POTENTIAL FOR METHICILLIN-RESISTANT *STAPHYLOCOCCUS AUREUS* CONTAMINATION AND TRANSMISSION

Food is also an important source of antimicrobial resistance by means of residues of antibiotics in food, through the transfer of antimicrobial resistant foodborne pathogens or through the ingestion of antimicrobial resistant strains of the original food microflora and subsequent resistance transfer to pathogenic microorganisms (Khan et al., 2000; Pesavento et al., 2007). Most animals may be colonized with *S. aureus*, but only a bit more than a decade ago it was when methicillin-resistant *Staphylococcus aureus* (MRSA) strains of a particular clonal lineage (belonging to clonal complex (CC) 398) were isolated from several food production animals, including pigs, cattle, chicken, turkey, and other animals (de Neeling et al., 2007; Huijsdens et al., 2006), subsequently called livestock-associated (LA) MRSA. Pigs in particular, and also pig farmers and their families, were found colonized with MRSA, and contact with pigs is now recognized as a risk factor for MRSA carriage in humans (Van Duijkeren et al., 2007). An association between the emergence of MRSA strains in pigs and the use of antibiotics in pig farming has been suggested (de Neeling et al., 2007; Wulf and Voss, 2008). For further details on MRSA in livestock and at the animal-to-human interference, please refer to other chapters of this book.

During slaughtering of MRSA-positive animals, contamination of carcasses and the environment with MRSA may occur and consequently meat from these animals may become contaminated. MRSA strains have been detected in different foods, including bovine milk and cheese (Normanno et al., 2007), meat products (Van Loo et al., 2007; de Boer et al., 2008), raw chicken and turkey meat (Kitai et al., 2005; Kwon et al., 2006; Kraushaar et al., 2016), and raw wild boar meat (Kraushaar and Fetsch, 2014).

Regarding the involvement of MRSA in SFP, Jones et al. (2002) reported for the first time an outbreak of gastrointestinal illness caused by community-acquired MRSA. In this outbreak, various *S. aureus* strains were isolated from food remnants, affected people, and food handlers. Among these strains, one produced SE C and was identified as being MRSA. This isolate was resistant to penicillin and oxacillin but sensitive to all other antibiotics tested. To our knowledge, only few data are available on the involvement of MRSA in SFP. In a study performed in France on foods incriminated in SFPOs, Kerouanton et al. (2007) highlighted 2 MRSA strains of the 33 tested. They concluded that with reference to human clinical isolates, the SFPO strains were more susceptible to antibiotics (except for two that were resistant to methicillin). LA-MRSA have so far not been directly linked to an SFPO most likely because SE encoding genes are not very common in strains found in animal-derived foods (Argudin et al., 2011: Kraushaar et al., 2016).

4. FACTORS INFLUENCING THE GROWTH OF *STAPHYLOCOCCUS AUREUS* IN FOOD

Microorganisms in foods are affected by a multiplicity of parameters described as intrinsic factors (i.e., factors related to the food itself), extrinsic factors (i.e., factors related to the environment in which the food is stored), implicit factors (i.e., factors related to the microorganisms themselves), and finally processing factors (i.e., factors affecting the composition of the food and also the types and numbers of microorganisms that remain in the food after treatment) (Hamad, 2012). Some of these parameters will be discussed below. However, it should be stressed that, in complex media such as foods, these factors interact to a great extent. Many of the data presented here were derived from laboratory experiments in which all other conditions beside the factor to be tested were ideal. Table 7.2 summarizes some of the factors affecting growth and SE production by *S. aureus*.

Table 7.2 Factors Affecting Growth and Enterotoxin Production by *Staphylococcus aureus* (Tatini, 1973)

	S. aureus Growth		Staphylococcal Enterotoxin Production	
Factor	**Optimum**	**Range**	**Optimum**	**Range**
Temperature (°C)	37	7–48	37–45	10–45
pH	6–7	4–10	7–8	4–9,6
Water activity (a_w)	0.98	0.83–0.99[1]	0.98	0.85–0.99[2]
NaCl (%)	0	0–20	0	0–10
Redox potential (E_h)	>+ 200 mV	<−200 to >+200 mV	>+ 200 mV	<−100 to >+200 mV
Atmosphere	Aerobic	anaerobic—aerobic	aerobic	anaerobic—aerobic

[1]*Aerobic (anaerobic 0.90→0.99).*
[2]*Aerobic (anaerobic 0.92→0.99).*

Table 7.3 Reporting of Staphylococcal Food Poisoning Outbreaks in the European Union, Years 2012–15

	Outbreaks (N)	Human Cases (N)	Hospitalizations (N)	References
2012	346 35 SE 311 WE	2532 497 SE 2035 WE	288 88 S 200 WE	EFSA and ECDC (2014)
2013	386 94 SE 292 WE	3203 1304 SE 1899 WE	210 52 SE 158 WE	EFSA and ECDC (2015a)
2014	393 31 SE 362 WE	2952 498 SE 2454 WE	264 86 SE 178 WE	EFSA and ECDC (2015b)
2015	434 39 SE 395 WE	3630 758 SE 2872 WE	316 113 SE 203 WE	EFSA and ECDC (2016)

N, *number;* SE, *strong evidence;* WE, *week evidence.*

4.1 WATER ACTIVITY (a_W)

Regarding *S. aureus* and other staphylococci, water activity (a_w) is of great importance because these bacteria are able to grow over a much wider a_w range than other food-associated pathogens. As shown in Table 7.3, *S. aureus* can grow at a minimum a_w of 0.83 (equivalent to about 20% NaCl) (Tatini, 1973), provided that all other conditions are optimal. The optimum a_w is >0.99 (Tatini, 1973). The a_w conditions for SE production are somewhat different than those for bacterial growth, depending on the type of toxin. SEA and staphylococcal enterotoxin D (SED) production occurs under nearly all a_w conditions allowing growth of *S. aureus* as long as all other conditions are optimal. Production of staphylococcal enterotoxin B (SEB) is very sensitive to reductions in a_w and hardly any is produced at $a_w \leq 0.93$ despite extensive growth. The effect of a_w on SEC production follows the same pattern as SEB production (Ewald and Notermans, 1988; Qi and Miller, 2000). Thota et al. (1973) found staphylococcal enterotoxin E (SEE) production in media containing 10% NaCl (according to Troller and Stinson, 1975, this concentration corresponds to a_w 0.92). Important factors affecting growth and SE production are also the humectant used to lower the a_w, the pH, the atmospheric composition, and also the incubation temperature. Thus, conditions for growth and SE production in laboratory media and in food, respectively, may differ to some extent. Studies on the osmoadaptive strategies of *S. aureus* have revealed that when cells are grown in a low a_w medium, they respond by accumulating certain low-molecular-weight compounds termed compatible solutes. Glycine betaine, carnitine, and proline have been shown to be principal compatible solutes accumulated within osmotically stressed *S. aureus* cells, and their accumulation results from sodium-dependent transport systems (Gutierrez et al., 1995; Qi and Miller, 2000). There is strong evidence that compatible solutes not only stimulate growth but also enterotoxin synthesis. For example, SEB production was significantly stimulated at low a_w when proline was available in broth (Qi and Miller, 2000).

4.2 pH

Most staphylococcal strains grow at pH values between 4 and 10, with the optimum being 6–7. When the other cultural parameters became nonoptimal, the pH range tolerated is reduced. For example, the lowest pH that permitted growth and SE production by aerobically cultured *S. aureus* strains was 4.0, whereas the lowest pH values that supported growth and SE production in anaerobic cultures were 4.6 and 5.3 (Smith et al., 1983). Other important parameters influencing the response of *S. aureus* to pH are the size of inoculum, the type of growth medium, the NaCl concentration (a_w), the temperature, and the atmosphere (Genigeorgis, 1989). The majority of *S. aureus* strains tested produced detectable amounts of SE aerobically at a pH of 5.1. However, in anaerobic conditions most strains failed to produce detectable SE below pH 5.7 (Tatini, 1973; Bergdoll, 1989; Smith et al., 1983).

4.3 REDOX POTENTIAL

Optimum redox potential and ranges for growth and SE formation are also provided in Table 7.2. *S. aureus* is a facultative anaerobic bacterium, which grows best in the presence of oxygen. Under anaerobic conditions, however, growth is much slower, and even after several days, cell numbers do not reach those attained under aerobic conditions. Thus, aerated cultures produced approximately tenfold more SEB than cultures incubated in an atmosphere of 95% N_2+5% CO_2. Similarly, greatly increased SEA, SEB, and SEC productions were observed in shaken as compared to static cultures. The level of dissolved oxygen plays a very important role (Bergdoll, 1989; Genigeorgis, 1989). Under strict anaerobic conditions, the growth of *S. aureus* was slower than when cultivated aerobically. In broth incubated at 37°C the anaerobic generation time was 80 min compared with 35 min for aerobic culture. With slower anaerobic growth, relatively less SEA was produced than under aerobic conditions, but in both cases toxin was detected after 120 min of incubation (Belay and Rasooly, 2002). It has already been mentioned that minimum a_w and minimum pH for growth as well as for SE formation are influenced by the atmosphere.

4.4 TEMPERATURE

S. aureus grows between 7 and 48°C, temperature being optimal at around 37°C. The effect of temperature depends on the strain tested and on the type of the growth medium. In an extensive study (Schmitt et al., 1990) using 77 strains isolated from different foods, the optimum growth temperature generally did not vary much within the range of 35–40°C. The minimum growth temperatures were irregularly distributed between 7 and 13°C and the maximum between 40 and 48°C. The minimum temperatures for SE production varied quite irregularly over a broad range between 15 and 38°C and the maximum temperatures from 35 to 45°C. For the lower temperature limit for SE production, production of low amounts of enterorotoxin has been observed after 3–4 days. Moreover, SE formation at 10°C was reported by Tatini (1973) without indicating the detailed experimental conditions.

One of the most effective measures for inactivating *S. aureus* in food is heating. The bacterium is killed in milk if proper heat treatment is applied. *S. aureus* was completely inactivated in milk after application of the following temperature/time conditions: 57.2°C/80 min, 60.0°C/24 min, 62.8°C/6.8 min, 65.6°C/1.9 min, and 71.7°C/0.14 min (Bergdoll, 1989). In the case of heat inactivation in other dairy products, however, one should keep in mind that staphylococci probably become more heat resistant as the a_w is lowered until at an a_w between 0.70 and 0.80, resistance begins to decline (Troller, 1986).

4.5 NUTRITIONAL FACTORS AND BACTERIAL ANTAGONISM

Growth of *S. aureus* and SE production is also influenced by nutritional factors. *S. aureus* does not grow well in the presence of a competitive flora. Its inhibition is mainly because of acidic products, lowering of the pH, production of H_2O_2, or other inhibitory substances such as antibiotics, volatile compounds, or nutritional competition (Haines and Harmon, 1973; Genigeorgis, 1989). Important factors affecting the degree of inhibition are the ratio of the numbers of competitors to the number of *S. aureus* as well as the temperature (Smith et al., 1983; Genigeorgis, 1989).

Starter cultures used in the production of fermented milk products such as cheese, yoghurt, buttermilk, and others can effectively prevent growth of *S. aureus* and SE formation. In the case of a failure of these cultures, however, the pathogen will not be inhibited and the product may be hazardous.

5. STAPHYLOCOCCAL FOOD POISONING OUTBREAKS: SYMPTOMATOLOGY; REPORTING SYSTEM INCLUDING THE US AND EU CONTROL; MONITORING SCHEMES AND OCCURRENCE

5.1 SYMPTOMATOLOGY AND TOXIC DOSE

The incubation period and severity of symptoms observed during an SFPO depend on the type and number of SEs ingested and the susceptibility of each person. Initial symptoms, nausea followed by incoercible characteristic vomiting (in spurts), appear within 30 min–8 h (3 h on average) after ingesting the contaminated food. Other commonly described symptoms are abdominal pain, diarrhea, dizziness, shivering, and general weakness, sometimes associated with a moderate fever. In the most severe cases, headaches, prostration, and low blood pressure have been reported. In the majority of cases, recovery occurs within 24–48 h without specific treatment, whereas diarrhea and general weakness can last 24 h or longer. Death is rare (0.02‰ according to Mead et al., 1999), occurring in the most susceptible people to dehydration such as infants and the elderly (Do Carmo et al., 2004) and people affected by an underlying illness.

Regarding the toxin dose, most of the studies are referred to SEA. Notermans et al. (1991) demonstrated the feasibility of a reference material containing about 0.5 μg of SEA because it had been suggested that this dose can cause symptoms such as vomiting (Bergdoll, 1989). Mossel et al. (1995) cited an emetic dose 50 value of about 0.2 μg SE per kg of human body weight. They concluded that an adult would need to ingest about 10–20 μg of SE to suffer symptoms. Other authors (Martin et al., 2001) considered that less than 1 μg of SE may cause food-poisoning symptoms in susceptible individuals. Evenson et al. (1988) estimated that the amount of SEA needed to cause vomiting and diarrhea was 0.144 μg, the amount recovered from a half-pint (~0.28 L) carton of a 2% chocolate milk. In SFP in Japan, the total intake of SEA in low-fat milk per capita was estimated mostly at approximately 20–100 ng (Asao et al., 2003; Ikeda et al., 2005). In an SFPO involving "coconut pearls" (a Chinese dessert based on tapioca), Hennekinne et al. (2009) estimated the total intake of SEA per body at around 100 ng. Also Ostyn et al. (2010) investigated SFPOs because of SEE and estimated that the total intake of SEE per body was 90 ng, a dose in accordance with those previously mentioned. However, these concentrations were taken from a limited number of food outbreaks. As quantitative date on SEs in food is scarce, establishing a dose response relationship must fail. That is why very recently the benchmark dose (BMD) approach was used to establish a dose response relationship for SEs in food, i.e., dose levels corresponding to specific response levels based on the systematic investigations carried out during recent years in France (Guillier et al., 2016). For this purpose, data of 63 European SPFOs were collected. The study focused on enterotoxins SEA, SEB, SEC, and SED, either present alone or in combination in the suspected foods. SEA was by far the most often detected SE type as it was solely detected in 36 outbreaks and as associated SE type in 18 outbreaks. SEA was also the only type of SE for which sufficient data were available for dose response modeling. Using this approach, the BMD lower limit ($BMDL_{10}$), which reflects the lower 95% confidence interval of the dose that induces effects in 10% of the exposed population, was estimated at 6.1 ng of SEA. This estimated value of $BMDL_{10}$ provides a basis for, e.g., determining the detection limit that should reach enterotoxin detection methods. For example, according to the available data under the assumption of a 100 g serving size, the limit of detection for qualitative methods should be lower than 0.06 ng/g for SEA (Guillier et al., 2016).

5.2 REPORTING SYSTEM AND OCCURRENCE OF STAPHYLOCOCCAL FOOD POISONING OUTBREAKS IN THE EUROPEAN UNION

The reporting of foodborne outbreaks has been mandatory for the European Union Member States (EU MSs) since 2005. Moreover, since 2007, new harmonized specifications on the reporting of these outbreaks at community level have come into force (Anon., 2007). However, the foodborne outbreak investigation and reporting systems at national level are not harmonized within the EU. Therefore, differences in the number of reported outbreaks, the types of outbreaks, and causative agents do not necessarily reflect different levels of food safety between EU MSs'; however, the high number of reported outbreaks may reflect the increasing efficiency of the EU MSs' systems in investigating and identifying the outbreaks.

The European Food Safety Authority (EFSA) is responsible for examining the data on zoonoses, antimicrobial resistance, and foodborne outbreaks submitted by member states in accordance with Directive 2003/99/EC (Anon., 2003) and for preparing the Community Summary Report from the results. Data are produced in collaboration with the European Centre for Disease Control (ECDC), which provides the information on zoonosis cases in humans; the Zoonoses Collaboration at the National Food Institute of the Technical University of Denmark assists EFSA and ECDC in preparing this yearly published report.

Microbiological safety in the food industry requires sanitary procedures and regular decontamination of production and storage equipment. Because consumers expect that the foods they purchase and consume will be safe, there is a need for close monitoring of pathogens. This trend is amplified both by Western countries population aging, leading to an increase in immunocompromised individuals, and by changes in eating patterns, favoring the emergence of new pathogens. For this purpose, the EU created the EFSA, defined general principles and requirements of the food law, and laid down procedures to ensure food safety especially in the EC regulation 178/2002 article 143. Basically, foods should not be placed on the market if they are unsafe (injurious to health or unfit for human consumption). The EU also defined rules to declare food-poisoning events and, in 2005, the reporting of foodborne outbreaks became mandatory for EU MS. During this first year for reporting of foodborne outbreaks in the EU, only seven member states reported foodborne outbreaks (n=36) caused by SEs.

In 2014, new harmonized specifications on the reporting of these outbreaks at EU level came into force. To harmonize data collected, the former classification "verified" and "possible" was abandoned and replaced by "strong" and "weak" evidences. When a foodborne outbreak is suspected, the Member State health system needs to coordinate activities to establish the degree of relationship between suspected foods and laboratory results. If medical, epidemiological, and laboratory data are consistent, a foodborne outbreak investigation can be considered well conducted and strong causal evidence can be highlighted. In the opposite case, or in the case where no particular food vehicle is suspected, the outbreak may be classified as weak evidence outbreak.

An overview of the reported data of SFPOs in the EU in the last years is provided in Table 7.3. Over the years, the number of foodborne outbreaks with staphylococcal toxins as causative agents are increasing in the EU. In 2015, 9.9% of all outbreaks in the EU were caused by staphylococcal toxins.

5.3 REPORTING SYSTEM AND OCCURRENCE IN THE UNITED STATES

Foodborne diseases because of known pathogens are estimated to cause about 9.4 million illnesses each year in the United States (Scallan et al., 2011). Although relatively few of these illnesses occur in the setting of a recognized outbreak, data collected during outbreak investigations provide insights into the pathogens and foods that cause illness. Foodborne disease outbreaks are a nationally notifiable condition. Center of disease control conducts surveillance of foodborne disease outbreaks in the United States through the foodborne disease outbreak surveillance system. Public health agencies in all 50 states, the District of Columbia, and US territories voluntarily submit reports of outbreaks investigated by their agencies using a web-based reporting platform, the National Outbreak Reporting System (NORS) (http://www.cdc.gov/nors/). NORS also collects reports of enteric disease outbreaks caused by other transmission modes, including water, animal contact, person-to-person contact, environmental contamination, and unknown modes of transmission.

Agencies use a standard form (http://www.cdc.gov/nors/pdf/NORS_CDC_5213.pdf) to report foodborne disease outbreaks. Data requested for each outbreak include the reporting state; date of first illness onset; number of illnesses, hospitalizations, and deaths; etiology; implicated food and ingredients; locations of food preparation; and factors contributing to food contamination. Foods reported in multistate outbreaks are further classified as confirmed or suspected sources.

An overview of the reported data of SFPOs in the United States over the last years is provided in Table 7.4.

Numbers of reported SFPOs in the EU and the United States strongly differ from each other. Over the 2012–14 period, EU reported by year around 370 outbreaks corresponding to 3000 human cases

Table 7.4 Reporting of Staphylococcal Food Poisoning Outbreaks in the United States, Years 2012–14

	Outbreaks (N)	Human Cases (N)	Hospitalizations (N)	References
2012	5 2 C 3 S	149 26 C 133 S	1 1 C 3 S	CDC (2014)
2013	10 6 C 4 S	263 221 C 42 S	27 25 C 2 S	CDC (2015)
2014	17 9 C 8 S	566 504 C 62 S	10 9 C 1 S	CDC (2016)

C, *confirmed;* N, *number;* S, *suspected.*

and 250 hospitalizations, whereas United States reported 10 outbreaks involving 330 cases and 13 hospitalizations, only. This could be due to the differences between reporting systems. Although US outbreaks are reported in a limited number, they are generally more characterized than the European SFPOs. It would be interesting to perform an in-depth analysis of the two reporting systems to improve and harmonize monitoring at EU and US levels; this would also allow for a better comparison of data.

6. CONCLUSION

S. aureus is an opportunistic pathogen of humans and warm-blooded animals. Some strains of *S. aureus* can produce SEs affecting human health when ingested. SFPO is one of the most common causes of foodborne disease worldwide. Outbreak investigations have suggested that improper handling of cooked or processed food by healthy carriers is the main source of *S. aureus*. Lack of maintaining cold chain allows growth of *S. aureus* and subsequent production of SEs. Although *S. aureus* can be eliminated by heat treatment and competition with other bacterial flora, SEs produced by *S. aureus* are still capable to cause SFPO because of their heat resistance. This fact should be considered in risk assessment and devising appropriate public health interventions. Thus, prevention of *S. aureus* contamination from farm to fork is crucial. Rapid surveillance in the event of SFPO outbreak and ongoing surveillance for the routine investigation of *S. aureus* and SEs implicated in food products could help to decrease the number of food-poisoning events in the next years.

REFERENCES

Alves, P.D.D., McCulloch, J.A., Even, S., Le Marechal, C., Thierry, A., Grosset, N., Azevedo, V., Rosa, C.A., Vautor, E., Le Loir, Y., 2009. Molecular characterisation of *Staphylococcus aureus* strains isolated from small and large ruminants reveals a host rather than tissue specificity. Vet. Microbiol. 137, 190–195.

Anon., 1968. CDC MMWR 17, 109–110.

Anon., 1976. CDC MMWR 25, 317–318.

Anon., 1983a. CDC MMWR 32 (14), 183–184.

Anon., 1983b. CDC MMWR 32 (22), 294–295.

Anon., 1986. CDC MMWR 35 (46), 715–716.

Anon., 1989. CDC MMWR 38 (24), 417–418.

Anon., 1992. Foodborne Pathogenic Microorganisms and Natural Toxins Handbook: *Staphylococcus aureus*. U.S. Food and Drug Administration. The Center for Food Safety and Applied Nutrition. (US FDA/CFSAN, 1992).

Anon., 1997. CDC MMWR 46 (50), 1189–1191.

Anon., 2003. Directive 2003/99/EC of the European Parliament and of the Council of 17 November 2003 on the monitoring of zoonoses and zoonotic agents, amending Council Decision 90/424/EEC and repealing Council Directive 92/117/EEC. Off. J. Eur. Union L325, 31–40.

Anon., 2007. Report of the Task Force on Zoonoses Data Collection on harmonising the reporting of food-borne outbreaks through the Community reporting system in accordance with Directive 2003/99/EC. EFSA J. 123, 1–16.

Argudin, M.A., Tenhagen, B.A., Fetsch, A., et al., 2011. Virulence and resistance determinants of German *Staphylococcus aureus* ST398 isolates from non human sources. Appl. Environ. Microbiol. 77, 3052–3060.

Asao, T., Kumeda, Y., Kawai, T., Shibata, T., Oda, H., Haruki, K., Nakazawa, H., Kozaki, S., 2003. An extensive outbreak of staphylococcal food poisoning due to low-fat milk in Japan: estimation of enterotoxin A in the incriminated milk and powdered skim milk. Epidemiol. Infect. 130, 33–40.

Baerthlein, K., 1922. Ueber: uogcdehnte Wurstvergiftungen, bedingt durch *Bacillus proteus* vulgaris. Med Wochenschr 69, Munch, pp. 155–156.

Balaban, N., Rasooly, A., 2000. Staphylococcal enterotoxins. Int. J. Food Microbiol. 61, 1–10.

Barber, M.A., 1914. Milk poisoning due to a type of *Staphylococcus albus* occurring in the udder of a healthy cow. Philipp. J. Sci. 9, 515–519.

Becker, K., Keller, B., von Eiff, C., Bruck, M., Lubritz, G., Etienne, J., Peters, G., 2001. Enterotoxigenic potential of *Staphylococcus intermedius*. Appl. Environ. Microbiol. 67 (12), 5551–5557.

Belay, N., Rasooly, A., 2002. *Staphylococcus aureus* growth and enterotoxin A production in an anaerobic environment. J. Food Prot. 65, 199–204.

Bennett, R.W., 1996. Atypical toxigenic *Staphylococcus* and non-*Staphylococcus aureus* species on the horizon? An update. J. Food Prot. 59, 1123–1126.

Bergdoll, M.S., 1989. In: Doyle, M.P. (Ed.), *Staphylococcus aureus*. Foodborne Bacterial Pathogens. Marcel Dekker Inc., New York, Basel, pp. 463–523.

Bone, F.J., Bogie, D., Morgan-Jone, S.C., 1989. Staphylococcal food poisoning from sheep milk cheese. Epidemiol. Infect 103, 449–458.

Bonnetain, F., Carbonel, S., Stoll, J., Legros, D., 2003. Toxi-infection alimentaire collective due à *Staphylococcus aureus*, Longevelle-sur-le-Doubs, juillet 2003. BEH 47, 231–232.

Brink, E.L., Van Metter, C.T., 1960. A Study of an Epidemic of Staphylococcal Enterotoxin Food Poisoning, Ad 419937, Defense Documentation Center for Technical Information. Cameron Station, Alexandria, Virginia. Contract No. DA 18-064-Cml-2733: 10 October 1960.

Centers for Disease Control and Prevention (CDC). Outbreak of staphylococcal food poisoning from a military unit lunch party - United States, July 2012. MMWR Morb. Mortal. Wkly. Rep. 2013 Dec 20 62 (50), 1026–1028 PMID:24352066.

Centers for Disease Control and Prevention (CDC), 2014. Surveillance for Foodborne Disease Outbreaks, United States, 2012, Annual Report. US Department of Health and Human Services, CDC, Atlanta, Georgia.

Centers for Disease Control and Prevention (CDC), 2015. Surveillance for Foodborne Disease Outbreaks, United States, 2013, Annual Report. US Department of Health and Human Services, CDC, Atlanta, Georgia.

Centers for Disease Control and Prevention (CDC), 2016. Surveillance for Foodborne Disease Outbreaks, United States, 2014, Annual Report. US Department of Health and Human Services, CDC, Atlanta, Georgia.

Chiou, C.S., Wei, H.L., Yang, L.C., 2000. Comparison of pulsed-field gel electrophoresis and coagulase gene restriction profile analysis techniques in the molecular typing of *Staphylococcus aureus*. J. Clin. Microbiol. 38, 2186–2190.

Dack, G.M., Cary, W.E., Woolpert, O., Wiggers, H., 1930. An outbreak of food poisoning proved to be due to a yellow hemolytic *staphylococcus*. J. Prev. Med. 4, 167–175.

de Boer, E., Zwartkruis-Nahuis, J.T., Wit, B., Huijsdens, X.W., de Neeling, A.J., Bosch, T., van Oosterom, R.A., Vila, A., Heuvelink, A.E., 2008. Prevalence of methicillin-resistant *Staphylococcus aureus* in meat. Int. J. Food Microbiol. 134, 52–56.

De Buyser, M.L., Dilasser, F., Hummel, R., Bergdoll, M.S., 1987. Enterotoxin and toxic shock syndrome toxin-1 production by staphylococci isolated from goat's milk. Int. J. Food Microbiol. 5, 301–309.

De Buyser, M.L., Dufour, B., Maire, M., Lafarge, V., 2001. Implication of milk and milk products in food-borne diseases in France and indifferent industrialized countries. Int. J. Food Microbiol. 67, 1–17.

de Neeling, A.J., van den Broek, M.J., Spalburg, E.C., van Santen-Verheuvel, M.G., Dam-Deisz, W.D., Boshuizen, H.C., van de Giessen, A.W., van Duijkeren, E., Huijsdens, X.W., 2007. High prevalence of methicillin resistant *Staphylococcus aureus* in pigs. Vet. Microbiol. 122, 366–372.

DeLay, P.D., 1944. Staphylococcal enterotoxin in bread pudding. Bull. U.S. Army Med. Dep. 72, 72–73.

Denison, G.A., 1936. Epidemiology and symptomatology of *staphylococcus* food poisoning. A report of recent outbreaks. Am. J. Public Nation Health 26, 1168–1175.

Denys, J., 1894. Présence de Staphylocoque dans une viande qui a déterminé des cas d'empoisonnement. Bull. Acad. Roy. Med. Belg. 8, 496.

Devriese, L.A., 1984. A simplified system for biotyping *Staphylococcus aureus* strains isolated from different animal species. J. Appl. Bacteriol. 56, 215–220.

Devriese, L.A., Yde, M., Godard, C., Isigidi, B.K., 1985. Use of biotyping to trace the origin of *Staphylococcus aureus* in foods. Int. J. Food Microbiol. 2, 365–369.

Do Carmo, L.S., Cummings, C., Linardi, V.R., Souza Diaz, R., De Souza, J.M., De Sena, M.J., Dos Santos, D.A., Shupp, J.W., Peres Pereira, R.K., Jett, M., 2004. A case study of a massive staphylococcal food poisoning incident. Foodborne Pathog. Dis. 1, 241–246.

Dolman, C.E., 1934. Ingestion of *staphylococcus* exotoxin by human volunteers, with special reference to staphylococci food poisoning. J. Infect. Dis. 55, 172–183.

EFSA (European Food Safety Authority) and ECDC (European Centre for Disease Prevention and Control), 2014. The European union summary report on trends and sources of zoonoses, zoonotic agents and food-borne outbreaks in 2012. EFSA J. 12 (2), 3547. http://dx.doi.org/10.2903/j.efsa.2014.3547. 312 pp.

EFSA and ECDC (European Food Safety Authority and European Centre for Disease Prevention and Control), 2015a. The European union summary report on trends and sources of zoonoses, zoonotic agents and food-borne outbreaks in 2013. EFSA J. 13 (1), 3991. http://dx.doi.org/10.2903/j.efsa.2015.3991. 162 pp.

EFSA (European Food Safety Authority) and ECDC (European Centre for Disease Prevention and Control), 2015b. The European Union summary report on trends and sources of zoonoses, zoonotic agents and food-borne outbreaks in 2014. EFSA J. 13 (12), 4329. http://dx.doi.org/10.2903/j.efsa.2015.4329. 191 pp.

EFSA (European Food Safety Authority) and ECDC (European Centre for Disease Prevention and Control), 2016. The European union summary report on trends and sources of zoonoses, zoonotic agents and food-borne outbreaks in 2015. EFSA J. 14 (12), 4634. http://dx.doi.org/10.2903/j.efsa.2016.4634. 231 pp.

Eisenberg, M.S., Gaarslev, K., Brown, W., Horwitz, D., 1975. Hill staphylococcal food poisoning aboard a commercial aircraft. Lancet 2, 595–599.

Even, S., Leroy, S., Charlier, C., Ben Zakour, N., Chacornac, J.P., Lebert, I., Jamet, E., Desmonts, M.H., Coton, E., Pochet, S., Donnio, P.Y., Gautier, M., Talon, R., Le Loir, Y., 2010. Low occurrence of safety hazards in coagulase negative staphylococci isolated from fermented foodstuffs. Int. J. Food Microbiol. 139, 87–95.

Evenson, M.L., Hinds, M.W., Bernstein, R.S., Bergdoll, M.S., 1988. Estimation of human dose of staphylococcal enterotoxin A from a large outbreak of staphylococcal food poisoning involving chocolate milk. Int. J. Food Microbiol. 7 (4).

Ewald, S., Notermans, S., 1988. Effect of water activity on growth and enterotoxin D production of *Staphylococcus aureus*. Int. J. Food Microbiol. 6, 25–30.

Fitz-James, I., Botteldoorn, N., In't Veld, P., Dierick, C., September 2, 2008. Joined investigation of a large outbreak involving *Staphylococcus aureus*. In: Proceeding FoodMicro 2008 Aberdeen, UK.
Fetsch, A., Contzen, M., Hartelt, K., Kleiser, A., Maassen, S., Rau, J., Kraushaar, B., Layer, F., Strommenger, B., September 18, 2014. *Staphylococcus aureus* food-poisoning outbreak associated with the consumption of ice-cream. Int. J. Food Microbiol. 187, 1–6. http://dx.doi.org/10.1016/j.ijfoodmicro.2014.06.017. 25033424.
Gallina, S., Bianchi, D.M., Bellio, A., Nogarol, C., Macori, G., Zaccaria, T., Biorci, F., Carraro, E., Decastelli, L., December 2013. Staphylococcal poisoning foodborne outbreak: epidemiological investigation and strain genotyping. J. Food Prot. 76 (12), 2093–2098. http://dx.doi.org/10.4315/0362-028X.JFP-13-190. 24290688.
Genigeorgis, C.A., 1989. Present state of knowledge on staphylococcal intoxication. Int. J. Food Microbiol. 9, 327–360.
Guillier, L., Bergis, B., Guillier, F., Noel, V., Auvray, F., Hennekinne, J.-A., 2016. Dose-response modelling of staphylococcal enterotoxins using outbreak data. Procedia Food Sci. 7, 129–132.
Gumbo, A., Bangure, D., Gombe, N.T., Mungati, M., Tshimanga, M., Hwalima, Z., Dube, I., September 2015. *Staphylococcus aureus* food poisoning among Bulawayo city council employees, Zimbabwe, 2014. BMC Res. Notes 28 (8), 485. http://dx.doi.org/10.1186/s13104-015-1490-4. PubMed Central PMCID: PMC4587832 26416028.
Gutierrez, C., Abee, T., Booth, I.R., 1995. Physiology of the osmotic stress response in microorganisms. Int. J. Food Microbiol. 28, 233–244.
Haines, W.C., Harmon, L.G., 1973. Effect of selected lactic acid bacteria on growth of *Staphylococcus aureus* and production of enterotoxin. Appl. Microbiol. 25, 436–441.
Hamad, S.H., 2012. Factors affecting the growth of microorganisms in food. In: Bhat, R., Karim Alias, A., Paliyath, G. (Eds.), Progress in Food Preservation. Wiley-Blackwell, Oxford, UK. http://dx.doi.org/10.1002/9781119962045.ch20.
Hennekinne, J.A., Kérouanton, A., Brisabois, A., De Buyser, M.L., 2003. Discrimination of *Staphylococcus aureus* biotypes by pulsed-field gel electrophoresis of DNA macro-restriction fragments. J. Appl. Microbiol. 94, 321–329.
Hennekinne, J.A., Brun, V., De Buyser, M.L., Dupuis, A., Ostyn, A., Dragacci, S., 2009. Innovative contribution of mass spectrometry to characterise staphylococcal enterotoxins involved in food outbreaks. Appl. Environ. Microbiol. 75, 882–884.
Hirooka, E.Y., Muller, E.E., Freitas, J.C., Vicente, E., Yoshimoto, Y., Bergdoll, M.S., 1988. Enterotoxigenicity of *Staphylococcus intermedius* of canine origin. Int. J. Food Microbiol. 7, 185–191.
Huijsdens, X.W., van Dijke, B.J., Spalburg, E., van Santen-Verheuvel, M.G., Heck, M.E., Pluister, G.N., Voss, A., Wannet, W.J., de Neeling, A.J., 2006. Community-acquired MRSA and pig-farming. Ann. Clin. Microbiol. Antimicrob. 5, 26.
Ikeda, T., Tamate, N., Yamaguchi, K., Makino, S., 2005. Mass outbreak of food poisoning disease caused by small amounts of staphylococcal enterotoxins A and H. Appl. Environ. Microbiol. 71, 2793–2795.
Isigidi, B.K., Devriese, L.A., Godard, C., Van Hoof, J., 1990. Characteristics of *Staphylococcus aureus* associated with meat products and meat workers. Lett. Appl. Microbiol. 11, 145–147.
Jablonski, L.M., Bohach, G.A., 1997. In: Doyle, M.P., Beuchat, L.R., Montville, T.J. (Eds.), *Staphylococcus aureus*. Food Microbiology Fundamentals and Frontiers. American Society for Microbiology Press, Washington, DC, pp. 353–357.
Johler, S., Weder, D., Bridy, C., Huguenin, M.C., Robert, L., Hummerjohann, J., Stephan, R., May 2015. Outbreak of staphylococcal food poisoning among children and staff at a Swiss boarding school due to soft cheese made from raw milk. J. Dairy Sci. 98 (5), 2944–2948. http://dx.doi.org/10.3168/jds.2014-9123. 25726108.
Jones, T.F., Kellum, M.E., Porter, S.S., Bell, M., Schaffner, W., 2002. An outbreak of community-acquired foodborne illness caused by methicillin-resistant *Staphylococcus aureus*. Emerg. Infect. Dis. 8, 82–84.
Jordan, E.O., 1930. The production by staphylococci of a substance causing food poisoning. JAMA 94, 1648.
Jordan, E.O., Burrows, W., 1934. Further observations on *staphylococcus* food poisoning. Am. J. Hyg. 20, 604.
Kerouanton, A., Hennekinne, J.A., Letertre, C., Petit, L., Chesneau, O., Brisabois, A., De Buyser, M.L., 2007. Characterization of *Staphylococcus aureus* strains associated with food poisoning outbreaks in France. Int. J. Food Microbiol. 115, 369–375.

Khambaty, F.M., Bennett, R.W., Shah, D.B., 1994. Application of pulse field gel electrophoresis to the epidemiological characterisation of *Staphylococcus intermedius* implicated in a food-related outbreak. Epidemiol. Infect 113, 75–81.

Khan, S.A., Nawaz, M.S., Khan, A.A., Cerniglia, C.E., 2000. Transfer of erythromycin resistance from poultry to human clinical strains of *Staphylococcus aureus*. J. Clin. Microbiol. 38, 1832–1838.

Kourtis, C., Payani, P., May 25, 2016. Staphylococcal food poisoning outbreaks in Cyprus, 2015. In: 10th NRL for CPS Workshop, Anses, Maisons-Alfort, France.

Kitai, S., Shimizu, A., Kawano, J., Sato, E., Nakano, C., Uji, T., Kitagawa, H., 2005. Characterization of methicillin-resistant *Staphylococcus aureus* isolated from retail raw chicken meat in Japan. J. Vet. Med. 67, 107–110.

Kitamoto, M., Kito, K., Niimi, Y., et al., 2009. Food poisoning by *Staphylococcus aureus* at a university festival. Jpn. J. Infect. Dis. 62, 242–243.

Kraushaar, B., Ballhausen, B., Leeser, D., Tenhagen, B.A., Kaesbohrer, A., Fetsch, A., 2016. Antimicrobial resistances and virulence markers in methicillin-resistant *Staphylococcus aureus* from broiler and Turkey: a molecular view from farm to fork. Vet. Microbiol. (16), S0378–S1135 30144–4.

Kraushaar, B., Fetsch, A., 2014. First description of PVL-positive methicillin-resistant *Staphylococcus aureus* (MRSA) in wild boar meat. Int. J. Food Microbiol. 186, 68–73.

Kwon, N.H., Park, K.T., Jung, W.K., Youn, H.Y., Lee, Y., Kim, S.H., Bae, W., Lim, J.Y., Kim, J.Y., Kim, J.M., Hong, S.K., Park, H.Y., 2006. Characteristics of methicillin-resistant *Staphylococcus aureus* isolated from chicken meat and hospitalized dogs in Korea and their epidemiological relatedness. Vet. Microbiol. 117, 304–312.

Le Loir, Y., Baron, F., Gautier, M., 2003. *Staphylococcus aureus* and food poisoning. Genet. Mol. Res. 2, 63–76.

Martin, S.E., Myers, E.R., Iandolo, J.J., 2001. *Staphylococcus aureus*. In: Hui, Y.H., Pierson, M.D., Gorham, J.R. (Eds.), Foodborne Disease Handbook. In: Bacterial Pathogens, vol. 1. Marcel Dekker Inc., New York, Basel, pp. 345–381.

Mead, P.S., Slutsker, L., Dietz, V., McCaig, L.F., Bresee, J.S., Shapiro, C., Griffin, P.M., Tauxe, R.V., 1999. Food-related illness and death in the United States. Emerg. Infect. Dis. 5, 607–625.

Minor, T.E., Marth, E.H., 1972. *Staphylococcus aureus* and staphylococcal food intoxications. A review. IV. *Staphylococci* in meat, bakery products and other foods. J. Milk Food Technol. 35, 228–241.

Morris, C.A., Conway, H.D., Everall, P.H., 1972. Food-poisoning due to staphylococcal enterotoxin E. Lancet 300, 1375–1376.

Mossel, D.A.A., Corry, J.E.L., Struijk, C.B., Baird, R.M., 1995. Essentials of the Microbiology of Foods. A Textbook for Advanced Studies. Wiley John & Sons, Chichester, England, pp. 146–150.

Mossong, J., Decruyenaere, F., Moris, G., Ragimbeau, C., Olinger, C.M., Johler, S., Perrin, M., Hau, P., Weicherding, P., 2015. Investigation of a staphylococcal food poisoning outbreak combining case-control, traditional typing and whole genome sequencing methods, Luxembourg, June 2014. Euro Surveill. 20 (45). http://dx.doi.org/10.2807/1560-7917.ES.2015.20.45.30059. 26608881.

Normanno, G., Corrente, M., La Salandra, G., Dambrosio, A., Quaglia, N.C., Parisi, A., Greco, G., Bellacicco, A.L., Virgilio, S., Celano, G.V., 2007. Methicillin-resistant *Staphylococcus aureus* (MRSA) in foods of animal origin product in Italy. Int. J. Food Microbiol. 117, 219–222.

Notermans, S., Dufrenne, J., In't Veld, P., 1991. Feasibility of a reference material for staphylococcal enterotoxin A. Int. J. Food Microbiol. 14, 325–331.

Ostyn, A., De Buyser, M.L., Guillier, F., Groult, J., Félix, B., Salah, S., Delmas, G., Hennekinne, J.A., 2010. First evidence of a food-poisoning due to staphylococcal enterotoxin type E in France. Eurosurveillance 15, 19528.

Pereira, M.L., Do Carmo, L., Dos Santos, E.J., Pereira, J.L., Bergdoll, M.S., 1996. Enterotoxin H in staphylococcal food poisoning. J. Food Prot. 59, 559–561.

Pesavento, G., Ducci, B., Comodo, N., Nostro, A.L., 2007. Antimicrobial resistance profile of *Staphylococcus aureus* isolated from raw meat: a research for methicillin resistant *Staphylococcus aureus* (MRSA). Food Control 18, 196–200.

Pillsbury, A., Chiew, M., Bates, J., Sheppeard, V., June 30, 2013. An outbreak of staphylococcal food poisoning in a commercially catered buffet. Commun. Dis. Intell. Q. Rep. 37 (2), E144–E148. 24168088.

Qi, Y., Miller, K.J., 2000. Effect of low water activity on staphylococcal enterotoxin A and B biosynthesis. J. Food Prot. 63, 473–478.

Rosec, J.P., Guiraud, J.P., Dalet, C., Richard, N., 1997. Enterotoxin production by staphylococci isolated from foods in France. Int. J. Food Microbiol. 35, 213–221.

Scallan, E., Hoekstra, R.M., Angulo, F.J., et al., 2011. Foodborne illness acquired in the United States — major pathogens. Emerg. Infect. Dis. 17, 7–15.

Schmid, D., Fretz, R., Winter, P., et al., 2009. Outbreak of staphylococcal food intoxication after consumption of pasteurized milk products, June 2007, Austria. Wien. Klin. Wochenschr. 121, 125–131.

Schmitt, M., Schuler-Schmidt, U., Schmidt-Lorenz, W., 1990. Temperature limits of growth, TNase and enterotoxin production of *Staphylococcus aureus* strains isolated from foods. Int. J. Food Microbiol. 11, 1–20.

Shimizu, A., Fujita, M., Igarashi, H., Takagi, M., Nagase, N., Sasaki, A., Kawano, J., 2000. Characterization of *Staphylococcus aureus* coagulase type VII isolates from staphylococcal food poisoning outbreaks (1980-1995) in Tokyo, Japan, by pulsed-field gel electrophoresis. J. Clin. Microbiol. 38, 3746–3749.

Smith, J.L., Buchanan, R.L., Palumbo, S.A., 1983. Effect of food environment on staphylococcal enterotoxin synthesis: a review. J. Food Prot. 46, 545–555.

Tatini, S.R., 1973. Influence of food environments on growth of *Staphylococcus aureus* and production of various enterotoxins. J. Milk Food Technol. 36, 559–563.

Thaikruea, L., Pataraarechachai, J., Savanpunyalert, P., Naluponjiragul, U., 1995. An unusual outbreak of food poisoning. Southeast Asian J. Trop. Med. Public Health 26, 78–85.

Thota, H., Tatini, S.R., Bennett, R.W., 1973. Effects of temperature, pH and NaCl on production of staphylococcal enterotoxins E and F. Abstr. Ann. Meet. Am. Soc. Microbiol. 1, 11.

Todd, E., Szabo, R., Gardiner, M.A., et al., 1981. Intoxication staphylococcique liée à du caillé de fromagerie – Québec. Rapp. Hebd. des Mal. du Can. 7, 171–172.

Troller, J.A., Stinson, J.V., 1975. Influence of water activity on growth and enterotoxin formation by *Staphylococcus aureus* in foods. J. Food Sci. 40, 802–804.

Troller, J.A., 1986. The water relations of foodborne bacterial pathogens – an updated review. J. Food Prot. 49, 656–670.

Van Duijkeren, E., Ikawaty, R., Broekhuizen-Stins, M.J., Jansen, M.D., Spalburg, E.C., de Neeling, A.J., Allaart, J.G., van Nes, A., Wagenaar, J.A., Fluit, A.C., 2007. Transmission of methicillin-resistant *Staphylococcus aureus* strains between different kinds of pig farms. Vet. Microbiol. 126, 383–389.

Van Loo, I.H.M., Diederen, B.M.W., Savelkoul, P.H.M., Woudenberg, J.H.C., Roosendaal, R., van Belkum, A., Lemmens-den Toom, N., Verhulst, C., van Keulen, P.H.J., Kluytmans, J.A.J.W., 2007. Methicillin-resistant *Staphylococcus aureus* in meat products, The Netherlands. Emerg. Infect. Dis. 13, 1753–1755.

Vernozy-Rozand, C., Mazuy, C., Prevost, G., Lapeyre, C., Bes, M., Brun, Y., Fleurette, J., 1996. Enterotoxin production by coagulase negative staphylococci isolated from goats' milk and cheese. Int. J. Food Microbiol. 30, 271–280.

Wei, H.L., Chiou, C.S., 2002. Molecular subtyping of *Staphylococcus aureus* from an outbreak associated with a food handler. Epidemiol. Infect 128, 15–20.

Woolaway, M.C., Bartlett, C.L.R., Wieneke, A.A., Gilbert, R.J., Murell, H.C., Aureli, P., 1986. International outbreak of staphylococcal food poisoning caused by contaminated lasagne. J. Hyg. 96, 67–73.

Wulf, M., Voss, A., 2008. MRSA in livestock animals — an epidemic waiting to happen? Clin. Microbiol. Infect 14, 519–521.

Zell, C., Resch, M., Rosenstein, R., Albrecht, T., Hertel, C., Gotz, F., 2008. Characterization of toxin production of coagulase negative staphylococci isolated from food and starter cultures. Int. J. Food Microbiol. 49, 1577–1593.

CHAPTER

8 STAPHYLOCOCCUS AUREUS FROM FARM TO FORK: IMPACT FROM A VETERINARY PUBLIC HEALTH PERSPECTIVE

Catherine M. Logue, Claire B. Andreasen
College of Veterinary Medicine, Iowa State University, Ames, IA, United States

1. INTRODUCTION

Staphylococcus aureus has long been recognized as a human pathogen of foodborne illness and clinical disease, but it also results in significant losses for the livestock and poultry industries due to poor health, poor weight gain, disease treatment expenses, and loss of food production via carcass condemnation. From a veterinary and public health perspective, *S. aureus* is a major concern, especially where animal-derived food products and humans intersect including consumers and animal production. The discovery of drug resistant and methicillin-resistant *S. aureus* (MRSA) has led to many avenues of research investigating the role of livestock in the distribution and potential impact of livestock-associated (LA) *S. aureus* in community-associated (CA) *S. aureus* and clinically associated disease. *S. aureus* is a major concern in public health, especially where animal-derived foods and human beings intersect. The major areas of impact include: (1) livestock-derived food products contaminated with *S. aureus*; (2) exposure of animal health workers to strains of *S. aureus* (see Chapter 11 of this book for further details); and (3) the potential selection of antimicrobial resistant *S. aureus* due to antibiotic use in food, animals, and human beings; the most significant being recognized as MRSA.

The goal of this chapter is to provide an overview of *S. aureus* and MRSA in the livestock industry by describing *S. aureus* as a cause of disease in animals and the relationship between animals used for food production and humans, and its impact from a veterinary public health perspective.

2. *STAPHYLOCOCCUS AUREUS* IN ANIMALS USED FOR FOOD PRODUCTION

2.1 GENERAL OVERVIEW

Staphylococcus spp. especially *S. aureus,* is not only a human pathogen and a recognized cause of pyodermas and abscesses in domesticated species, including companion (dogs and cats) and zoo animals (Anon., 2016; Wang et al., 2012; Zhang et al., 2011) but is also a significant cause of disease in production animals. Disease in livestock animals includes mastitis in cattle and dairy producing animals such as sheep and goats (Doyle et al., 2012; Fluit, 2012; Leonard and Markey, 2008; Peton and Le Loir, 2014); septicemia, "bumblefoot," and chondronecrosis in chickens and other poultry species

Staphylococcus aureus. http://dx.doi.org/10.1016/B978-0-12-809671-0.00008-5

(Andreasen, 2013; McNamee and Smyth, 2000); exudative epidermitis (EE) in swine often due to *Staphylococcus hycius* (Park et al., 2013; Slifierz et al., 2014) and lesions in farmed meat rabbits (Viana et al., 2007).

However, not all animal hosts become diseased and in a considerable number of cases, animals can be healthy carriers of *S. aureus*. Carriage frequency rates reported in some animal hosts are 90% in poultry, 42% in swine, 14%–23% in cattle (Nagase et al., 2002), and 29% in sheep (Vautor et al., 2005). In addition to being a source of *S. aureus* more recent studies have also recognized that livestock has been identified as a significant source of MRSA (see later sections of this chapter) with multiple researchers identifying the organism in a range of hosts (Frana et al., 2013; Graveland et al., 2011; Kraushaar et al., 2016; Mulders et al., 2010; Vanderhaeghen et al., 2010) and research has highlighted the relationship between the host and strain types associated with humans either as a cause of disease or carriage (Doyle et al., 2012; Fetsch et al., 2017; Frana et al., 2013; Neela et al., 2009; van Cleef et al., 2011; Weese, 2010).

Molecular typing has proven useful in helping to classify the types of *S. aureus* associated with livestock. In general, multilocus sequence typing has been used to identify over 2200 sequence types (STs), which are then grouped into clonal complexes (CC) (Robinson and Enright, 2004). Studies have indicated that most LA STs belong to a small number of animal-associated clones. LA *S. aureus* are commonly associated as follows: bovine infections commonly have CC97, ST151, CC130, and CC126; sheep or goats have CC133 and mastitis associated with CC133 and ST522; and poultry have ST5 and ST9. For comparison, human infections are often linked with ST1, ST8, CC5, ST121, and ST398 (Eriksson et al., 2013; Fitzgerald, 2012). Human strains of *S. aureus* also have been found in livestock (e.g., ST5; ST398), including in the nares and farm dust (Frana et al., 2013; Kadariya et al., 2014) suggesting there is overlap between the hosts and potential sources.

For the farm to fork aspect, this section will focus on the most common animal species used for food production.

2.2 *STAPHYLOCOCCUS AUREUS* IN POULTRY

Staphylococcus infections are common in poultry and these are primarily due to *S. aureus,* which result in significant economic losses due to morbidity, mortality, or condemnation at processing (Andreasen, 2013). The most frequent sites of entry are the bones, tendon sheaths, and joints, with fewer infections of the sternal bursa, yolk sac, heart, vertebrae, eyelid, testis, liver, and lungs also being reported (Andreasen, 2013). The economic impact to the poultry industry results in decreased weight gain in meat birds, loss of egg production, lameness, and bird loss from osteomyelitis, septicemia, and carcass condemnation at slaughter. *Staphylococcus* spp. are ubiquitous and a common contaminant of the skin and mucous membranes as well as the environments of hatcheries, rearing units, or processing plants (Andreasen, 2013; Notermans et al., 1982; Rodgers et al., 2003; Thompson et al., 1980). Morbidity and mortality associated with *Staphylococcus* is relatively low, with *S. aureus* being the most common source of disease associated with leg and joint infections (Argudin et al., 2013). From a public health perspective, ~50% of typical and atypical *S. aureus* strains produce enterotoxins that can cause food poisoning in human beings, primarily because of the contamination of carcasses at processing (Notermans et al., 1982). *S. aureus* strains from processed poultry are thought to be endemic to the processing plant or transmitted from the hands of workers in the plant. The literature varies as to the origin of processing plant strains, in that biotyping indicates passage of human staphylococcal strains to poultry in processing plants, and plasmid profiling indicates that endemic strains in the processing plant are primarily introduced by incoming birds (Dodd et al., 1987). In addition, a report by Lowder and Fitzgerald supports evidence that pathogenic *S. aureus* CC5 strains of poultry may have a human origin (Lowder and Fitzgerald, 2010; Lowder et al., 2009) and that the "human to poultry" jump may have occurred in the last 40 years or so.

2.3 *STAPHYLOCOCCUS AUREUS* IN SWINE

Staphylococcal infections of economic importance in swine are most often not due to *S. aureus*, as these are usually sporadic and individual because of infections via wounds. The most significant staphylococcal swine disease is caused by *Staphylococcus hyicus*, the agent associated with EE also commonly known as "greasy pig disease," which is a skin disease of pigs under 3 months of age that results in dermal damage, lesions, exudation, erosion, dehydration, and perivascular cellular infiltrations in the dermis. Significant mortality (5%–90%) and morbidity (10%–90%) rates are associated with EE worldwide (Chen et al., 2007; Park et al., 2013; Slifierz et al., 2014). A recent study by Chen et al. (2007) has also recognized another *Staphylococcus* species *S. sciuri* as an agent of EE in young pigs that also showed high morbidity (33%–68%) and mortality rates of 13%–84%. Aside from disease linked with *S. hycius* and *S. scuiri*, swine can be healthy carriers of *Staphylococcus* species including *S. aureus*, *S. rostri*, and coagulase-negative staphylococci (*epidermidis*, *haemolyticus*, *warnerii*, etc.) and they are often found associated with the nares, skin surfaces, and in fecal shedding (Linhares et al., 2015; Stegmann and Perreten, 2010; Tulinski et al., 2012) with prevalence rates at various anatomical sites ranging from 40% to 67%. Of particular concern with some of these microflora is their potential carriage of resistance to antimicrobials and heavy metals (Argudin et al., 2016; Simeoni et al., 2008), which could be transferred to pathogenic strains.

2.4 *STAPHYLOCOCCUS AUREUS* IN CATTLE AND SMALL RUMINANTS

In production animals, infection of dairy cattle with *S. aureus* causes mastitis which is an inflammation of the udder tissue resulting in abnormalities in milk production (Vanderhaeghen et al., 2014). *S. aureus*–associated mastitis is also recognized in other milk-producing ruminants including sheep and goats (Bergonier et al., 2014; Cortimiglia et al., 2015; Merz et al., 2016; Zhao et al., 2015). Of interest, Merz et al. (2016) suggest that strains of *S. aureus* causing mastitis in sheep and in goats may be more closely related than those implicated in mastitis of dairy cattle, forming a distinct population especially when studied from a single region.

In dairy cattle, the economic impact of *S. aureus*–associated mastitis can also be a major cause of recurring mastitis, as well as spreading mastitis within herds (Sommerhauser et al., 2003), and is often a significant focus of antimicrobial treatment. *S. aureus* is the most common bacterial agent of mastitis with bacterial colonization of the teat skin, udder, and muzzle, while extra-mammary sites including the vagina and hair coat are recognized (De Vliegher et al., 2003; White et al., 1989), with infected cows serving as the major source of continuing farm infections (Capurro et al., 2010). *S. aureus* mastitis is extremely difficult to control by antibiotic treatment alone and prevention of new infections and culling of infected animals is often employed. In addition to other cows, spread of infection can occur through workers' hands, washcloths, teat cup liners, and flies. In small ruminants, *S. aureus* enterotoxins have been isolated from goat's milk, udder, and teats, as a potential foodborne pathogen (Foschino et al., 2002; Valle et al., 1990). Mastitis in dairy cattle (goats and sheep) can also pose a risk for consumers should contaminated milk enter the food chain (Merz et al., 2016; Oliver et al., 2009). Contaminated milk and dairy products have been implicated in human disease and in most cases linked with the consumption of raw milk products; however, pasteurized products have also been linked with illness due to intoxication (Asao et al., 2003; Johler et al., 2015; Schmid et al., 2009).

Also of importance in recent years has been the emergence of mastitis associated in dairy animals linked with coagulase-negative staphylococci, both non–Methicillin-resistant coagulase-negative staphylococci (MRCoNS) and MRCoNS strains (El-Jakee et al., 2013; Luthje and Schwarz, 2006).

3. ANTIMICROBIAL RESISTANCE IN *STAPHYLOCOCCUS AUREUS* AND THE EMERGENCE OF METHICILLIN-RESISTANT *STAPHYLOCOCCUS AUREUS*

3.1 GENERAL OVERVIEW

Methicillin-resistant *S. aureus* (MRSA) infections accounted for an estimated 80,000 invasive infections in the United States annually with an estimated 11,285 related deaths (Anon., 2014); while numbers in the European Union suggest MRSA accounts for >25% of human bacteremia, with 170,000 cases annually (Johnson, 2011; Kock et al., 2010, 2011). Prevalence levels in Germany suggest rates as high as 18%–20% of all *S. aureus* cultures examined are MRSA (Kock et al., 2011). MRSA is also being diagnosed more often in people with no hospital connection—many of them young, healthy students, or professional athletes which supports the association between CA strains and individuals that are not exposed to the typical clinical situation. Health care–associated strains are more often being replaced by CA-MRSA due to the expanding community reservoir of these organisms and the increasing influx into the hospital of individuals harboring CA-MRSA (Nimmo et al., 2013). The source of these CA strain types remains underexplored and could be a combination of sources such as LA and clinical. This is particularly true over the past decade or so as MRSA primarily of clonal lineages CC398, also called "livestock-associated" (LA)-MRSA have been frequently detected among livestock but also in the clinical setting.

Antibiotic use in livestock has included disease treatment, prevention, and control, and since the mid-1950s, growth promotion (Dibner and Richards, 2005; Hao et al., 2014). It was discovered that small subtherapeutic doses of antibiotics could enhance the feed to weight efficiency in poultry, swine, and beef cattle (Dibner and Richards, 2005; Hao et al., 2014). However, later it was found that farms using antibiotics for growth promotion had more antibiotic resistant bacteria in the intestinal flora of farm animals and farm workers than farms that did not use antibiotics for animal growth promotion (Gorbach, 2001). The link to human strains of antibiotic resistant bacteria due to nontherapeutic antibiotic use in livestock still often remains circumstantial in studies, because often authors do not report statistics on the farm use of antibiotics (Marshall and Levy, 2011). Antibiotics used to treat individual livestock for diseases tend to control antimicrobial resistant strains due to the short-term treatment and small number of treated animals (Marshall and Levy, 2011). Understanding the "farm to fork" relationship with antimicrobial use in livestock has become critical to provide knowledge for risk management and public health due to the continuous emergence of drug-resistant bacterial infections in humans and animals. There is a concern that these uses have contributed to antibiotic-resistant bacterial strains that may evolve on farms and may help explain the emergence of LA-MRSA.

The restrictions on the use of antibiotics in food-producing animals vary with species and country. Agricultural use of antibiotics in food animals includes many that are used for human health including tetracycline, macrolides, penicillin, sulfonamides, and fluoroquinolones with the potential generation of antimicrobial resistance (Smith, 2015). Resistant antibiotic strains may then spread to the general human population by handling or ingestion of contaminated meat products; occupational contact (farmers, meat packers, butchers, etc.) with potential secondary spread into the larger community from those who are occupationally exposed; entry into and transmission via hospitals or other health care facilities; or spread via environmental routes including air, water, or manure in areas in proximity to animal farms or crop farms where manure has been used as a fertilizer (Carrel et al., 2014; Casey et al., 2013; Ferguson et al., 2016; Friese et al., 2013; Smith et al., 2013; Wardyn et al., 2015).

Several species of meat producing livestock, including pigs, poultry, and cattle have been implicated as sources of MRSA in humans in Europe, Asia, North and South America, and Australia (Smith, 2015). In particular, swine are recognized as a significant source of LA-MRSA colonization and infection in humans (Graveland et al., 2011; Weese, 2010; Wendlandt et al., 2013b) through work exposure (at the farm or slaughterhouse). Swine-associated MRSA from farm to slaughter (Lassok and Tenhagen, 2013) has been the focus of significant public health research (Doyle et al., 2012; Fluit, 2012; Lassok and Tenhagen, 2013; Smith, 2015; Weese, 2010). These findings have led to further examination of the former paradigm of host species-specific *S. aureus*. In addition, CC398 strains have been found in pork and other raw meat from chicken, turkey, and veal calves (Buyukcangaz et al., 2013; de Boer et al., 2009; Hanson et al., 2011; O'Brien et al., 2012; Velasco et al., 2014); however, MRSA transmission from the farm to the general human population is not as clearly defined and there are still knowledge gaps in tracing the bacterial point of origin to the product, food handlers, and consumers.

In the following Sections 3.2–3.4, the prevalence and veterinary public health implications of MRSA found in the major food-producing animals are described.

3.2 METHICILLIN-RESISTANT *STAPHYLOCOCCUS AUREUS* ASSOCIATED WITH POULTRY

In poultry, the association of *S. aureus* with poultry disease has been recognized; however, poultry has also been recognized as a source of MRSA (Hanson et al., 2011) in turkeys as well as broilers and layers (Agunos et al., 2016). MRSA has been detected in poultry in several countries, but our understanding of its prevalence and significance for human health is currently incomplete. Studies by some researchers have demonstrated transmission of LA-MRSA to humans by food-producing animals especially in persons who have close contact with animals and estimated its associated risk (Graveland et al., 2011; Weese, 2010; Wendlandt et al., 2013a) while there appear to be limited number of studies that have identified poultry as a potential source of transmission to humans (Fetsch et al., 2017; Mulders et al., 2010; Richter et al., 2012; Wendlandt et al., 2013b). LA-MRSA strains have, however, been identified in both diseased and healthy flocks and their associated meat (Fessler et al., 2011; Fetsch et al., 2017; Monecke et al., 2013). MRSA was detected in 2%–35% of chicken flocks in a number of studies from Europe, and some reports suggest that the prevalence may be higher on broiler than layer farms (Nemeghaire et al., 2013). A German national monitoring scheme found MRSA in 20% of turkeys, and regional studies in Germany have reported that up to 25%–90% of turkey flocks may be colonized with *S. aureus*. The strain type *S. aureus* CC398 (also known as

LA-MRSA) was often detected in these studies, although other lineages e.g., ST9 and ST5 were also found (Kraushaar et al., 2016; Lowder et al., 2009; Mulders et al., 2010; Schwarz et al., 2008). A higher prevalence of MRSA was noted in turkey (33%) than chicken (5.1%) or duck (1.3%); whereas the comparison of live (6.9%) versus carcasses (9.4%) at processing found that the prevalence was relatively similar (Agunos et al., 2016). Poultry as a potential zoonotic source of *Staphylococcus* and MRSA to the food chain and personnel associated with poultry production warrants attention, in particular, prevalence studies would suggest that broilers and turkeys may hold greater risks of transmission than previously considered.

3.3 METHICILLIN-RESISTANT *STAPHYLOCOCCUS AUREUS* ASSOCIATED WITH SWINE

Swine are recognized as a significant source of *S. aureus* including multidrug and methicillin-resistant *S. aureus* (MRSA), and the unique association with the strain type ST398 which has been linked with carriage and or human colonization (Graveland et al., 2011; van Cleef et al., 2011, 2015; Weese, 2010; Wendlandt et al., 2013b) with the prevailing view being that production swine may be a LA source of MRSA to humans through work exposure (at the farm or slaughterhouse), meats, and other such factors. This view, however, remains contentious and current research suggests that this is no longer the case and that humans were probably the source of MRSA for swine (Deiters et al., 2015; Lekkerkerk et al., 2015). Of particular concern is the strain type identified as MRSA ST398, which was originally considered an animal clone, but phylogenetic analysis would appear to suggest that this clonal lineage originated in man as *S. aureus* and has since spread to animals where it subsequently acquired tetracycline and methicillin resistance (Price et al., 2012). Additional work has demonstrated that most human originating strains carry the *scn* gene (a staphylococcal complement inhibitor gene) while those that are livestock originated lack this gene (Sung et al., 2008).

In general, MRSA associated with swine has been recognized in the sense of healthy carriage by the host; however, disease linked with MRSA has also been recognized. In the Netherlands, MRSA ST398 was implicated in EE disease of piglets as the organism was isolated from the lesions of piglets diagnosed with EE and no evidence of *S. hycius*, which is the usual agent of this disease, was evident suggesting that MRSA can cause disease under appropriate circumstances (van Duijkeren et al., 2007). Since this time other researchers have also reported disease in swine linked with MRSA including endometritis associated with a sow and dermatitis in her piglets in Serbia (Zutic et al., 2012) as well as urinary tract infections, metritis, and mastitis in swine in Germany (Schwarz et al., 2008).

3.4 METHICILLIN-RESISTANT *STAPHYLOCOCCUS AUREUS* ASSOCIATED WITH CATTLE

In cattle, the association between animal host and source of MRSA appears to follow similar lines to those reported with poultry and swine, with dairy and beef cattle implicated as sources of MRSA that are also linked with milk produced from dairy cattle as well as meat from beef (Cortimiglia et al., 2016; Ge et al., 2017; Kreausukon et al., 2012; Nemeghaire et al., 2014; Tenhagen et al., 2014; Weese et al., 2010). In addition, MRSA-associated mastitis has been identified in dairy herds worldwide (Fessler et al., 2010; Luini et al., 2015; Spohr et al., 2011; Wang et al., 2015; Zhang et al., 2016).

Bulk tank milk has also been identified as a source of MRSA demonstrating the potential food safety risk for contaminated milk and dairy products entering the human food chain (Al-Ashmawy et al., 2016; Kreausukon et al., 2012; Oliveira et al., 2016; Paterson et al., 2014b; Tavakol et al., 2012).

In cattle, some MRSA strains seem to be of human origin, and MRSA CC398 also has been found; however, there also seem to be some bovine-associated strains. Similarly, human-adapted MRSA, CC398, and isolates that might be host-adapted have been reported in both small and large ruminants (Silva et al., 2014; Tavakol et al., 2012; Velasco et al., 2015).

Recently, MRSA harboring a new methicillin-encoding gene called *mecC* gene have been described and appears that this gene has been first identified in dairy cattle and bulk milk (Garcia-Alvarez et al., 2011; Paterson et al., 2014b).

3.5 METHICILLIN-RESISTANT *STAPHYLOCOCCUS AUREUS* IN LIVESTOCK—FACTORS CONTRIBUTING TO SELECTION AND SPREAD

In addition to traits associated with methicillin resistance and the identification of MRSA, the emergence of multidrug resistance in *S. aureus* not linked with methicillin is also common with strains displaying multidrug resistance and, resistance to heavy metals and disinfection agents, such as hydrogen peroxide, chlorhexidine, and formaldehyde (Aarestrup and Hasman, 2004; Argudin et al., 2016; Wardyn et al., 2015). Argudin et al. (2016) noted that LA-MRSA in animal production and its emergence is not fully understood and drivers of selection may include factors besides antimicrobial compounds such as metal containing compounds. In their study, heavy metal containing resistance genes such *arsA* (arsenic compounds); *cadD* (cadmium); *copB* (copper); and *crcZ* (zinc/cadmium) were reported at a prevalence level of 4.8%, 0.2%, 24.3%, and 71.5%, respectively, in LA-MRSA CC398 strains. Nair et al. (2014) also noted heavy metal resistance in MRSA of humans and LA-MRSA to cadmium and zinc with 79% of LA-MRSA and 10% of human strains and 36% of retail meat *S. aureus* positive for *crcZ*. The authors suggested prolonged exposure to zinc through livestock feeding may propagate resistance.

The potential impact of heavy metals on selection of LA-MRSA, particularly in swine, is also supported by others who report that selection of LA-MRSA in swine production is directly linked with zinc supplementation likely increasing the prevalence and persistence of MRSA in swine (Aarestrup et al., 2010; Cavaco et al., 2011; Slifierz et al., 2015a,b).

Regardless of how we view antimicrobial resistance in *S. aureus* and the emergence of LA-MRSA, the role of feeding regimes (especially supplementation of feed) and disinfection protocols may have a significant role in the selection of MRSA.

4. SUMMARY AND CONCLUSION

LA *S. aureus* research has focused more on carriers than transmission and infection. It was recently suggested that bidirectional infection between animals and farm workers is not rare (McNamee and Smyth, 2000; Smith, 2015; Viana et al., 2007; Weese, 2010) and our understanding of the movement of these strains is continuing to evolve (Larsen et al., 2016; Sung et al., 2008). There are currently some limitations to determine if a strain of *S. aureus* is definitively a human or livestock adapted strain due to the lack of validation of many marker genes or single nucleotide polymorphism for various lineages on a large scale (Stegger et al., 2013). Livestock and human transmission of antibiotic resistance currently requires a more defined understanding of the genetic interactions, transmission, and the environment that includes commensal and environmental bacteria.

Public health conclusions are difficult in the absence of a national farm and worker sampling and surveillance program that would characterize isolates and the associated epidemiology. A program of this type could be difficult to implement on a national scale due to various private/corporate, state, and national regulations, but there is a need to determine antimicrobial resistance at all stages of the transmission chain. The lack of these data limits direct evidence of the strain origin or specific gene associations, molecular tools that can now provide the means to augment a risk management knowledge base. There has been some progress via national monitoring schemes such as the National Antimicrobial Resistance Monitoring System at https://www.cdc.gov/narms/ in the United States, which monitors microbial resistance trends in human isolates, while food animal isolates are monitored by the US Department of Agriculture https://www.ars.usda.gov/southeast-area/athens-ga/us-national-poultry-research-center/bacterial-epidemiology-antimicrobial-resistance-research/docs/narms-national-antimicrobial-resistance-monitoring-system-animal-isolates/, and the Food and Drugs Administration monitors retail meat isolates http://www.fda.gov/AnimalVeterinary/SafetyHealth/AntimicrobialResistance/NationalAntimicrobialResistanceMonitoringSystem/; while similar such schemes are also used in Europe though the European Food Safety Authority https://www.efsa.europa.eu/en/topics/topic/antimicrobial-resistance and additional monitoring programs are used at the various country and state levels, however, there are still knowledge gaps that need to be addressed that will require continued partnership with the industry to monitor and identify emerging trends and resistances.

Predictive modeling is another option; however, a weakness in these models is the lack of knowledge related to microbial loads and antibiotic use at each stage of the "farm to fork" process (Marshall and Levy, 2011). Collaborations between animal, human, and public health sectors will be needed to further define prevention and solutions.

REFERENCES

Aarestrup, F.M., Cavaco, L., Hasman, H., 2010. Decreased susceptibility to zinc chloride is associated with methicillin resistant *Staphylococcus aureus* CC398 in Danish swine. Vet. Microbiol. 142, 455–457.

Aarestrup, F.M., Hasman, H., 2004. Susceptibility of different bacterial species isolated from food animals to copper sulphate, zinc chloride and antimicrobial substances used for disinfection. Vet. Microbiol. 100, 83–89.

Agunos, A., Pierson, F.W., Lungu, B., Dunn, P.A., Tablante, N., 2016. Review of nonfoodborne zoonotic and potentially zoonotic poultry diseases. Avian Dis. 60, 553–575.

Al-Ashmawy, M.A., Sallam, K.I., Abd-Elghany, S.M., Elhadidy, M., Tamura, T., 2016. Prevalence, molecular characterization, and antimicrobial susceptibility of methicillin-resistant *Staphylococcus aureus* isolated from milk and dairy products. Foodborne Pathog. Dis. 13, 156–162.

Andreasen, C.B., 2013. Staphylococcosis. In: Swayne, D.E., Glisson, J.R., McDougals, L.R., Nolan, L.K., Suarez, D.L., Nair, V. (Eds.), Diseases of Poultry. Wiley-Blackwell, Ames, IA, pp. 971–977.

Anon., 2014. In: President, O.O.T. (Ed.), National Strategy for Combating Antibiotic-Resistant Bacteria. White House, Washington, DC.

Anon., 2016. Methicillin Resistant *Staphylococcus aureus*. The Center for Food Security and Public Health.

Argudin, M.A., Cariou, N., Salandre, O., Le Guennec, J., Nemeghaire, S., Butaye, P., 2013. Genotyping and antimicrobial resistance of *Staphylococcus aureus* isolates from diseased turkeys. Avian Pathol. 42, 572–580.

Argudin, M.A., Lauzat, B., Kraushaar, B., Alba, P., Agerso, Y., Cavaco, L., Butaye, P., Porrero, M.C., Battisti, A., Tenhagen, B.A., Fetsch, A., Guerra, B., 2016. Heavy metal and disinfectant resistance genes among livestock-associated methicillin-resistant *Staphylococcus aureus* isolates. Vet. Microbiol. 191, 88–95.

Asao, T., Kumeda, Y., Kawai, T., Shibata, T., Oda, H., Haruki, K., Nakazawa, H., Kozaki, S., 2003. An extensive outbreak of staphylococcal food poisoning due to low-fat milk in Japan: estimation of enterotoxin A in the incriminated milk and powdered skim milk. Epidemiol. Infect. 130, 33–40.

Bergonier, D., Sobral, D., Fessler, A.T., Jacquet, E., Gilbert, F.B., Schwarz, S., Treilles, M., Bouloc, P., Pourcel, C., Vergnaud, G., 2014. *Staphylococcus aureus* from 152 cases of bovine, ovine and caprine mastitis investigated by multiple-locus variable number of tandem repeat analysis (MLVA). Vet. Res. 45, 97.

Buyukcangaz, E., Velasco, V., Sherwood, J.S., Stepan, R.M., Koslofsky, R.J., Logue, C.M., 2013. Molecular typing of *Staphylococcus aureus* and methicillin-resistant *S. aureus* (MRSA) isolated from animals and retail meat in North Dakota, United States. Foodborne Pathog. Dis. 10, 608–617.

Capurro, A., Aspan, A., Ericsson Unnerstad, H., Persson Waller, K., Artursson, K., 2010. Identification of potential sources of *Staphylococcus aureus* in herds with mastitis problems. J. Dairy Sci. 93, 180–191.

Carrel, M., Schweizer, M.L., Sarrazin, M.V., Smith, T.C., Perencevich, E.N., 2014. Residential proximity to large numbers of swine in feeding operations is associated with increased risk of methicillin-resistant *Staphylococcus aureus* colonization at time of hospital admission in rural Iowa veterans. Infect. Control Hosp. Epidemiol. 35, 190–193.

Casey, J.A., Curriero, F.C., Cosgrove, S.E., Nachman, K.E., Schwartz, B.S., 2013. High-density livestock operations, crop field application of manure, and risk of community-associated methicillin-resistant *Staphylococcus aureus* infection in Pennsylvania. JAMA Intern. Med. 173, 1980–1990.

Cavaco, L.M., Hasman, H., Aarestrup, F.M., 2011. Zinc resistance of *Staphylococcus aureus* of animal origin is strongly associated with methicillin resistance. Vet. Microbiol. 150, 344–348.

Chen, S., Wang, Y., Chen, F., Yang, H., Gan, M., Zheng, S.J., 2007. A highly pathogenic strain of *Staphylococcus sciuri* caused fatal exudative epidermitis in piglets. PLoS One 2, e147.

Cortimiglia, C., Bianchini, V., Franco, A., Caprioli, A., Battisti, A., Colombo, L., Stradiotto, K., Vezzoli, F., Luini, M., 2015. Short communication: prevalence of *Staphylococcus aureus* and methicillin-resistant *S. aureus* in bulk tank milk from dairy goat farms in Northern Italy. J. Dairy Sci. 98, 2307–2311.

Cortimiglia, C., Luini, M., Bianchini, V., Marzagalli, L., Vezzoli, F., Avisani, D., Bertoletti, M., Ianzano, A., Franco, A., Battisti, A., 2016. Prevalence of *Staphylococcus aureus* and of methicillin-resistant *S. aureus* clonal complexes in bulk tank milk from dairy cattle herds in Lombardy region (Northern Italy). Epidemiol. Infect. 144, 3046–3051.

de Boer, E., Zwartkruis-Nahuis, J.T., Wit, B., Huijsdens, X.W., de Neeling, A.J., Bosch, T., van Oosterom, R.A., Vila, A., Heuvelink, A.E., 2009. Prevalence of methicillin-resistant *Staphylococcus aureus* in meat. Int. J. Food Microbiol. 134, 52–56.

De Vliegher, S., Laevens, H., Devriese, L.A., Opsomer, G., Leroy, J.L., Barkema, H.W., de Kruif, A., 2003. Prepartum teat apex colonization with *Staphylococcus* chromogenes in dairy heifers is associated with low somatic cell count in early lactation. Vet. Microbiol. 92, 245–252.

Deiters, C., Gunnewig, V., Friedrich, A.W., Mellmann, A., Kock, R., 2015. Are cases of Methicillin-resistant *Staphylococcus aureus* clonal complex (CC) 398 among humans still livestock-associated? Int. J. Med. Microbiol. 305, 110–113.

Dibner, J.J., Richards, J.D., 2005. Antibiotic growth promoters in agriculture: history and mode of action. Poult. Sci. 84, 634–643.

Dodd, C.E., Adams, B.W., Mead, G.C., Waites, W.M., 1987. Use of plasmid profiles to detect changes in strains of *Staphylococcus aureus* during poultry processing. J. Appl. Bacteriol. 63, 417–425.

Doyle, M.E., Hartmann, F.A., Lee Wong, A.C., 2012. Methicillin-resistant staphylococci: implications for our food supply? Anim. Health Res. Rev. 13, 157–180.

El-Jakee, J.K., Aref, N.E., Gomaa, A., El-Hariri, M.D., Galal, H.M., Omar, S.A., Samir, A., 2013. Emerging of coagulase negative staphylococci as a cause of mastitis in dairy animals: an environmental hazard. Int. J. Vet. Sci. Med. 1, 74–78.

Eriksson, J., Espinosa-Gongora, C., Stamphoj, I., Larsen, A.R., Guardabassi, L., 2013. Carriage frequency, diversity and methicillin resistance of *Staphylococcus aureus* in Danish small ruminants. Vet. Microbiol. 163, 110–115.

Ferguson, D.D., Smith, T.C., Hanson, B.M., Wardyn, S.E., Donham, K.J., 2016. Detection of airborne methicillin-resistant *Staphylococcus aureus* inside and downwind of a swine building, and in animal feed: potential occupational, animal health, and environmental implications. J. Agromedicine 21, 149–153.

Fessler, A., Scott, C., Kadlec, K., Ehricht, R., Monecke, S., Schwarz, S., 2010. Characterization of methicillin-resistant *Staphylococcus aureus* ST398 from cases of bovine mastitis. J. Antimicrob. Chemother. 65, 619–625.

Fessler, A.T., Kadlec, K., Hassel, M., Hauschild, T., Eidam, C., Ehricht, R., Monecke, S., Schwarz, S., 2011. Characterization of methicillin-resistant *Staphylococcus aureus* isolates from food and food products of poultry origin in Germany. Appl. Environ. Microbiol. 77, 7151–7157.

Fetsch, A., Kraushaar, B., Kasbohrer, A., Hammerl, J.A., 2017. Turkey meat as source of CC9/CC398 methicillin-resistant *Staphylococcus aureus* in humans? Clin. Infect. Dis. 64, 102–103.

Fitzgerald, J.R., 2012. Livestock-associated *Staphylococcus aureus*: origin, evolution and public health threat. Trends Microbiol. 20, 192–198.

Fluit, A.C., 2012. Livestock-associated *Staphylococcus aureus*. Clin. Microbiol. Infect. 18, 735–744.

Foschino, R., Invernizzi, A., Barucco, R., Stradiotto, K., 2002. Microbial composition, including the incidence of pathogens, of goat milk from the bergamo region of Italy during a lactation year. J. Dairy Res. 69, 213–225.

Frana, T.S., Beahm, A.R., Hanson, B.M., Kinyon, J.M., Layman, L.L., Karriker, L.A., Ramirez, A., Smith, T.C., 2013. Isolation and characterization of methicillin-resistant *Staphylococcus aureus* from pork farms and visiting veterinary students. PLoS One 8, e53738.

Friese, A., Schulz, J., Zimmermann, K., Tenhagen, B.A., Fetsch, A., Hartung, J., Rosler, U., 2013. Occurrence of livestock-associated methicillin-resistant *Staphylococcus aureus* in Turkey and broiler barns and contamination of air and soil surfaces in their vicinity. Appl. Environ. Microbiol. 79, 2759–2766.

Garcia-Alvarez, L., Holden, M.T., Lindsay, H., Webb, C.R., Brown, D.F., Curran, M.D., Walpole, E., Brooks, K., Pickard, D.J., Teale, C., Parkhill, J., Bentley, S.D., Edwards, G.F., Girvan, E.K., Kearns, A.M., Pichon, B., Hill, R.L., Larsen, A.R., Skov, R.L., Peacock, S.J., Maskell, D.J., Holmes, M.A., 2011. Meticillin-resistant *Staphylococcus aureus* with a novel mecA homologue in human and bovine populations in the UK and Denmark: a descriptive study. Lancet Infect. Dis. 11, 595–603.

Ge, B., Mukherjee, S., Hsu, C.H., Davis, J.A., Tran, T.T., Yang, Q., Abbott, J.W., Ayers, S.L., Young, S.R., Crarey, E.T., Womack, N.A., Zhao, S., McDermott, P.F., 2017. MRSA and multidrug-resistant *Staphylococcus aureus* in U.S. retail meats, 2010–2011. Food Microbiol. 62, 289–297.

Gorbach, S.L., 2001. Antimicrobial use in animal feed–time to stop. N. Engl. J. Med. 345, 1202–1203.

Graveland, H., Duim, B., van Duijkeren, E., Heederik, D., Wagenaar, J.A., 2011. Livestock-associated methicillin-resistant *Staphylococcus aureus* in animals and humans. Int. J. Med. Microbiol. 301, 630–634.

Hanson, B.M., Dressler, A.E., Harper, A.L., Scheibel, R.P., Wardyn, S.E., Roberts, L.K., Kroeger, J.S., Smith, T.C., 2011. Prevalence of *Staphylococcus aureus* and methicillin-resistant *Staphylococcus aureus* (MRSA) on retail meat in Iowa. J. Infect. Public Health 4, 169–174.

Hao, H., Cheng, G., Iqbal, Z., Ai, X., Hussain, H.I., Huang, L., Dai, M., Wang, Y., Liu, Z., Yuan, Z., 2014. Benefits and risks of antimicrobial use in food-producing animals. Front. Microbiol. 5, 288.

Johler, S., Weder, D., Bridy, C., Huguenin, M.C., Robert, L., Hummerjohann, J., Stephan, R., 2015. Outbreak of staphylococcal food poisoning among children and staff at a Swiss boarding school due to soft cheese made from raw milk. J. Dairy Sci. 98, 2944–2948.

Johnson, A.P., 2011. Methicillin-resistant *Staphylococcus aureus*: the European landscape. J. Antimicrob. Chemother. 66 (Suppl. 4), iv43–iv48.

Kadariya, J., Smith, T.C., Thapaliya, D., 2014. *Staphylococcus aureus* and staphylococcal food-borne disease: an ongoing challenge in public health. BioMed Res. Int. 2014, 827965.

Kock, R., Becker, K., Cookson, B., van Gemert-Pijnen, J.E., Harbarth, S., Kluytmans, J., Mielke, M., Peters, G., Skov, R.L., Struelens, M.J., Tacconelli, E., Navarro Torne, A., Witte, W., Friedrich, A.W., 2010. Methicillin-resistant *Staphylococcus aureus* (MRSA): burden of disease and control challenges in Europe. Euro Surveill. 15, 19688.

Kock, R., Mellmann, A., Schaumburg, F., Friedrich, A.W., Kipp, F., Becker, K., 2011. The epidemiology of methicillin-resistant *Staphylococcus aureus* (MRSA) in Germany. Dtsch. Arztebl. Int. 108, 761–767.

Kraushaar, B., Ballhausen, B., Leeser, D., Tenhagen, B.A., Kasbohrer, A., Fetsch, A., 2016. Antimicrobial resistances and virulence markers in Methicillin-resistant *Staphylococcus aureus* from broiler and Turkey: a molecular view from farm to fork. Vet. Microbiol. 200.

Kreausukon, K., Fetsch, A., Kraushaar, B., Alt, K., Muller, K., Kromker, V., Zessin, K.H., Kasbohrer, A., Tenhagen, B.A., 2012. Prevalence, antimicrobial resistance, and molecular characterization of methicillin-resistant *Staphylococcus aureus* from bulk tank milk of dairy herds. J. Dairy Sci. 95, 4382–4388.

Larsen, J., Stegger, M., Andersen, P.S., Petersen, A., Larsen, A.R., Westh, H., Agerso, Y., Fetsch, A., Kraushaar, B., Kasbohrer, A., Febetaler, A.T., Schwarz, S., Cuny, C., Witte, W., Butaye, P., Denis, O., Haenni, M., Madec, J.Y., Jouy, E., Laurent, F., Battisti, A., Franco, A., Alba, P., Mammina, C., Pantosti, A., Monaco, M., Wagenaar, J.A., de Boer, E., van Duijkeren, E., Heck, M., Dominguez, L., Torres, C., Zarazaga, M., Price, L.B., Skov, R.L., 2016. Evidence for human adaptation and foodborne transmission of livestock-associated methicillin-resistant *Staphylococcus aureus*. Clin. Infect. Dis. 63, 1349–1352.

Lassok, B., Tenhagen, B.A., 2013. From pig to pork: methicillin-resistant *Staphylococcus aureus* in the pork production chain. J. Food Prot. 76, 1095–1108.

Lekkerkerk, W.S., van Wamel, W.J., Snijders, S.V., Willems, R.J., van Duijkeren, E., Broens, E.M., Wagenaar, J.A., Lindsay, J.A., Vos, M.C., 2015. What is the origin of livestock-associated methicillin-resistant *Staphylococcus aureus* clonal complex 398 isolates from humans without livestock Contact? An epidemiological and genetic analysis. J. Clin. Microbiol. 53, 1836–1841.

Leonard, F.C., Markey, B.K., 2008. Meticillin-resistant *Staphylococcus aureus* in animals: a review. Vet. J. 175, 27–36.

Linhares, L.L., Yang, M., Sreevatsan, S., Munoz-Zanzi, C.A., Torremorell, M., Davies, P.R., 2015. The effect of anatomic site and age on detection of *Staphylococcus aureus* in pigs. J. Vet. Diagn. Invest. 27, 55–60.

Lowder, B.V., Fitzgerald, J.R., 2010. Human origin for avian pathogenic *Staphylococcus aureus*. Virulence 1, 283–284.

Lowder, B.V., Guinane, C.M., Ben Zakour, N.L., Weinert, L.A., Conway-Morris, A., Cartwright, R.A., Simpson, A.J., Rambaut, A., Nubel, U., Fitzgerald, J.R., 2009. Recent human-to-poultry host jump, adaptation, and pandemic spread of *Staphylococcus aureus*. Proc. Natl. Acad. Sci. U.S.A. 106, 19545–19550.

Luini, M., Cremonesi, P., Magro, G., Bianchini, V., Minozzi, G., Castiglioni, B., Piccinini, R., 2015. Methicillin-resistant *Staphylococcus aureus* (MRSA) is associated with low within-herd prevalence of intra-mammary infections in dairy cows: genotyping of isolates. Vet. Microbiol. 178, 270–274.

Luthje, P., Schwarz, S., 2006. Antimicrobial resistance of coagulase-negative staphylococci from bovine subclinical mastitis with particular reference to macrolide-lincosamide resistance phenotypes and genotypes. J. Antimicrob. Chemother. 57, 966–969.

Marshall, B.M., Levy, S.B., 2011. Food animals and antimicrobials: impacts on human health. Clin. Microbiol. Rev. 24, 718–733.

McNamee, P.T., Smyth, J.A., 2000. Bacterial chondronecrosis with osteomyelitis ('femoral head necrosis') of broiler chickens: a review. Avian Pathol. 29, 253–270.

Merz, A., Stephan, R., Johler, S., 2016. *Staphylococcus aureus* isolates from goat and sheep milk seem to be closely related and differ from isolates detected from bovine milk. Front. Microbiol. 7, 319.

Monecke, S., Ruppelt, A., Wendlandt, S., Schwarz, S., Slickers, P., Ehricht, R., Jackel, S.C., 2013. Genotyping of *Staphylococcus aureus* isolates from diseased poultry. Vet. Microbiol. 162, 806–812.

Mulders, M.N., Haenen, A.P., Geenen, P.L., Vesseur, P.C., Poldervaart, E.S., Bosch, T., Huijsdens, X.W., Hengeveld, P.D., Dam-Deisz, W.D., Graat, E.A., Mevius, D., Voss, A., Van De Giessen, A.W., 2010. Prevalence of livestock-associated MRSA in broiler flocks and risk factors for slaughterhouse personnel in The Netherlands. Epidemiol. Infect. 138, 743–755.

Nagase, N., Sasaki, A., Yamashita, K., Shimizu, A., Wakita, Y., Kitai, S., Kawano, J., 2002. Isolation and species distribution of staphylococci from animal and human skin. J. Vet. Med. Sci. 64, 245–250.

Nair, R., Thapaliya, D., Su, Y., Smith, T.C., 2014. Resistance to zinc and cadmium in *Staphylococcus aureus* of human and animal origin. Infect. Control Hosp. Epidemiol. 35 (Suppl. 3), S32–S39.

Neela, V., Mohd Zafrul, A., Mariana, N.S., van Belkum, A., Liew, Y.K., Rad, E.G., 2009. Prevalence of ST9 methicillin-resistant *Staphylococcus aureus* among pigs and pig handlers in Malaysia. J. Clin. Microbiol. 47, 4138–4140.

Nemeghaire, S., Argudin, M.A., Haesebrouck, F., Butaye, P., 2014. Epidemiology and molecular characterization of methicillin-resistant *Staphylococcus aureus* nasal carriage isolates from bovines. BMC Vet. Res. 10, 153.

Nemeghaire, S., Roelandt, S., Argudin, M.A., Haesebrouck, F., Butaye, P., 2013. Characterization of methicillin-resistant *Staphylococcus aureus* from healthy carrier chickens. Avian Pathol. 42, 342–346.

Nimmo, G.R., Bergh, H., Nakos, J., Whiley, D., Marquess, J., Huygens, F., Paterson, D.L., 2013. Replacement of healthcare-associated MRSA by community-associated MRSA in Queensland: confirmation by genotyping. J. Infect. 67, 439–447.

Notermans, S., Dufrenne, J., van Leeuwen, W.J., 1982. Contamination of broiler chickens by *Staphylococcus aureus* during processing; incidence and origin. J. Appl. Bacteriol. 52, 275–280.

O'Brien, A.M., Hanson, B.M., Farina, S.A., Wu, J.Y., Simmering, J.E., Wardyn, S.E., Forshey, B.M., Kulick, M.E., Wallinga, D.B., Smith, T.C., 2012. MRSA in conventional and alternative retail pork products. PLoS One 7, e30092.

Oliveira, C.J., Tiao, N., de Sousa, F.G., de Moura, J.F., Santos Filho, L., Gebreyes, W.A., 2016. Methicillin-resistant *Staphylococcus aureus* from Brazilian dairy farms and identification of novel sequence types. Zoonoses Public Health 63, 97–105.

Oliver, S.P., Boor, K.J., Murphy, S.C., Murinda, S.E., 2009. Food safety hazards associated with consumption of raw milk. Foodborne Pathog. Dis. 6, 793–806.

Park, J., Friendship, R.M., Poljak, Z., Weese, J.S., Dewey, C.E., 2013. An investigation of exudative epidermitis (greasy pig disease) and antimicrobial resistance patterns of Staphylococcus hyicus and *Staphylococcus aureus* isolated from clinical cases. Can. Vet. J. 54, 139–144.

Paterson, G.K., Morgan, F.J., Harrison, E.M., Peacock, S.J., Parkhill, J., Zadoks, R.N., Holmes, M.A., 2014b. Prevalence and properties of mecC methicillin-resistant *Staphylococcus aureus* (MRSA) in bovine bulk tank milk in Great Britain. J. Antimicrob. Chemother. 69, 598–602.

Peton, V., Le Loir, Y., 2014. *Staphylococcus aureus* in veterinary medicine. Infect. Genet. Evol. 21, 602–615.

Price, L.B., Stegger, M., Hasman, H., Aziz, M., Larsen, J., Andersen, P.S., Pearson, T., Waters, A.E., Foster, J.T., Schupp, J., Gillece, J., Driebe, E., Liu, C.M., Springer, B., Zdovc, I., Battisti, A., Franco, A., Zmudzki, J., Schwarz, S., Butaye, P., Jouy, E., Pomba, C., Porrero, M.C., Ruimy, R., Smith, T.C., Robinson, D.A., Weese, J.S., Arriola, C.S., Yu, F., Laurent, F., Keim, P., Skov, R., Aarestrup, F.M., 2012. *Staphylococcus aureus* CC398: host adaptation and emergence of methicillin resistance in livestock. MBio 3.

Richter, A., Sting, R., Popp, C., Rau, J., Tenhagen, B.A., Guerra, B., Hafez, H.M., Fetsch, A., 2012. Prevalence of types of methicillin-resistant *Staphylococcus aureus* in Turkey flocks and personnel attending the animals. Epidemiol. Infect. 140, 2223–2232.

Robinson, D.A., Enright, M.C., 2004. Multilocus sequence typing and the evolution of methicillin-resistant *Staphylococcus aureus*. Clin. Microbiol. Infect. 10, 92–97.

Rodgers, J.D., McCullagh, J.J., McNamee, P.T., Bell, C., Brice, N., Smyth, J.A., Ball, H.J., 2003. Recovery of pathogenic *Staphylococcus aureus* from broiler hatchery air samples. Vet. Rec. 153, 656–657.

Schmid, D., Fretz, R., Winter, P., Mann, M., Hoger, G., Stoger, A., Ruppitsch, W., Ladstatter, J., Mayer, N., de Martin, A., Allerberger, F., 2009. Outbreak of staphylococcal food intoxication after consumption of pasteurized milk products, June 2007, Austria. Wien. Klin. Wochenschr. 121, 125–131.

Schwarz, S., Kadlec, K., Strommenger, B., 2008. Methicillin-resistant *Staphylococcus aureus* and *Staphylococcus pseudintermedius* detected in the BfT-GermVet monitoring programme 2004–2006 in Germany. J. Antimicrob. Chemother. 61, 282–285.

Silva, N.C., Guimaraes, F.F., Manzi, M.P., Junior, A.F., Gomez-Sanz, E., Gomez, P., Langoni, H., Rall, V.L., Torres, C., 2014. Methicillin-resistant *Staphylococcus aureus* of lineage ST398 as cause of mastitis in cows. Lett. Appl. Microbiol. 59, 665–669.

Simeoni, D., Rizzotti, L., Cocconcelli, P., Gazzola, S., Dellaglio, F., Torriani, S., 2008. Antibiotic resistance genes and identification of staphylococci collected from the production chain of swine meat commodities. Food Microbiol. 25, 196–201.

Slifierz, M.J., Friendship, R., Weese, J.S., 2015a. Zinc oxide therapy increases prevalence and persistence of methicillin-resistant *Staphylococcus aureus* in pigs: a randomized controlled trial. Zoonoses Public Health 62, 301–308.

Slifierz, M.J., Friendship, R.M., Weese, J.S., 2015b. Methicillin-resistant *Staphylococcus aureus* in commercial swine herds is associated with disinfectant and zinc usage. Appl. Environ. Microbiol. 81, 2690–2695.

Slifierz, M.J., Park, J., Friendship, R.M., Weese, J.S., 2014. Zinc-resistance gene CzrC identified in methicillin-resistant Staphylococcus hyicus isolated from pigs with exudative epidermitis. Can. Vet. J. 55, 489–490.

Smith, T.C., 2015. Livestock-associated *Staphylococcus aureus*: the United States experience. PLoS Pathog. 11, e1004564.

Smith, T.C., Gebreyes, W.A., Abley, M.J., Harper, A.L., Forshey, B.M., Male, M.J., Martin, H.W., Molla, B.Z., Sreevatsan, S., Thakur, S., Thiruvengadam, M., Davies, P.R., 2013. Methicillin-resistant *Staphylococcus aureus* in pigs and farm workers on conventional and antibiotic-free swine farms in the USA. PLoS One 8, e63704.

Sommerhauser, J., Kloppert, B., Wolter, W., Zschock, M., Sobiraj, A., Failing, K., 2003. The epidemiology of *Staphylococcus aureus* infections from subclinical mastitis in dairy cows during a control programme. Vet. Microbiol. 96, 91–102.

Spohr, M., Rau, J., Friedrich, A., Klittich, G., Fetsch, A., Guerra, B., Hammerl, J.A., Tenhagen, B.A., 2011. Methicillin-resistant *Staphylococcus aureus* (MRSA) in three dairy herds in southwest Germany. Zoonoses Public Health 58, 252–261.

Stegger, M., Liu, C.M., Larsen, J., Soldanova, K., Aziz, M., Contente-Cuomo, T., Petersen, A., Vandendriessche, S., Jimenez, J.N., Mammina, C., van Belkum, A., Salmenlinna, S., Laurent, F., Skov, R.L., Larsen, A.R., Andersen, P.S., Price, L.B., 2013. Rapid differentiation between livestock-associated and livestock-independent *Staphylococcus aureus* CC398 clades. PLoS One 8, e79645.

Stegmann, R., Perreten, V., 2010. Antibiotic resistance profile of *Staphylococcus rostri*, a new species isolated from healthy pigs. Vet. Microbiol. 145, 165–171.

Sung, J.M., Lloyd, D.H., Lindsay, J.A., 2008. *Staphylococcus aureus* host specificity: comparative genomics of human versus animal isolates by multi-strain microarray. Microbiology 154, 1949–1959.

Tavakol, M., Riekerink, R.G., Sampimon, O.C., van Wamel, W.J., van Belkum, A., Lam, T.J., 2012. Bovine-associated MRSA ST398 in The Netherlands. Acta Vet. Scand. 54, 28.

Tenhagen, B.A., Vossenkuhl, B., Kasbohrer, A., Alt, K., Kraushaar, B., Guerra, B., Schroeter, A., Fetsch, A., 2014. Methicillin-resistant *Staphylococcus aureus* in cattle food chains – prevalence, diversity, and antimicrobial resistance in Germany. J. Anim. Sci. 92, 2741–2751.

Thompson, J.K., Gibbs, P.A., Patterson, J.T., 1980. *Staphylococcus aureus* in commercial laying flocks: incidence and characteristics of strains isolated from chicks, pullets and hens in an integrated commercial enterprise. Br. Poult. Sci. 21, 315–330.

Tulinski, P., Fluit, A.C., Wagenaar, J.A., Mevius, D., van de Vijver, L., Duim, B., 2012. Methicillin-resistant coagulase-negative staphylococci on pig farms as a reservoir of heterogeneous staphylococcal cassette chromosome mec elements. Appl. Environ. Microbiol. 78, 299–304.

Valle, J., Gomez-Lucia, E., Piriz, S., Goyache, J., Orden, J.A., Vadillo, S., 1990. Enterotoxin production by staphylococci isolated from healthy goats. Appl. Environ. Microbiol. 56, 1323–1326.

van Cleef, B.A., Graveland, H., Haenen, A.P., van de Giessen, A.W., Heederik, D., Wagenaar, J.A., Kluytmans, J.A., 2011. Persistence of livestock-associated methicillin-resistant *Staphylococcus aureus* in field workers after short-term occupational exposure to pigs and veal calves. J. Clin. Microbiol. 49, 1030–1033.

van Cleef, B.A., van Benthem, B.H., Verkade, E.J., van Rijen, M.M., Kluytmans-van den Bergh, M.F., Graveland, H., Bosch, T., Verstappen, K.M., Wagenaar, J.A., Bos, M.E., Heederik, D., Kluytmans, J.A., 2015. Livestock-associated MRSA in household members of pig farmers: transmission and dynamics of carriage, a prospective cohort study. PLoS One 10, e0127190.

van Duijkeren, E., Jansen, M.D., Flemming, S.C., de Neeling, H., Wagenaar, J.A., Schoormans, A.H., van Nes, A., Fluit, A.C., 2007. Methicillin-resistant *Staphylococcus aureus* in pigs with exudative epidermitis. Emerg. Infect. Dis. 13, 1408–1410.

Vanderhaeghen, W., Hermans, K., Haesebrouck, F., Butaye, P., 2010. Methicillin-resistant *Staphylococcus aureus* (MRSA) in food production animals. Epidemiol. Infect. 138, 606–625.

Vanderhaeghen, W., Piepers, S., Leroy, F., Van Coillie, E., Haesebrouck, F., De Vliegher, S., 2014. Invited review: effect, persistence, and virulence of coagulase-negative Staphylococcus species associated with ruminant udder health. J. Dairy Sci. 97, 5275–5293.

Vautor, E., Abadie, G., Guibert, J.M., Chevalier, N., Pepin, M., 2005. Nasal carriage of *Staphylococcus aureus* in dairy sheep. Vet. Microbiol. 106, 235–239.

Velasco, V., Buyukcangaz, E., Sherwood, J.S., Stepan, R.M., Koslofsky, R.J., Logue, C.M., 2015. Characterization of *Staphylococcus aureus* from humans and a comparison with isolates of animal origin, in North Dakota, United States. PLoS One 10, e0140497.

Velasco, V., Sherwood, J.S., Rojas-Garcia, P.P., Logue, C.M., 2014. Multiplex real-time PCR for detection of *Staphylococcus aureus*, mecA and Panton-Valentine Leukocidin (PVL) genes from selective enrichments from animals and retail meat. PLoS One 9, e97617.

Viana, D., Selva, L., Segura, P., Penades, J.R., Corpa, J.M., 2007. Genotypic characterization of *Staphylococcus aureus* strains isolated from rabbit lesions. Vet. Microbiol. 121, 288–298.

Wang, D., Wang, Z., Yan, Z., Wu, J., Ali, T., Li, J., Lv, Y., Han, B., 2015. Bovine mastitis *Staphylococcus aureus*: antibiotic susceptibility profile, resistance genes and molecular typing of methicillin-resistant and methicillin-sensitive strains in China. Infect. Genet. Evol. 31, 9–16.

Wang, Y., Yang, J., Logue, C.M., Liu, K., Cao, X., Zhang, W., Shen, J., Wu, C., 2012. Methicillin-resistant *Staphylococcus pseudintermedius* isolated from canine pyoderma in North China. J. Appl. Microbiol. 112, 623–630.

Wardyn, S.E., Forshey, B.M., Farina, S.A., Kates, A.E., Nair, R., Quick, M.K., Wu, J.Y., Hanson, B.M., O'Malley, S.M., Shows, H.W., Heywood, E.M., Beane-Freeman, L.E., Lynch, C.F., Carrel, M., Smith, T.C., 2015. Swine farming is a risk factor for infection with and high prevalence of carriage of multidrug-resistant *Staphylococcus aureus*. Clin. Infect. Dis. 61, 59–66.

Weese, J.S., 2010. Methicillin-resistant *Staphylococcus aureus* in animals. ILAR J. 51, 233–244.

Weese, J.S., Avery, B.P., Reid-Smith, R.J., 2010. Detection and quantification of methicillin-resistant *Staphylococcus aureus* (MRSA) clones in retail meat products. Lett. Appl. Microbiol. 51, 338–342.

Wendlandt, S., Kadlec, K., Fessler, A.T., Mevius, D., van Essen-Zandbergen, A., Hengeveld, P.D., Bosch, T., Schouls, L., Schwarz, S., van Duijkeren, E., 2013a. Transmission of methicillin-resistant *Staphylococcus aureus* isolates on broiler farms. Vet. Microbiol. 167, 632–637.

Wendlandt, S., Schwarz, S., Silley, P., 2013b. Methicillin-resistant *Staphylococcus aureus*: a food-borne pathogen? Annu. Rev. Food Sci. Technol. 4, 117–139.

White, D.G., Harmon, R.J., Matos, J.E., Langlois, B.E., 1989. Isolation and identification of coagulase-negative Staphylococcus species from bovine body sites and streak canals of nulliparous heifers. J. Dairy Sci. 72, 1886–1892.

Zhang, L., Li, Y., Bao, H., Wei, R., Zhou, Y., Zhang, H., Wang, R., 2016. Population structure and antimicrobial profile of *Staphylococcus aureus* strains associated with bovine mastitis in China. Microb. Pathog. 97, 103–109.

Zhang, W., Hao, Z., Wang, Y., Cao, X., Logue, C.M., Wang, B., Yang, J., Shen, J., Wu, C., 2011. Molecular characterization of methicillin-resistant *Staphylococcus aureus* strains from pet animals and veterinary staff in China. Vet. J. 190, e125–129.

Zhao, Y., Liu, H., Zhao, X., Gao, Y., Zhang, M., Chen, D., 2015. Prevalence and pathogens of subclinical mastitis in dairy goats in China. Trop. Anim. Health Prod. 47, 429–435.

Zutic, M., Cirkovic, I., Pavlovic, L., Asanin, J., Jovanovic, S., Zutic, J., Asanin, R., 2012. First isolation of methicillin-resistant *Staphylococcus aureus* from pigs' clinical samples in Serbia. Acta Vet. Brno 81, 225–227.

[illegible]

[illegible]

[illegible]

[illegible]

CHAPTER

9 *STAPHYLOCOCCUS AUREUS* AND METHICILLIN-RESISTANT *STAPHYLOCOCCUS AUREUS* IN WORKERS IN THE FOOD INDUSTRY

Birgit Strommenger, Franziska Layer, Guido Werner
Robert Koch Institute, Wernigerode, Germany

1. INTRODUCTION

Staphylococcus aureus impresses with an enormous versatility. It is well known as part of the common flora on the skin and mucous membranes of mammals and approximately 20%–30% of humans are persistently colonized with *S. aureus*, mainly by mostly susceptible human-adapted isolates. In contrast, colonization with methicillin-resistant *S. aureus* (MRSA) is rare, predominantly transient and associated with prior contact to the health care system. Additionally, in recent years, community and livestock-associated *S. aureus* clones contributed to colonization in humans, latter especially in those working in close contact to farm animals.

On the other hand, *S. aureus* can cause a variety of different diseases leading from uncomplicated skin and soft tissue infections to life-threatening diseases such as septicemia. Additionally, toxin-mediated diseases such as toxic shock syndrome, staphylococcal scalded skin syndrome, and staphylococcal food poisoning (SFP) contribute to its burden of disease (Lowy, 1998).

SFP is a common foodborne disease, mediated by the ingestion of enterotoxins produced by enterotoxigenic strains of *S. aureus*. A considerable percentage of colonizing *S. aureus* isolates is equipped with enterotoxin genes. Humans carrying enterotoxigenic isolates represent a contamination source when handling food, thus generating a continuous risk of *S. aureus* food intoxication. While the majority of pathogens causing foodborne diseases infect humans (including workers in the food industry) and lead to symptomatic disease, staphylococci are resident skin inhabitants, putatively constituting a part of the skin microbiome. As such they largely remain unrecognized, until they become visible during the course of outbreak investigations related to epidemiologically linked cases of staphylococcal disease such as SFP (Todd et al., 2008).

Molecular characterization of isolates colonizing humans and obtained from food, respectively, enables the tracing of food-related outbreaks back to the source of food intoxication. However, because of the widespread occurrence of enterotoxigenic strains as human colonizers and the often transient nature of colonization the source of contamination cannot always be identified unambiguously. Therefore, compliance with hygiene measures is the most important requirement to prevent food contamination by both, human colonization and environmental *S. aureus* reservoirs.

Staphylococcus aureus. http://dx.doi.org/10.1016/B978-0-12-809671-0.00009-7

In this chapter we will summarize current knowledge about the *S. aureus* population colonizing humans, including those in close contact to animals and food, respectively. Additionally, we will review on the molecular characterization of *S. aureus* isolates related to staphylococcal foodborne disease and the elucidation of staphylococcal foodborne outbreaks.

2. *STAPHYLOCOCCUS AUREUS* AND METHICILLIN-RESISTANT *STAPHYLOCOCCUS AUREUS* COLONIZATION IN HUMANS

As a commensal bacterium *S. aureus* colonizes the skin and the mucous membranes of humans worldwide (Williams, 1963; Wertheim et al., 2005; Verhoeven et al., 2014), and it is well established that colonization predisposes for *S. aureus* generalization and subsequent endogenous infection (von Eiff et al., 2001; Wertheim et al., 2005; Paling et al., 2016). The most common site of carriage is the nasal vestibule (Williams, 1963; Kaspar et al., 2016); additionally, *S. aureus* can be present at mucosa of the oropharynx, the skin, and the perineum but also the gastrointestinal tract (Acton et al., 2009), vagina (Guinan et al., 1982), and axillae are colonized in lower frequency (Williams, 1963). However, nasal carriage is regarded as the source of colonization of secondary body sites implicating that *S. aureus* transmission via colonized hands can be reduced by adherence to commonly recommended hand hygiene measures (see Chapter 12 for further details).

Historically, *S. aureus* carriage has been separated into three classes: persistent carriers, intermittent, and noncarriers based on the presence of *S. aureus* in nasal swab cultures of an individual over time (Nouwen et al., 2004). However, in a more recent study carriage states were reclassified based on staphylococcal antibody profiles and *S. aureus* elimination from the nose (van Belkum et al., 2009) leading to only two carriage types: persistent and "other" carriers. Approximately 20%–30% of humans are presumed to be persistent carriers, whereas the remainders of the population are potentially intermittent carriers. It can be considered that most individuals are exposed to the bacterium transiently throughout lifetime (Mulcahy and McLoughlin, 2016). A number of studies have shown that persistent carriage can be associated with clonally diverse *S. aureus* strains (VandenBergh et al., 1999; Cespedes et al., 2005; Bloemendaal et al., 2009; Miller et al., 2014). Persistent carriers are characterized by an elevated nasal bacterial load and shed a higher number of bacteria into the environment, thus increasing the risk of cross-transmission within households and consequently also into the food chain if workers in the food industry are affected (Davis et al., 2012).

2.1 *STAPHYLOCOCCUS AUREUS* CARRIAGE

As reflected by the growing number of studies on *S. aureus* carriage over the recent years, carriage decreased in the general population most probably because of improvements in personal hygiene and general living conditions (Verhoeven et al., 2014; Wertheim et al., 2005). These studies also indicated that prevalence rates of *S. aureus* nasal carriage and its antimicrobial drug resistance vary slightly from country to country and in dependence of the study population.

For Central Europe carriage rates reported from the general population were in the range of 20%–35% (Wertheim and Verbrugh, 2006; Sakwinska et al., 2009; Sangvik et al., 2011; den Heijer et al., 2013; Holtfreter et al., 2016). In the US prevalences around 30% were documented (Kuehnert et al., 2006; Mainous et al., 2006; Gorwitz et al., 2008). Similar rates were found among the healthy Japanese population (Uemura et al., 2004), whereas the nasal carriage rate in a Chinese study was reported to be only 23% (Ma et al., 2011). For South and South East Asia quite variable carriage rates have been found, ranging from below 10% to more than 50% (Saxena and Panhotra, 2003; Anwar et al., 2004; Erdenizmenli et al., 2004; Lu et al., 2005; Choi et al., 2006; Severin et al., 2008; Chatterjee et al., 2009). For Australia a nasal colonization rate around 30% has been reported (Vlack et al., 2006; Munckhof et al., 2008). Recent reports from Africa also showed variable colonization rates ranging from around 10% to 30% (Adesida et al., 2007; Ben Slama et al., 2011; Ateba Ngoa et al., 2012; Omuse et al., 2012; Ouedraogo et al., 2016). In South America rather high carriage rates between 30% and 60% were reported (Hamdan-Partida et al., 2010; Ruimy et al., 2010). In several studies nasal carriage was associated with male sex and younger age (Skramm et al., 2011; Gorwitz et al., 2008; Holtfreter et al., 2016); however, in general, several different factors contribute to *S. aureus* carriage including those associated with the host, the pathogen itself, and the environment (Mulcahy and McLoughlin, 2016; Kluytmans et al., 1997).

2.2 METHICILLIN-RESISTANT *STAPHYLOCOCCUS AUREUS* IN GENERAL AND HOSPITAL-ASSOCIATED METHICILLIN-RESISTANT *STAPHYLOCOCCUS AUREUS*

MRSA colonizes the same body niches as *S. aureus*, however, in contrast to *S. aureus*, MRSA carriage, e.g., hospital-associated methicillin-resistant *S. aureus* (HA-MRSA) carriage in nonrisk populations remains rare. Depending on geographic region and study design 0%–35% of individuals are colonized with MRSA. For nine countries in Central Europe den Heijer et al. (2013) reported low MRSA prevalence in the healthy community from 0.0% (Sweden) to 2.1% (Belgium). Other studies from Europe showed similar colonization rates (Sakwinska et al., 2009; Lozano et al., 2011; Mehraj et al., 2014; Holtfreter et al., 2016). Similar or slightly elevated carriage rates were also reported in studies from Australia and several Asian countries (Munckhof et al., 2008; Chatterjee et al., 2009; Ma et al., 2011; Chen et al., 2013). Significantly higher rates were recorded in Southern European countries and the United States (up to 35%, Tavares et al., 2013; Champion et al., 2014). MRSA carriage rates are usually higher in children (Creech et al., 2005; Braga et al., 2014; Rodriguez et al., 2014); additionally, certain risk factors related to contact to health care institutions increase the probability of MRSA, particularly HA-MRSA carriage; these include contact to HA-MRSA positive patients, antibiotic therapy, hospitalization, or even contact with the health care system either as visitor or employee (Hardy et al., 2004; Coia et al., 2006). HA-MRSA, frequently exhibit multidrug-resistant phenotypes.

2.3 COMMUNITY-ASSOCIATED METHICILLIN-RESISTANT *STAPHYLOCOCCUS AUREUS*

Since the 1990s colonization and infection with community-associated methicillin-resistant *S. aureus* (CA-MRSA) became increasingly prevalent in community settings worldwide (Mediavilla et al., 2012). CA-MRSA colonize individuals independent of risk factors mentioned before; they cause a growing number of uncomplicated to severe skin and soft tissue infections but can also lead to serious invasive infections such as necrotizing pneumonia or fasciitis. Although CA-MRSA are still rare in the general population in Europe (Sangvik et al., 2011; Lozano et al., 2011; Sakwinska et al., 2009; Monecke et al., 2009; Ruimy et al., 2009), they represent a major threat to human health in the United States, where they became endemic by now (Seybold et al., 2006; Maree et al., 2007). However, elevated rates of CA-MRSA colonization were also reported from Southern Europe and the Middle East [11.4% (Tavares et al., 2013)]. The reason for the increased virulence potential of CA-MRSA is still controversially discussed; the frequent occurrence of Panton–Valentine leukocidin (PVL) in CA-MRSA strains is supposed to be a contributing factor but others are under discussion (Voyich et al., 2006; Labandeira-Rey et al., 2007; Wang et al., 2007). Most probably, the acquisition of new virulence and fitness determinants together with an altered gene expression of common virulence genes and alterations in protein sequence that increase fitness are involved (Thurlow et al., 2012). CA-MRSA are usually only resistant to few antibiotics.

2.4 LIVESTOCK-ASSOCIATED METHICILLIN-RESISTANT *STAPHYLOCOCCUS AUREUS*

During the last decade livestock has been described as an additional reservoir of MRSA beyond the hospital environment and the newly described community reservoir. Livestock-associated methicillin-resistant *S. aureus* (LA-MRSA) have been found in several food producing animals and also frequently as colonizers in humans in close contact with these animals, such as farmers and veterinarians (Cuny et al., 2015). Thus, occupational contact to livestock became an important risk factor for carriage of MRSA, i.e., LA-MRSA.

LA-MRSA, such as HA-MRSA, frequently exhibit multidrug-resistant phenotypes. The spread of these *S. aureus* isolates outside hospitals poses a substantial risk to the community because it might lead to an increase in community-acquired infections, which are difficult to treat.

3. MOLECULAR EPIDEMIOLOGY OF COLONIZING *STAPHYLOCOCCUS AUREUS* ISOLATES

3.1 CLONALITY OF *STAPHYLOCOCCUS AUREUS*

Although the *S. aureus* population as such is rather clonal, *S. aureus* carriage isolates from community-based studies have been shown to be much more diverse than MRSA (Day et al., 2001; Feil et al., 2003; Kuehnert et al., 2006; Holtfreter et al., 2016). However, some clonal lineages, such as clonal complex (CC) CC30, 45, 15, 5, 121, and 8, are more common in carriage isolates and appear to be globally disseminated. Among these clonal lineages CC30 has been reported to be predominant in a variety of studies from all over the world (Melles et al., 2008; Ko et al., 2008; Sakwinska et al., 2009; Ruimy et al., 2009) suggesting its ecological success and overall transmissibility. Less pronounced than in MRSA, the composition of CCs representing the population of nasal colonizers can vary geographically, as demonstrated in studies investigating carriage isolates from several different continents (Ruimy et al., 2008, 2009; Schaumburg et al., 2011).

3.2 CLONALITY OF HOSPITAL-ASSOCIATED METHICILLIN-RESISTANT *STAPHYLOCOCCUS AUREUS*

Some of the lineages present as frequent methicillin-sensitive *S. aureus* (MSSA) nasal colonizers have acquired staphylococcal cassette chromosome *mec* (SCC*mec*), a mobile genetic element of staphylococci carrying the *mecA* gene and leading to the emergence of MRSA within the same clonal lineage; however, the global HA-MRSA population can be assigned to a limited number of CCs, which are distributed worldwide; these include CCs CC5, 8, 22, 30, and 45 (Enright et al., 2002).

Within these CCs several lineages evolved independently over time at different geographic locations. As mentioned above the composition of clonal lineages within the HA-MRSA population in a given geographic region varies considerably. For example, ST22 and ST36 dominate in the United Kingdom, ST22 and ST225 in Germany, ST239 in Asia and Australia, and ST5 in North America, Japan, and Korea (Deurenberg and Stobberingh, 2008; Grundmann et al., 2010; Chatterjee and Otto, 2013; Bal et al., 2016). However, the distribution of clones is subject to a constant dynamic resulting in a continuously changing epidemiology (Witte et al., 2001; Wyllie et al., 2011).

3.3 CLONALITY OF COMMUNITY-ASSOCIATED METHICILLIN-RESISTANT *STAPHYLOCOCCUS AUREUS*

CA-MRSA are phylogenetically distinct from traditional HA-MRSA clones. Although they share the same clonal lineage in some instances, they evolved independently from hospital-adapted strains. The most prevalent CCs in the community are CCs 1, 8, 30, 59, and 80. MRSA strains from CCs 8 and 30 are pandemic both in the hospitals and in the community and are a common cause of infections. Similar to HA-MRSA the distribution of predominant clones differs geographically: sequence type (ST)80 is most prevalent in Europe and Northern Africa, ST59 in the Far East, ST93 and ST1 in Australia, and ST1 and ST8 in the United States (Chatterjee and Otto, 2013; Bal et al., 2016). Finally, there are some regional clones, such as ST772, which is prevalent in Bangladesh and India, ST72 strains in South Korea and Portugal, ST152 in the Balkan States, and ST88 strains in Africa and Asia (Kim et al., 2007; Shambat et al., 2012; Tavares et al., 2014; Dermota et al., 2015; Abdulgader et al., 2015).

3.4 CLONALITY OF LIVESTOCK-ASSOCIATED METHICILLIN-RESISTANT *STAPHYLOCOCCUS AUREUS*

As for CA-MRSA, LA-MRSA evolved independently from common HA- and CA-MRSA clones. LA-MRSA colonization in livestock and individuals in occupational contact in Europe is predominantly due to clonal lineage ST398 clone, whereas ST9 dominates in Asia. However, a substantial number of additional clonal lineages show apparent promiscuity presumably enabling both, long-term colonization in animals and colonization and infection of humans (Price et al., 2012). For example, MRSA isolates from poultry are often associated with ST5, which belongs to the most prevalent HA-MRSA lineages; additional MRSA lineages with supposed promiscuity are CC1, CC97, CC130, and ST425 (Fitzgerald, 2012).

4. COLONIZING *STAPHYLOCOCCUS AUREUS* AND METHICILLIN-RESISTANT *STAPHYLOCOCCUS AUREUS*: THEIR ENTEROTOXIGENIC POTENTIAL AND INVOLVEMENT IN STAPHYLOCOCCAL FOODBORNE OUTBREAKS

4.1 ENTEROTOXIN GENE CONTENT OF COLONIZING *STAPHYLOCOCCUS AUREUS* AND METHICILLIN-RESISTANT *STAPHYLOCOCCUS AUREUS*

The majority of virulence and resistance genes are located on mobile genetic elements, thus constituting a major part of the so-called "accessory genome" (Lindsay et al., 2006). It has been shown previously that elements of the accessory genome, such as enterotoxin genes, are frequently associated with particular clonal lineages (Melconian et al., 1983; Isigidi et al., 1992; Peacock et al., 2002; van Belkum et al., 2006; Lindsay et al., 2006). Several studies report on screening of *S. aureus* nasal carriage isolates for enterotoxin and enterotoxin-like genes (Becker et al., 2003; Omoe et al., 2005; Bania et al., 2006; Boerema et al., 2006; Nashev et al., 2007; Collery et al., 2008; Lozano et al., 2011). In these studies 75%–100% of investigated isolates carry enterotoxin (like) genes with 30%–67% of them being positive for the classical enterotoxin genes *sea* to *see*. The *egc* locus (complete or incomplete) was detected in 38%–84% of all isolates. Single toxin genes as well as combinations of different toxins were detected in all studies. The predominant classical enterotoxin gene varies considerably from country to country. Comparative investigation of MSSA and MRSA isolates or isolates from carriage and invasive infections revealed association of enterotoxin gene distribution with clonal lineage rather than with methicillin resistance or invasiveness.

The association between clonal lineage and toxin gene distribution was also described by Holtfreter et al. (2007) who investigated more than 200 *S. aureus* isolates from nasal colonization and from bacteremia cases for the presence of superantigen genes including enterotoxin genes in dependence of clonal lineage. Although they observed remarkably variable virulence profiles within each *S. aureus* CC, each lineage was characterized by a typical repertoire of superantigen toxin genes. For example, the *egc*-cluster enterotoxins, which cluster on the *S. aureus* genomic island vSAβ, were present in all CC5, CC22, and CC45 isolates but completely absent from CC8, CC12, CC15, and CC395. Moreover, an *egc* variant was almost exclusively linked to the CC30 background. Other enterotoxin genes with very strong linkage to certain CCs were *sec-sel*, and *sed-sej-ser*, they are colocalized on the pathogenicity island SaPI3 and were detected mainly in CC45 isolates, whereas the plasmid-borne enterotoxin genes *sed-sej-ser* were usually found in CC8. Enterotoxins with a broader distribution were the phage-borne *sea*, which was occasionally detected in CC8, -30, -45, and -395, and *seb*, on SaPI, which was infrequently found in CC5, -8, -12, -25, and -45. Similar gene distributions were also described by van Trijp et al. (2010) for strains from carriers with autologous invasive infection and by Monecke et al. (2011) for an *S. aureus* collection encompassing strains from a variety of different clonal lineages. Similar to MSSA and HA-MRSA lineages, common CA-MRSA harbor a lineage specific repertoire of enterotoxin genes; additionally, CA-MRSA lineages frequently harbor the genes for PVL (Diep et al., 2006; Tenover et al., 2006; El Garch et al., 2009; Shukla et al., 2010; Monecke et al., 2011; Portillo et al., 2013; Shore et al., 2014).

Although almost all human-associated MSSA and MRSA isolates carry enterotoxin and/or leukotoxin genes, several studies demonstrated that LA-MRSA ST398 rarely carry these genes (Köck et al., 2009; Gomez-Sanz et al., 2010; Huber et al., 2010; Hallin et al., 2011; Williamson et al., 2014; Moon et al., 2015; Normanno et al., 2015). However, only limited data are available for alternative LA-MRSA lineages. For LA-MRSA ST9 and ST5 from poultry the presence of the *egc*-cluster was demonstrated (Fessler et al., 2011; Wendlandt et al., 2013a; Monecke et al., 2013; Kraushaar et al., 2016) indicating that particular LA-MRSA lineages also might have enterotoxigenic properties (Johler et al., 2015). Beyond LA-MRSA, MSSA from different animal species have been shown to carry specific repertoires of enterotoxin genes (Fluit, 2012).

4.2 COLONIZING ENTEROTOXIGENIC *STAPHYLOCOCCUS AUREUS* AS CAUSE OF STAPHYLOCOCCAL FOODBORNE OUTBREAKS

Several studies compared isolates associated with SFPs to isolates from human carriage and/or human clinical infections to investigate the role of *S. aureus* colonizing and infecting humans as a possible source of SFP. For example, Wattinger et al. (2012) found that SFP isolates could be assigned to the same common clonal lineages (mainly CC15, CC30, CC45) and showed highly similar virulence gene profiles as isolates from nasal colonization or invasive infection indicating that contamination of foodstuff with *S. aureus* colonizing and infecting food handlers represents a main source of SFP. Interestingly, the same group found no overlap in clonal lineages and virulence gene pattern for SFP isolates and isolates obtained from milk or pork, suggesting that pork or bovine mastitis milk do not represent the most common sources of SFP (Johler et al., 2011; Baumgartner et al., 2014). In contrast, several other studies identified enterotoxigenic *S. aureus* strains from animal sources, i.e., from bovine sources, as potential cause of primary contamination (Kerouanton et al., 2007; Bianchi et al., 2014).

Regarding the high prevalence of enterotoxigenic *S. aureus* as nasal colonizer in the population worldwide and the knowledge about human carriage as a major source of food contamination, the relatively low number of notified SFP outbreaks is striking. There might be different reasons for this discrepancy:

- Although there is no doubt that nasal and hand contamination is a major source of food contamination, mere carriage is not sufficient to initiate an outbreak if proper care is taken to prevent contamination of foods including the exclusion of workers with open wounds from preparing and handling food (al Bustan et al., 1996). Many of the studies cited previously in this chapter could link high carriage rates to a lack of personal and kitchen hygiene. Especially in developing countries training in food hygiene and safety often was shown to be insufficient.
- The real incidence of SFP is probably underestimated, which is due to several reasons, including unreported minor outbreaks, improper sample collection, and laboratory examination (Argudin et al., 2010).

Besides the low number of SFP outbreaks reported, only in a limited proportion of SFP outbreaks it is possible to trace back the causing strain to its source in the nose or on the hand of a food handler. In 2013, Johler et al. investigated an SFP outbreak caused by an enterotoxin A and *egc*-cluster positive strain of CC30 that was also found in the nose of a food handler. Another SFP outbreak reported recently was caused by an enterotoxin A and I producing strain, which was introduced most likely by a food handler (Gallina et al., 2013). Similarly, Wei et al. were able to trace back enterotoxigenic outbreak strains carrying *sea* and *seb*, respectively, to the hand lesion of a food handler in a cafeteria (Wei and Chiou, 2002). However, in a large percentage of outbreaks the initial source of contamination remains unknown. This might be due to the fact that carriage is often transient and might not be detected at the time of epidemiological investigations, which take place sometime after contamination of food.

5. ARE WORKERS IN THE FOOD INDUSTRY DIFFERENT?

Generally, workers in the food industry are colonized by *S. aureus* and MRSA with the same probability as the residual population. As such they mainly carry strains of those clonal lineages, which are common nasal colonizers in the respective geographic region.

However, because of their work environment and/or occupational behavior they might be at additional risk for acquisition of *S. aureus* or MRSA of specific clonal lineages as transient or persistent colonizer. The probability of acquisition, the strain type acquired, the risk of endogenous infection with multidrug resistant *S. aureus* acquired but also the risk of transmission of enterotoxigenic *S. aureus* to food is associated with the stage of the food production chain (production, processing, distribution, or preparation, http://www.cdc.gov/foodsafety/outbreaks/investigating-outbreaks/production-chain.html) they are working at. To assess colonization rates and associated risks of infection and transmission of enterotoxigenic or multidrug-resistant isolates, a number of studies were conducted during recent years.

5.1 WORKERS EXPOSED TO LIVING ANIMALS AT SLAUGHTERHOUSES

A multitude of studies has investigated MRSA colonization in individuals in close contact with livestock, in especially farmers and veterinarians (van Cleef et al., 2014) as well as MRSA contamination of meat (de Boer et al., 2009; Kluytmans, 2010; Wendlandt et al., 2013b; Tenhagen et al., 2014; Kraushaar and Fetsch, 2014; Fetsch et al., 2017); however, although LA-MRSA from livestock can be easily transmitted to the abattoir environment and to abattoir workers, data on MRSA prevalence among abattoir workers are scarce.

Two Dutch studies documented an increased nasal MRSA carriage rate (3.2%–5.6%) in pig slaughterhouse workers in The Netherlands (Van Cleef et al., 2010; Gilbert et al., 2012). Colonization was predominantly due to LA-MRSA of CC398, which is the predominant LA-MRSA lineage among Dutch pigs. The colonization rates reported were significantly higher than the general population prevalence in The Netherlands (Wertheim et al., 2004), although they were much lower than reported from pig farmers or veterinarians with daily exposure to livestock (Cuny et al., 2009). Working with live pigs was shown to be the most important risk factor for colonization, although environmental contamination might also have played a role in the acquisition of MRSA. Mulders et al. (2010) found very similar results for the poultry food chain in The Netherlands. However, in contrast to the studies previously mentioned, LA-MRSA of the poultry-associated ST9 were frequently obtained from animals and workers in addition to CC398. LA-MRSA CC398 and ST9 were also isolated from broiler chickens and abattoir workers in Germany (Wendlandt et al., 2013a). Normanno et al. (2015) found nasal MRSA carriage in ~9% of abattoir workers screened in two industrial abattoirs in Southern Italy; MRSA strains isolated were more heterogeneous, but reflected the MRSA population isolated from slaughtered pigs during the same study, and included LA-MRSA isolates from CC398, CC1, and CC8.

Moon et al. (2015) collected data on MRSA carriage of workers in different slaughterhouses in Korea. The prevalence of MRSA in workers was 6.9% in chicken slaughterhouse workers, but no MRSA was detected in pig and cattle slaughterhouse workers. Besides LA-MRSA ST398 and ST692, isolates of Korean CA-MRSA lineage ST72 were identified in workers.

Finally, Leibler et al. (2016) recently investigated the *S. aureus* nasal carriage among beef-packing workers in Nebraska and did not find any indication of worker carriage with MRSA lineages previously associated with livestock. *S. aureus* carriage rates in general were concordant with data mentioned earlier in this chapter, and a slightly elevated MRSA carriage rate was due to human adapted isolates, mainly CA-MRSA lineage USA300.

5.2 BUTCHERS AND FOOD HANDLERS EXPOSED TO RAW MEAT

Gilbert et al. (2012) demonstrated that occupational exposure to MRSA decreased along the slaughter line, accompanied by a reduced risk of carriage for workers employed. However, on the other hand high MRSA contamination levels of retail meat were reported from several geographic regions (de Boer et al., 2009; Kluytmans, 2010; Wendlandt et al., 2013b; Tenhagen et al., 2014; Kraushaar and Fetsch, 2014; Fetsch et al., 2017). To determine to what extent butchers and other food handlers exposed to raw meat are at increased risk of colonization with *S. aureus* and MRSA, several studies were conducted focusing on these individuals. Boost et al. (2013) investigated nasal colonization in 300 pork butchers at traditional "wet markets" throughout Hong Kong, dealing with fresh meat as well as carcasses, which may increase the risk of contamination, especially if protective clothing is not worn. They found an MRSA colonization rate of 5.6% in butchers, consisting of 2.3% of individuals colonized by non-LA-MRSA and 3.3% by LA-MRSA, which was considerably higher than in the general population in Hong Kong. The majority of LA-MRSA belonged to CC9, which was previously reported from pig carcasses in Hong Kong (Guardabassi et al., 2009; Ho et al., 2012), suggesting cross-contamination from carcasses to fresh meat and workers. In contrast, de Jonge et al. (2010) found no MRSA among 95 employees working in the Dutch cold meat processing industry and in institutional kitchens, although 31 participants (33%) were colonized with MSSA. Boost et al. attributed this difference to the higher level of automation associated with improved hygiene standards in the Dutch meat processing industry in comparison to the traditional butchering process at Hong Kong wet markets.

In a follow-up study the same group (Ho et al., 2014) investigated *S. aureus*/MRSA colonization in more than 400 food handlers from six large catering establishments in Hong Kong to determine whether individuals regularly in contact with raw meat exhibit an increased risk for carriage. Overall 22.8% of food handlers were colonized, but colonization rate was significantly higher in workers handling raw meat (30%), indicating an increased risk for colonization depending on regular exposition to raw meat. However, colonization rate was still in a range previously reported for the general public in Hong Kong and elsewhere (Zhang et al., 2011). A diverse range of clonal lineages was detected among MSSA isolates, also including clonal lineages previously associated with livestock (ST5, ST1, ST9, ST130, ST97, ST398); however, the majority of isolates belonged to clonal lineages common in human nasal colonization. Only five out of 99 *S. aureus* isolates were MRSA resulting in an MRSA colonization rate of 1.2%, which is comparable to the local MRSA colonization rate previously reported (Zhang et al., 2011); all MRSA isolates belonged to clonal lineage CC45, which represents the predominant HA-MRSA in Hong Kong (Ho et al., 2009; Gruteke et al., 2015) and was not previously associated with livestock indicating human rather than animal origin.

El Bayomi et al. (2016) recently investigated hand swaps from 30 raw chicken meat vendors in Egypt and found that 60% of food handlers were positive for *S. aureus* including 10% positive for MRSA. Within the study 31% of *S. aureus* isolates (and the majority of MRSA) harbored the PVL gene, whereas 10% were positive for the sea and *sed* genes each. Although most isolates were MSSA, they exhibited antibiotic resistance to several antibiotic compounds. Genotyping revealed relatedness between MRSA isolates of human and chicken meat origin, indicating cross-transmission.

In contrast to the previous studies Nnachi et al. (2014) reported an alarming MRSA carriage rate of 51% among raw meat handlers in Nigeria, which was most probably associated with insufficient hygiene measures during handling of raw meat.

5.3 FOOD HANDLERS NOT EXPOSED TO THE PREVIOUS RISK FACTORS

The studies mentioned above suggest that food handlers not exposed to living animals or carcasses likely do not have an increased risk to carry LA-MRSA. Similarly, they are only at low risk of carrying HA-MRSA as long as they are not exposed to respective risk factors described earlier in this chapter. However, during the last decades food workers have been implicated in the occurrence of staphylococcal foodborne diseases in many different settings around the world mainly by contaminating food with methicillin sensitive, enterotoxin producing isolates originating from the handlers nose or hands. For example, Holmberg and Blake (1984) reviewed more than 100 SFP outbreaks in the United States and found the disease-causing strains in food handlers in a considerable proportion of cases. Wieneke et al. (1993) obtained similar results reviewing more than 300 outbreaks of SFP from 1969 to 1990 in the United Kingdom. To quantify the risk of food contamination by multidrug resistant or toxinogenic *S. aureus*, several studies have been conducted to determine the percentage of individuals colonized by multidrug resistant and toxin producing *S. aureus* isolates. Among these studies large variations in colonization rates, resistance rates, and in the proportions of toxinogenic *S. aureus* isolates occur. This variation may be attributed not only to geographical location of different studies, as mentioned earlier in this chapter, but also to methodological differences in study design, sampling, culturing (preenrichment vs. direct plating), typing (as far as conducted), and toxin or toxin gene detection. This immense variation makes objective comparison of presented results quite difficult. Because the problem of foodborne illness is more widespread and serious in developing countries, a great number of cross-sectional studies originate from these geographical regions (Akhtar et al., 2014).

In the following, an overview of *S. aureus* and MRSA carriage in humans in different geographical regions—Africa, Latin America, Asia, Middle East, Europe, North America and Oceania—is provided.

5.3.1 Africa

Oteri and Ekanem (1989) investigated nasal swabs from 160 food handlers in two hospital settings in Laos, Nigeria and found a carriage rate of 24% along with alarmingly poor adherence to personal hygiene measures. In a very recent study from Nigeria hand swaps from 60 food handlers at school cafeterias in Benin City revealed an *S. aureus* carriage rate of 38% (Okareh and Erhahon, 2015).

A carriage rate of 21.6% was reported for 259 food handlers, including restaurant workers, bakers, store keepers, milk distributors, butchers, and vegetable sellers, from Sudan (Saeed and Hamid, 2010); unexpectedly, carriage rates in this study varied substantially in dependence of food handler group with the highest abundance in store keepers (44.6%) and the lowest in vegetable sellers (3.6%). A similar carriage rate (20.5%) was reported from food handlers working at a university cafeteria in Ethiopia (Dagnew et al., 2012) and in 2% of the employees nasal MRSA carriage was detected; the remaining MSSA isolates were partially resistant to several antibiotics. At the same location Andargie et al. (2008) found *S. aureus* in 16.5% of 127 fingernail contents collected from food handlers.

In a study from Botswana, where 200 food handlers were sampled, also a high percentage of antibiotic resistant *S. aureus*, including MRSA, was found (Loeto et al., 2007). With a carriage rate of 57.7% a similarly high percentage of all employees were tested positive for *S. aureus* in nasal and/or skin swaps. Of the 204 *S. aureus* strains isolated, only 21% were enterotoxigenic and enterotoxin A was the most prevalent toxin. Resistance to methicillin was encountered in 33 isolates and, most alarming, for nine of the isolates elevated minimum inhibitory concentrations for vancomycin, one of the critically important antimicrobials for human medicine according to WHO (2016), was reported.

In 2012 and 2013 El-Shenawy et al. investigated the prevalence of *S. aureus* nasal and skin carriage, respectively, among 200 food handlers each, working in three different food processing plants in Egypt (milk, meat, and vegetable processing plants, respectively El-Shenawy et al., 2013, 2014). Colonization was quite similar in either of the studies (31% nasal colonization vs. 38% skin colonization), ranging from 27% to 34%, respectively, in workers in the vegetable processing plant to 36% and 45%, respectively, in workers in the milk and dairy processing plant. Although in the first study 34% of the isolates produced enterotoxins A–D or a combination thereof, in the second study only 14% of isolates tested positive for enterotoxin production. However, in both studies enterotoxins A and C were most predominant. In contrast to these results Zeinhom et al. (2015) investigated skin swaps from 75 milk and cheese handlers in Egypt and reported a carriage rate among individuals of only 17% with 5% of food handlers carrying enterotoxin B positive isolates.

5.3.2 Latin America

Cross-sectional studies from Latin America were reported predominantly from Brazil and Chile. In 1997, Soares et al. investigated *S. aureus* isolates obtained from nasal and skin swabs of 196 food handlers in community or hospital-located kitchens. 46% of employees were colonized, of which approximately two-third were colonized nasally, whereas 17% were colonized on their hands, and 18% were colonized at both locations. Pulsed field gel electrophoresis revealed a high diversity within the strain set. Of 91 carriers, 28.6% were colonized by enterotoxigenic strains. The majority of enterotoxigenic strains produced enterotoxin C, but toxins A, B, D, and E or combinations thereof were also found. Most isolates were resistant to Penicillin and Ampicillin but susceptible to other antimicrobials agents tested; however, three MRSA (1.5% of individuals) isolates were detected, which revealed multidrug resistant and were affiliated to a clonal lineage highly prevalent as HA-MRSA in Brazil; two of them likely had been acquired in the hospital environment by workers employed in the hospital-located kitchens.

Rall et al. (2010) collected samples from the hands and anterior nares of 82 food handlers in three industrial kitchens in a small town in Brazil and found 21.2% of the individuals to be colonized by *S. aureus*; 95% of *S. aureus* strains carried one of the enterotoxin genes *sea* to *sei* of which *sea* was the most frequent. However, they also found ~47% of commensal coagulase-negative staphylococci to be putatively enterotoxigenic. In a follow-up study the group analyzed 35 *S. aureus* isolates obtained by *spa*-typing and by polymerase chain reaction for the presence of the immune evasion cluster (IEC), which was present in only 10 (28.6%) strains (Baptistao et al., 2016). According to *spa*-typing the isolates were quite diverse with clonal lineages ST1, ST5, ST30, and ST45 being most prevalent. However, with respect to the high percentage of IEC-negative isolates the authors concluded that the food handlers in part might have been contaminated by IEC-negative *S. aureus* strains through food, which was also supported by the identification of *S. aureus* exhibiting *spa*-types associated with both, *S. aureus* from animals and from humans.

A much higher colonization rate was reported by da Silva Sdos et al. (2015) after investigating swabs from the anterior nostrils and hands of 30 food handlers at a university restaurant in Northeast Brazil. Forty percent of all food handlers investigated were tested positive for *S. aureus*, however, only two of them were colonized by staphylococci on their hands. Seventy five percent of these isolates carried enterotoxin genes (*sea* to *sei*), with *sei* being most prevalent. Most isolates showed sensitivity to the tested antibiotics, except for penicillin. Similar to the previous study the authors found a high percentage of commensal coagulase-negative staphylococci carrying enterotoxin genes indicating the potential to produce toxins and cause foodborne infections.

In another study conducted in Brazil, Ferreira et al. (2014) sampled 146 food handlers from 14 public hospitals in northeastern Brazil. The results indicated that 50% and 29% of them had coagulase-positive staphylococci on their hands and in their anterior nares, respectively. The majority of isolates were resistant to Penicillin, and 29% of the isolates were phenotypically identified as MRSA; additionally, resistances to other clinically important antibiotics, including vancomycin, were reported by the authors.

In a study conducted in Chile, Soto et al. (1996) screened 87 food handlers working at a metropolitan university in Chile for *S. aureus* carriage by swabbing several body sites. They found a carriage rate of 65.5%, with 41% of food handlers carrying enterotoxigenic strains. Enterotoxin B was the predominant toxin type detected. In contrast, Figueroa et al. (2002) found *S. aureus* colonization in only 34% of 102 food handlers from 19 restaurants in Santiago with 19% of them carrying enterotoxin producing isolates, predominantly of enterotoxin type A.

In a single study conducted in Argentina nasal swabs were obtained from 88 food handlers (Jorda et al., 2012). A total of 37.5% food handlers were positive for *S. aureus* with four (4.5%) of them carrying MRSA. Enterotoxin genes were found in 14.7% of food handlers. Enterotoxin gene *sea* was most frequently detected.

5.3.3 Asia

In Hong Kong, Ho et al. (2015b) conducted a longitudinal study of nasal colonization and hand contamination of food handlers with *S. aureus* starting in 2002 and lasting until 2011. Within this time frame hygiene measures were strictly implemented in 2003 in Hong Kong because of the outbreak of severe acute respiratory syndrome (SARS). The study started in 2002 with samples from 619 food handlers at 15 catering establishments, including supermarkets, canteens, and centralized kitchens serving hospitals and sporting facilities. Fourteen of these premises were revisited in 2003 and a further 527 samples were taken; 499 participants were sampled at both occasions. In 2011, 434 food handlers were sampled from six large catering establishments, similar to those mentioned above. In 2002 the authors found a nasal carriage rate of 35% and a hand contamination rate of 41%; in 2003 both, nasal and hand colonization rates were significantly reduced to 24% and 12%, respectively. In 2011 the nasal carriage rate remained stable (23%), whereas the hand contamination rate was further reduced to 4%. Most hand contamination was attributable to nasal isolates of persistently colonized (co-) workers who had presumably contaminated the environment (Ho et al., 2015c). Handling of cooked and raw meat was shown to increase the risk for colonization, as already previously anticipated (Ho et al., 2014). MRSA carriage was found in 1.2% in 2011 compared with 0.6% and 0.8% in 2002 and 2003, respectively. *Spa*-typing of isolates revealed a high diversity with isolates of clonal lineages ST188, CC15, and CC30 predominating and indicated a certain strain dynamic over time. Although MRSA isolates were diverse in 2002 and 2003, they were dominated by isolates of t1081/CC45, which was also the most common HA-MRSA in Hong Kong (Gruteke et al., 2015; Ho et al., 2009). Typing also enabled to estimate the rate of persistent carriers (18%). The considerable decrease in colonization rates, which was mainly due to a reduction of transient carriers, was attributed to the enhanced hygiene practices implemented during the SARS epidemic. These included, for example, the mandatory use of gloves and masks and the increased emphasis on hand hygiene in food handlers. Carriage rates were similar to those reported from the general population over the same time span (O'Donoghue and Boost, 2004; Zhang et al., 2011). Isolates originating from the study were subsequently analyzed for the presence of enterotoxin genes *sea* to *sej*, *sek* to *ser* and *selu*, *ses*, and *set* (Ho et al., 2015a). Enterotoxin and enterotoxin-like genes were detected in 83%, 82%, and 80% of isolates, respectively. They were more often detected in isolates of persistent carriers, with *sea*, *seb*, and *sem* found to be statistically associated with persistent carriage. The most common enterotoxin genes were those associated with the *egc*-cluster; however, of the classical enterotoxin genes, *sea* and *see* were most prevalent. In contrast to other studies (Holtfreter et al., 2007) no association was found between clonal lineage and the presence of enterotoxin determinants.

Tan et al. (2014) investigated hand carriage rates in 85 food handlers at primary schools in Malaysia and found a mean carriage rate of 70%, depending on the time of sampling (before/during/after preparation of food). Of the *S. aureus* isolates obtained in this study 72% showed antibiotic resistance to at least one antibiotic tested, and a large proportion of isolates was resistant to ampicillin and penicillin; however, multidrug resistance was rare.

5.3.4 Middle East

In a cross-sectional study conducted in Anatolia and encompassing 299 food handlers Simsek et al. (2009) found 23.1% of individuals to be colonized with *S. aureus*. A second study performed in the same geographical area enrolled 30 food handlers who worked in catering services (Vatansever et al., 2016). Authors found that 36% of individuals showed nasal *S. aureus* carriage, whereas mouth and hand carriage were considerably lower (19% and 14%, respectively). Although no MRSA isolate was found, several isolates with resistance to critically important antimicrobials, including one multidrug-resistant isolate were found.

Al-Bustan et al. (1996) screened 500 restaurant workers in 100 restaurants of Kuwait City and found an *S. aureus* nasal carriage rate of 26.6%, which was low in comparison to the general public and the hospital population in the same region; they also looked for the strains ability of enterotoxin A–D production and found that 86.6% of isolates were able to produce either one or a combination of several toxins, with enterotoxin A and B being most prevalent. In a later study Udo et al. (2009) characterized 200 isolates obtained from nasal, skin, stool, and throat swaps from 250 food handlers working in 50 different restaurants in Kuwait City. Workers were sampled during mandatory screening or in context of outbreak associated epidemiological investigations. The authors found a much higher colonization rate (53.2%) than in the previous study; further (molecular) characterization revealed highly diverse strains and 92.5% of isolates expressing antibacterial resistance; two-third were unsusceptible to at least two different bacterial agents, and 23% were multidrug resistant, including one MRSA isolate. 71% of isolates were positive for enterotoxin genes (*sea* to *sei*), with *sei* being the most prevalent gene, and almost two-third of enterotoxigenic isolates carried more than one enterotoxin gene. Additionally, the authors found genes for PVL in 9% of the isolates, including the MRSA isolate indicating the presence of CA-MRSA.

In one of the most recent cross-sectional studies, conducted in Iran, 220 food handlers attending a public health center laboratory for annual checkup were examined by culturing nasal and fingernail content swaps. 65.4% and 46% of individuals were found to harbor *S. aureus* in their nostrils and in their fingernail content, respectively (Nasrolahei et al., 2016).

5.3.5 Europe, North America, and Oceania

Cross-sectional studies from Central Europe, North America, or Oceania focusing on *S. aureus* carriage in individuals working in the food industry are scarce. Screening of food handlers is conducted mostly regarding cases of SFP, as part of subsequent epidemiological investigations. In 2000, Finnish researchers reported about hand and nasal samples of flight catering staff taken between 1995 and 1997 (Hatakka et al., 2000). 29% of the nasal samples and 9% of the hand samples were positive for *S. aureus*. 46% of the strains were enterotoxigenic, resulting in 12% and 6% of individuals carrying enterotoxigenic isolates in their nose and on their hands, respectively. Enterotoxin B was the most prevalent toxin within the highly diverse collection of isolates.

Uzunović et al. performed a laboratory-based study on all consecutive, nonduplicate *S. aureus* strains isolated from nasal swabs of food handlers in Bosnia-Herzegovina. Swabs were taken between 2007 and 2009 twice a year as part of regulatory mandatory surveillance for all persons dealing with food. Only 189 nonduplicate *S. aureus* isolates (including three MRSA) were recovered out of 13,690 nasal swabs resulting in an unexpectedly low average nasal carriage rate of 1.4%; however, all isolates were *spa*-typed, thus reflecting at least a snapshot of the clonal diversity of nasal carriage. *Spa*-typing revealed a high diversity of the population with the common clonal lineages, CC15, CC22, CC7, CC30, and CC5 being most prevalent (Uzunovic et al., 2013).

Castro et al. sampled nose and hands of 162 volunteers from a food processing company in Portugal. Nasal and hand carriage were found in 19.85% and 11.1%, respectively, with 6.2% of individuals being colonized at both sites. 82% of the isolates were resistant to at least one antibiotic tested, but no MRSA were detected and multidrug-resistant *S. aureus* were rare (Castro et al., 2016). The majority of isolates detected in this study (71.9% of nasal and 56% of hand isolates) were enterotoxigenic, with *egc*-located *seg* and *sei* being the most prevalent enterotoxin genes.

Recently, Caggiano et al. (2016) published a study on nasal screening of 323 healthy workers employed in the pasta and pork industry in Italy. *S. aureus* and MRSA carriage rate was 26.3% and 2.2%, respectively. Workers employed in the pasta industry had higher carriage rates than those employed in the pork industry (61.2% and 3.7% vs. 38.8% and 0.6%). Two MRSA isolates from pasta industry workers were assigned to CC398.

6. SUMMARY AND CONCLUSION

The impact of *S. aureus* and MRSA in workers in the food industry is not only manifold but also highly variable. For instance, employees working at the beginning of the food chain are at increased risk for carrying MRSA, especially if they are handling living animal. MRSA carriage in these individuals is mostly associated with LA-MRSA lineages present in the respective animal population. However, LA-MRSA carriage rate is significantly lower in food industry workers in comparison to farmers, veterinarians, etc. Nevertheless, it implies a certain risk for endogenous MRSA infections and for spreading LA-MRSA to the worker's household members. Data on MRSA and MSSA carriage in butchers are controversial. The probability of MRSA colonization seems to be associated with the degree of automation within the slaughter process as increased automation also implies improved hygiene standards. MRSA lineages found in these individuals mostly include LA-MRSA, however, colonization with CA-MRSA was also reported. Carriage is most probably associated with insufficient hygiene measures during handling of raw meat and in personal hygiene. Additionally, a few studies also report a slightly increased risk for MSSA colonization.

In general, occupational exposure to MRSA decreases along the slaughter line, accompanied by a reduced risk of MRSA carriage for workers employed. Food handlers without occupational contact to living animals and not exposed to "classical" risk factors for MRSA colonization (e.g., working in the hospital environment) are not at increased risk for MRSA carriage, which is also reflected by the overall low MRSA carriage rates (mostly below 5%) reported in the studies summarized above.

As expected, MSSA colonization is common in workers in the food industry and MSSA carriage rates vary considerably (~20% to ~60%). This is in accordance to MSSA carriage in the general population, as outlined previously in this chapter. Also, the percentage of enterotoxigenic isolates (~20% to 90%) and the distribution of enterotoxin types (as far as investigated) is subject to an enormous variation, which is also seen in the general population. Although MRSA carriage is rare in most reports, the majority of isolates tested for their susceptibility exhibit antibiotic resistances to several antibiotics usually including penicillin.

Although controversially discussed in several studies, nasal colonization seems to be more common than contamination of hands. It is anticipated that persistently colonized individuals contaminate their hands as well as the environment and serve as the source for hand contamination of noncarriers. Unfortunately, most studies did not include comparable typing data for carriage isolates. However, existing typing data together with resistance and enterotoxin gene profiles indicate that MSSA carriage in food handling individuals is highly diverse and is due to the same clonal lineages as in the general population within a given geographic region. However, this implies that food handlers might also be colonized with CA-MRSA strains, which can be spread via the food chain because of improper personal and process hygiene. As a consequence, compliance with hygiene measures is the most important prerequisite to prevent food contamination by both, human and environmental *S. aureus* reservoirs.

REFERENCES

Abdulgader, S.M., Shittu, A.O., Nicol, M.P., Kaba, M., 2015. Molecular epidemiology of methicillin-resistant *Staphylococcus aureus* in Africa: a systematic review. Front. Microbiol. 6, 348.

Acton, D.S., Plat-Sinnige, M.J., van Wamel, W., de Groot, N., van Belkum, A., 2009. Intestinal carriage of *Staphylococcus aureus*: how does its frequency compare with that of nasal carriage and what is its clinical impact? Eur. J. Clin. Microbiol. Infect. Dis. 28, 115–127.

Adesida, S.A., Abioye, O.A., Bamiro, B.S., Brai, B.I., Smith, S.I., Amisu, K.O., Ehichioya, D.U., Ogunsola, F.T., Coker, A.O., 2007. Associated risk factors and pulsed field gel electrophoresis of nasal isolates of *Staphylococcus aureus* from medical students in a tertiary hospital in Lagos, Nigeria. Braz. J. Infect. Dis. 11, 63–69.

Akhtar, S., Sarker, M.R., Hossain, A., 2014. Microbiological food safety: a dilemma of developing societies. Crit. Rev. Microbiol. 40, 348–359.

al Bustan, M.A., Udo, E.E., Chugh, T.D., 1996. Nasal carriage of enterotoxin-producing *Staphylococcus aureus* among restaurant workers in Kuwait city. Epidemiol. Infect. 116, 319–322.

Andargie, G., Kassu, A., Moges, F., Tiruneh, M., Huruy, K., 2008. Prevalence of bacteria and intestinal parasites among food-handlers in Gondar town, northwest Ethiopia. J. Health Popul. Nutr. 26, 451–455.

Anwar, M.S., Jaffery, G., Rehman Bhatti, K.U., Tayyib, M., Bokhari, S.R., 2004. *Staphylococcus aureus* and MRSA nasal carriage in general population. J. Coll. Physicians Surg. Pak 14, 661–664.

Argudin, M.A., Mendoza, M.C., Rodicio, M.R., 2010. Food poisoning and *Staphylococcus aureus* enterotoxins. Toxins (Basel) 2, 1751–1773.

Ateba Ngoa, U., Schaumburg, F., Adegnika, A.A., Kosters, K., Moller, T., Fernandes, J.F., Alabi, A., Issifou, S., Becker, K., Grobusch, M.P., Kremsner, P.G., Lell, B., 2012. Epidemiology and population structure of *Staphylococcus aureus* in various population groups from a rural and semi urban area in Gabon, Central Africa. Acta Trop. 124, 42–47.

Bal, A.M., Coombs, G.W., Holden, M.T., Lindsay, J.A., Nimmo, G.R., Tattevin, P., Skov, R.L., 2016. Genomic insights into the emergence and spread of international clones of healthcare-, community- and livestock-associated meticillin-resistant *Staphylococcus aureus*: blurring of the traditional definitions. J. Glob. Antimicrob. Resist. 6, 95–101.

Bania, J., Dabrowska, A., Korzekwa, K., Zarczynska, A., Bystron, J., Chrzanowska, J., Molenda, J., 2006. The profiles of enterotoxin genes in *Staphylococcus aureus* from nasal carriers. Lett. Appl. Microbiol. 42, 315–320.

Baptistao, L.G., Silva, N.C., Bonsaglia, E.C., Rossi, B.F., Castilho, I.G., Fernandes Junior, A., Rall, V.L., 2016. Presence of immune evasion cluster and molecular typing of methicillin-susceptible *Staphylococcus aureus* isolated from food handlers. J. Food Prot. 79, 682–686.

Baumgartner, A., Niederhauser, I., Johler, S., 2014. Virulence and resistance gene profiles of *staphylococcus aureus* strains isolated from ready-to-eat foods. J. Food Prot. 77, 1232–1236.

Becker, K., Friedrich, A.W., Lubritz, G., Weilert, M., Peters, G., von Eiff, C., 2003. Prevalence of genes encoding pyrogenic toxin superantigens and exfoliative toxins among strains of *Staphylococcus aureus* isolated from blood and nasal specimens. J. Clin. Microbiol. 41, 1434–1439.

Ben Slama, K., Gharsa, H., Klibi, N., Jouini, A., Lozano, C., Gomez-Sanz, E., Zarazaga, M., Boudabous, A., Torres, C., 2011. Nasal carriage of *Staphylococcus aureus* in healthy humans with different levels of contact with animals in Tunisia: genetic lineages, methicillin resistance, and virulence factors. Eur. J. Clin. Microbiol. Infect. Dis. 30, 499–508.

Bianchi, D.M., Gallina, S., Bellio, A., Chiesa, F., Civera, T., Decastelli, L., 2014. Enterotoxin gene profiles of *Staphylococcus aureus* isolated from milk and dairy products in Italy. Lett. Appl. Microbiol. 58, 190–196.

Bloemendaal, A.L., Fluit, A.C., Jansen, W.T., Vriens, M.R., Ferry, T., Amorim, J.M., Pascual, A., Stefani, S., Papaparaskevas, J., Borel Rinkes, I.H., Verhoef, J., 2009. Colonization with multiple *Staphylococcus aureus* strains among patients in European intensive care units. Infect. Control Hosp. Epidemiol. 30, 918–920.

Boerema, J.A., Clemens, R., Brightwell, G., 2006. Evaluation of molecular methods to determine enterotoxigenic status and molecular genotype of bovine, ovine, human and food isolates of *Staphylococcus aureus*. Int. J. Food Microbiol. 107, 192–201.

Boost, M., Ho, J., Guardabassi, L., O'Donoghue, M., 2013. Colonization of butchers with livestock-associated methicillin-resistant *Staphylococcus aureus*. Zoonoses Public Health 60, 572–576.

Braga, E.D., Aguiar-Alves, F., de Freitas Mde, F., de e Silva, M.O., Correa, T.V., Snyder, R.E., de Araujo, V.A., Marlow, M.A., Riley, L.W., Setubal, S., Silva, L.E., Araujo Cardoso, C.A., 2014. High prevalence of *Staphylococcus aureus* and methicillin-resistant S. aureus colonization among healthy children attending public daycare centers in informal settlements in a large urban center in Brazil. BMC Infect. Dis. 14, 538.

Caggiano, G., Dambrosio, A., Ioanna, F., Balbino, S., Barbuti, G., De Giglio, O., Diella, G., Lovero, G., Rutigliano, S., Scarafile, G., Baldassarre, A., Vimercati, L., Musti, M., Montagna, M.T., 2016. Prevalence and characterization of methicillin-resistant *Staphylococcus aureus* isolates in food industry workers. Ann. Ig. 28, 8–14.

Castro, A., Santos, C., Meireles, H., Silva, J., Teixeira, P., 2016. Food handlers as potential sources of dissemination of virulent strains of *Staphylococcus aureus* in the community. J. Infect. Public Health 9, 153–160.

Cespedes, C., Said-Salim, B., Miller, M., Lo, S.H., Kreiswirth, B.N., Gordon, R.J., Vavagiakis, P., Klein, R.S., Lowy, F.D., 2005. The clonality of *Staphylococcus aureus* nasal carriage. J. Infect. Dis. 191, 444–452.

Champion, A.E., Goodwin, T.A., Brolinson, P.G., Werre, S.R., Prater, M.R., Inzana, T.J., 2014. Prevalence and characterization of methicillin-resistant *Staphylococcus aureus* isolates from healthy university student athletes. Ann. Clin. Microbiol. Antimicrob. 13, 33.

Chatterjee, S.S., Otto, M., 2013. Improved understanding of factors driving methicillin-resistant *Staphylococcus aureus* epidemic waves. Clin. Epidemiol. 5, 205–217.

Chatterjee, S.S., Ray, P., Aggarwal, A., Das, A., Sharma, M., 2009. A community-based study on nasal carriage of *Staphylococcus aureus*. Indian J. Med. Res. 130, 742–748.

Chen, C.J., Wang, S.C., Chang, H.Y., Huang, Y.C., 2013. Longitudinal analysis of methicillin-resistant and methicillin-susceptible *Staphylococcus aureus* carriage in healthy adolescents. J. Clin. Microbiol. 51, 2508–2514.

Choi, C.S., Yin, C.S., Bakar, A.A., Sakewi, Z., Naing, N.N., Jamal, F., Othman, N., 2006. Nasal carriage of *Staphylococcus aureus* among healthy adults. J. Microbiol. Immunol. Infect. 39, 458–464.

Coia, J.E., Duckworth, G.J., Edwards, D.I., Farrington, M., Fry, C., Humphreys, H., Mallaghan, C., Tucker, D.R., Chemotherapy Joint Working Party of the British Society of Antimicrobial, Society Hospital Infection, Association Infection Control Nurses, 2006. Guidelines for the control and prevention of meticillin-resistant *Staphylococcus aureus* (MRSA) in healthcare facilities. J. Hosp. Infect. 63 (Suppl. 1), S1–S44.

Collery, M.M., Smyth, D.S., Twohig, J.M., Shore, A.C., Coleman, D.C., Smyth, C.J., 2008. Molecular typing of nasal carriage isolates of *Staphylococcus aureus* from an Irish university student population based on toxin gene PCR, agr locus types and multiple locus, variable number tandem repeat analysis. J. Med. Microbiol. 57, 348–358.

Creech 2nd, C.B., Kernodle, D.S., Alsentzer, A., Wilson, C., Edwards, K.M., 2005. Increasing rates of nasal carriage of methicillin-resistant *Staphylococcus aureus* in healthy children. Pediatr. Infect. Dis. J. 24, 617–621.

Cuny, C., Nathaus, R., Layer, F., Strommenger, B., Altmann, D., Witte, W., 2009. Nasal colonization of humans with methicillin-resistant *Staphylococcus aureus* (MRSA) CC398 with and without exposure to pigs. PLoS One 4, e6800.

Cuny, C., Wieler, L.H., Witte, W., 2015. Livestock-associated MRSA: the impact on humans. Antibiotics (Basel) 4, 521–543.

da Silva Sdos, S., Cidral, T.A., Soares, M.J., de Melo, M.C., 2015. Enterotoxin-Encoding genes in *Staphylococcus* spp. from food handlers in a university restaurant. Foodborne Pathog. Dis. 12, 921–925.

Dagnew, M., Tiruneh, M., Moges, F., Tekeste, Z., 2012. Survey of nasal carriage of *Staphylococcus aureus* and intestinal parasites among food handlers working at Gondar University, Northwest Ethiopia. BMC Public Health 12, 837.

Davis, M.F., Iverson, S.A., Baron, P., Vasse, A., Silbergeld, E.K., Lautenbach, E., Morris, D.O., 2012. Household transmission of meticillin-resistant *Staphylococcus aureus* and other staphylococci. Lancet Infect. Dis. 12, 703–716.

Day, N.P., Moore, C.E., Enright, M.C., Berendt, A.R., Smith, J.M., Murphy, M.F., Peacock, S.J., Spratt, B.G., Feil, E.J., 2001. A link between virulence and ecological abundance in natural populations of *Staphylococcus aureus*. Science 292, 114–116.

de Boer, E., Zwartkruis-Nahuis, J.T., Wit, B., Huijsdens, X.W., de Neeling, A.J., Bosch, T., van Oosterom, R.A., Vila, A., Heuvelink, A.E., 2009. Prevalence of methicillin-resistant *Staphylococcus aureus* in meat. Int. J. Food Microbiol. 134, 52–56.

de Jonge, R., Verdier, J.E., Havelaar, A.H., 2010. Prevalence of meticillin-resistant *Staphylococcus aureus* amongst professional meat handlers in The Netherlands, March–July 2008. Euro Surveill. 15.

den Heijer, C.D., van Bijnen, E.M., Paget, W.J., Pringle, M., Goossens, H., Bruggeman, C.A., Schellevis, F.G., Stobberingh, E.E., Apres Study Team, 2013. Prevalence and resistance of commensal *Staphylococcus aureus*, including meticillin-resistant *S aureus*, in nine European countries: a cross-sectional study. Lancet Infect. Dis. 13, 409–415.

Dermota, U., Mueller-Premru, M., Svent-Kucina, N., Petrovic, Z., Ribic, H., Rupnik, M., Janezic, S., Zdovc, I., Grmek-Kosnik, I., 2015. Survey of community-associated-methicillin-Resistant *Staphylococcus aureus* in Slovenia: identification of community-associated and livestock-associated clones. Int. J. Med. Microbiol. 305, 505–510.

Deurenberg, R.H., Stobberingh, E.E., 2008. The evolution of *Staphylococcus aureus*. Infect. Genet. Evol. 8, 747–763.

Diep, B.A., Carleton, H.A., Chang, R.F., Sensabaugh, G.F., Perdreau-Remington, F., 2006. Roles of 34 virulence genes in the evolution of hospital- and community-associated strains of methicillin-resistant *Staphylococcus aureus*. J. Infect. Dis. 193, 1495–1503.

El-Shenawy, M., El-Hosseiny, L., Tawfeek, M., El-Shenawy, M., Baghdadi, H., Saleh, O., Mañes, J., Soriano, J.M., 2013. Nasal carriage of enterotoxigenic *Staphylococcus aureus* and risk factors among food handlers-Egypt. Food Public Health 3, 284–288.

El-Shenawy, M., Tawfeek, M., El-Hosseiny, L., El-Shenawy, M., Farag, A., Baghdadi, H., Saleh, O., Mañes, J., Soriano, J.M., 2014. Cross sectional study of skin carriage and enterotoxigenicity of *Staphylococcus aureus* among food handlers. Open J. Med. Microbiol. 16–22.

El Bayomi, R.M., Ahmed, H.A., Awadallah, M.A., Mohsen, R.A., Abd El-Ghafar, A.E., Abdelrahman, M.A., 2016. Occurrence, virulence factors, antimicrobial resistance, and genotyping of *Staphylococcus aureus* strains isolated from chicken products and humans. Vector Borne Zoonotic Dis. 16, 157–164.

El Garch, F., Hallin, M., De Mendonca, R., Denis, O., Lefort, A., Struelens, M.J., 2009. StaphVar-DNA microarray analysis of accessory genome elements of community-acquired methicillin-resistant *Staphylococcus aureus*. J. Antimicrob. Chemother. 63, 877–885.

Enright, M.C., Robinson, D.A., Randle, G., Feil, E.J., Grundmann, H., Spratt, B.G., 2002. The evolutionary history of methicillin-resistant *Staphylococcus aureus* (MRSA). Proc. Natl. Acad. Sci. U.S.A. 99, 7687–7692.

Erdenizmenli, M., Yapar, N., Senger, S.S., Ozdemir, S., Yuce, A., 2004. Investigation of colonization with methicillin-resistant and methicillin-susceptible *Staphylococcus aureus* in an outpatient population in Turkey. Jpn. J. Infect. Dis. 57, 172–175.

Feil, E.J., Cooper, J.E., Grundmann, H., Robinson, D.A., Enright, M.C., Berendt, T., Peacock, S.J., Smith, J.M., Murphy, M., Spratt, B.G., Moore, C.E., Day, N.P., 2003. How clonal is *Staphylococcus aureus*? J. Bacteriol. 185, 3307–3316.

Ferreiraa, J.S., Costab, W.L.R., Cerqueirac, E.S., Carvalhoc, J.S., Oliveirac, L.C., Almeidac, R.C.C., 2014. Food handler-associated methicillin-resistant *Staphylococcus aureus* in public hospitals in Salvador, Brazil. Food Control 37.

Fessler, A.T., Kadlec, K., Hassel, M., Hauschild, T., Eidam, C., Ehricht, R., Monecke, S., Schwarz, S., 2011. Characterization of methicillin-resistant *Staphylococcus aureus* isolates from food and food products of poultry origin in Germany. Appl. Environ. Microbiol. 77, 7151–7157.

Fetsch, A., Kraushaar, B., Kasbohrer, A., Hammerl, J.A., 2017. Turkey meat as source of CC9/CC398 methicillin-resistant *Staphylococcus aureus* in humans? Clin. Infect. Dis. 64, 102–103.

Figueroa, G., Navarrete, P., Caro, M., Troncoso, M., Faundez, G., 2002. Carriage of enterotoxigenic *Staphylococcus aureus* in food handlers. Rev. Med. Chil. 130, 859–864.

Fitzgerald, J.R., 2012. Human origin for livestock-associated methicillin-resistant *Staphylococcus aureus*. MBio 3 e00082–12.

Fluit, A.C., 2012. Livestock-associated *Staphylococcus aureus*. Clin. Microbiol. Infect. 18, 735–744.

Gallina, S., Bianchi, D.M., Bellio, A., Nogarol, C., Macori, G., Zaccaria, T., Biorci, F., Carraro, E., Decastelli, L., 2013. Staphylococcal poisoning foodborne outbreak: epidemiological investigation and strain genotyping. J. Food Prot. 76, 2093–2098.

Gilbert, M.J., Bos, M.E., Duim, B., Urlings, B.A., Heres, L., Wagenaar, J.A., Heederik, D.J., 2012. Livestock-associated MRSA ST398 carriage in pig slaughterhouse workers related to quantitative environmental exposure. Occup. Environ. Med. 69, 472–478.

Gomez-Sanz, E., Torres, C., Lozano, C., Fernandez-Perez, R., Aspiroz, C., Ruiz-Larrea, F., Zarazaga, M., 2010. Detection, molecular characterization, and clonal diversity of methicillin-resistant *Staphylococcus aureus* CC398 and CC97 in Spanish slaughter pigs of different age groups. Foodborne Pathog. Dis. 7, 1269–1277.

Gorwitz, R.J., Kruszon-Moran, D., McAllister, S.K., McQuillan, G., McDougal, L.K., Fosheim, G.E., Jensen, B.J., Killgore, G., Tenover, F.C., Kuehnert, M.J., 2008. Changes in the prevalence of nasal colonization with *Staphylococcus aureus* in the United States, 2001–2004. J. Infect. Dis. 197, 1226–1234.

Grundmann, H., Aanensen, D.M., van den Wijngaard, C.C., Spratt, B.G., Harmsen, D., Friedrich, A.W., Group European Staphylococcal Reference Laboratory Working, 2010. Geographic distribution of *Staphylococcus aureus* causing invasive infections in Europe: a molecular-epidemiological analysis. PLoS Med. 7, e1000215.

Gruteke, P., Ho, P.L., Haenen, A., Lo, W.U., Lin, C.H., de Neeling, A.J., 2015. MRSA spa t1081, a highly transmissible strain endemic to Hong Kong, China, in The Netherlands. Emerg. Infect. Dis. 21, 1074–1076.

Guardabassi, L., O, M., Donoghue, Moodley, A., Ho, J., Boost, M., 2009. Novel lineage of methicillin-resistant *Staphylococcus aureus*, Hong Kong. Emerg. Infect. Dis. 15, 1998–2000.

Guinan, M.E., Dan, B.B., Guidotti, R.J., Reingold, A.L., Schmid, G.P., Bettoli, E.J., Lossick, J.G., Shands, K.N., Kramer, M.A., Hargrett, N.T., Anderson, R.L., Broome, C.V., 1982. Vaginal colonization with *Staphylococcus aureus* in healthy women: a review of four studies. Ann. Intern. Med. 96, 944–947.

Hallin, M., De Mendonca, R., Denis, O., Lefort, A., El Garch, F., Butaye, P., Hermans, K., Struelens, M.J., 2011. Diversity of accessory genome of human and livestock-associated ST398 methicillin resistant *Staphylococcus aureus* strains. Infect. Genet. Evol. 11, 290–299.

Hamdan-Partida, A., Sainz-Espunes, T., Bustos-Martinez, J., 2010. Characterization and persistence of *Staphylococcus aureus* strains isolated from the anterior nares and throats of healthy carriers in a Mexican community. J. Clin. Microbiol. 48, 1701–1705.

Hardy, K.J., Hawkey, P.M., Gao, F., Oppenheim, B.A., 2004. Methicillin resistant *Staphylococcus aureus* in the critically ill. Br. J. Anaesth. 92, 121–130.

Hatakka, M., Bjorkroth, K.J., Asplund, K., Maki-Petays, N., Korkeala, H.J., 2000. Genotypes and enterotoxicity of *Staphylococcus aureus* isolated from the hands and nasal cavities of flight-catering employees. J. Food Prot. 63, 1487–1491.

Ho, J., Boost, M., O'Donoghue, M., 2015a. Prevalence of enterotoxin genes in *Staphylococcus aureus* colonising food handlers: does nasal carriage status matter? Eur. J. Clin. Microbiol. Infect. Dis. 34, 2177–2181.

Ho, J., Boost, M., O'Donoghue, M., 2015b. Sustainable reduction of nasal colonization and hand contamination with *Staphylococcus aureus* in food handlers, 2002–2011. Epidemiol. Infect. 143, 1751–1760.

Ho, J., Boost, M.V., O'Donoghue, M.M., 2015c. Tracking sources of *Staphylococcus aureus* hand contamination in food handlers by spa typing. Am. J. Infect. Control 43, 759–761.

Ho, J., O'Donoghue, M., Guardabassi, L., Moodley, A., Boost, M., 2012. Characterization of methicillin-resistant *Staphylococcus aureus* isolates from pig carcasses in Hong Kong. Zoonoses Public Health 59, 416–423.

Ho, J., O'Donoghue, M.M., Boost, M.V., 2014. Occupational exposure to raw meat: a newly-recognized risk factor for *Staphylococcus aureus* nasal colonization amongst food handlers. Int. J. Hyg. Environ. Health 217, 347–353.

Ho, P.L., Chow, K.H., Lo, P.Y., Lee, K.F., Lai, E.L., 2009. Changes in the epidemiology of methicillin-resistant *Staphylococcus aureus* associated with spread of the ST45 lineage in Hong Kong. Diagn. Microbiol. Infect. Dis. 64, 131–137.

Holmberg, S.D., Blake, P.A., 1984. Staphylococcal food poisoning in the United States. New facts and old misconceptions. JAMA 251, 487–489.

Holtfreter, S., Grumann, D., Balau, V., Barwich, A., Kolata, J., Goehler, A., Weiss, S., Holtfreter, B., Bauerfeind, S.S., Doring, P., Friebe, E., Haasler, N., Henselin, K., Kuhn, K., Nowotny, S., Radke, D., Schulz, K., Schulz, S.R., Trube, P., Hai Vu, C., Walther, B., Westphal, S., Cuny, C., Witte, W., Volzke, H., Grabe, H.J., Kocher, T., Steinmetz, I., Broker, B.M., 2016. Molecular epidemiology of *Staphylococcus aureus* in the general population in northeast Germany – results of the study of health in Pomerania (SHIP-TREND-0). J. Clin. Microbiol. 54.

Holtfreter, S., Grumann, D., Schmudde, M., Nguyen, H.T., Eichler, P., Strommenger, B., Kopron, K., Kolata, J., Giedrys-Kalemba, S., Steinmetz, I., Witte, W., Broker, B.M., 2007. Clonal distribution of superantigen genes in clinical *Staphylococcus aureus* isolates. J. Clin. Microbiol. 45, 2669–2680.

Huber, H., Koller, S., Giezendanner, N., Stephan, R., Zweifel, C., 2010. Prevalence and characteristics of meticillin-resistant *Staphylococcus aureus* in humans in contact with farm animals, in livestock, and in food of animal origin, Switzerland, 2009. Euro Surveill. 15.

Isigidi, B.K., Mathieu, A.M., Devriese, L.A., Godard, C., Van Hoof, J., 1992. Enterotoxin production in different *Staphylococcus aureus* biotypes isolated from food and meat plants. J. Appl. Bacteriol. 72, 16–20.

Johler, S., Giannini, P., Jermini, M., Hummerjohann, J., Baumgartner, A., Stephan, R., 2015. Further evidence for staphylococcal food poisoning outbreaks caused by egc-encoded enterotoxins. Toxins (Basel) 7, 997–1004.

Johler, S., Layer, F., Stephan, R., 2011. Comparison of virulence and antibiotic resistance genes of food poisoning outbreak isolates of *Staphylococcus aureus* with isolates obtained from bovine mastitis milk and pig carcasses. J. Food Prot. 74, 1852–1859.

Johler, S., Tichaczek-Dischinger, P.S., Rau, J., Sihto, H.M., Lehner, A., Adam, M., Stephan, R., 2013. Outbreak of Staphylococcal food poisoning due to SEA-producing *Staphylococcus aureus*. Foodborne Pathog. Dis. 10, 777–781.

Jorda, G.B., Marucci, R.S., Guida, A.M., Pires, P.S., Manfredi, E.A., 2012. Carriage and characterization of *Staphylococcus aureus* in food handlers. Rev. Argent. Microbiol. 44, 101–104.

Kaspar, U., Kriegeskorte, A., Schubert, T., Peters, G., Rudack, C., Pieper, D.H., Wos-Oxley, M., Becker, K., 2016. The culturome of the human nose habitats reveals individual bacterial fingerprint patterns. Environ. Microbiol. 18, 2130–2142.

Kerouanton, A., Hennekinne, J.A., Letertre, C., Petit, L., Chesneau, O., Brisabois, A., De Buyser, M.L., 2007. Characterization of *Staphylococcus aureus* strains associated with food poisoning outbreaks in France. Int. J. Food Microbiol. 115, 369–375.

Kim, E.S., Song, J.S., Lee, H.J., Choe, P.G., Park, K.H., Cho, J.H., Park, W.B., Kim, S.H., Bang, J.H., Kim, D.M., Park, K.U., Shin, S., Lee, M.S., Choi, H.J., Kim, N.J., Kim, E.C., Oh, M.D., Kim, H.B., Choe, K.W., 2007. A survey of community-associated methicillin-resistant *Staphylococcus aureus* in Korea. J. Antimicrob. Chemother. 60, 1108–1114.

Kluytmans, J.A., 2010. Methicillin-resistant *Staphylococcus aureus* in food products: cause for concern or case for complacency? Clin. Microbiol. Infect. 16, 11–15.

Kluytmans, J., van Belkum, A., Verbrugh, H., 1997. Nasal carriage of *Staphylococcus aureus*: epidemiology, underlying mechanisms, and associated risks. Clin. Microbiol. Rev. 10, 505–520.

Ko, K.S., Lee, J.Y., Baek, J.Y., Peck, K.R., Rhee, J.Y., Kwon, K.T., Heo, S.T., Ahn, K.M., Song, J.H., 2008. Characterization of *Staphylococcus aureus* nasal carriage from children attending an outpatient clinic in Seoul, Korea. Microb. Drug Resist. 14, 37–44.

Köck, R., Harlizius, J., Bressan, N., Laerberg, R., Wieler, L.H., Witte, W., Deurenberg, R.H., Voss, A., Becker, K., Friedrich, A.W., 2009. Prevalence and molecular characteristics of methicillin-resistant *Staphylococcus aureus* (MRSA) among pigs on German farms and import of livestock-related MRSA into hospitals. Eur. J. Clin. Microbiol. Infect. Dis. 28, 1375–1382.

Kraushaar, B., Ballhausen, B., Leeser, D., Tenhagen, B.A., Kasbohrer, A., Fetsch, A., 2016. Antimicrobial resistances and virulence markers in methicillin-resistant *Staphylococcus aureus* from broiler and Turkey: a molecular view from farm to fork. Vet. Microbiol. 200.

Kraushaar, B., Fetsch, A., 2014. First description of PVL-positive methicillin-resistant *Staphylococcus aureus* (MRSA) in wild boar meat. Int. J. Food Microbiol. 186, 68–73.

Kuehnert, M.J., Kruszon-Moran, D., Hill, H.A., McQuillan, G., McAllister, S.K., Fosheim, G., McDougal, L.K., Chaitram, J., Jensen, B., Fridkin, S.K., Killgore, G., Tenover, F.C., 2006. Prevalence of *Staphylococcus aureus* nasal colonization in the United States, 2001–2002. J. Infect. Dis. 193, 172–179.

Labandeira-Rey, M., Couzon, F., Boisset, S., Brown, E.L., Bes, M., Benito, Y., Barbu, E.M., Vazquez, V., Hook, M., Etienne, J., Vandenesch, F., Bowden, M.G., 2007. *Staphylococcus aureus* Panton-Valentine leukocidin causes necrotizing pneumonia. Science 315, 1130–1133.

Leibler, J.H., Jordan, J.A., Brownstein, K., Lander, L., Price, L.B., Perry, M.J., 2016. *Staphylococcus aureus* nasal carriage among beefpacking workers in a Midwestern United States slaughterhouse. PLoS One 11, e0148789.

Lindsay, J.A., Moore, C.E., Day, N.P., Peacock, S.J., Witney, A.A., Stabler, R.A., Husain, S.E., Butcher, P.D., Hinds, J., 2006. Microarrays reveal that each of the ten dominant lineages of *Staphylococcus aureus* has a unique combination of surface-associated and regulatory genes. J. Bacteriol. 188, 669–676.

Loeto, D., Matsheka, M.I., Gashe, B.A., 2007. Enterotoxigenic and antibiotic resistance determination of *Staphylococcus aureus* strains isolated from food handlers in Gaborone, Botswana. J. Food Prot. 70, 2764–2768.

Lowy, F.D., August 20, 1998. *Staphylococcus aureus* infections. N. Engl. J. Med. 339 (8), 520–532.

Lozano, C., Gomez-Sanz, E., Benito, D., Aspiroz, C., Zarazaga, M., Torres, C., 2011. *Staphylococcus aureus* nasal carriage, virulence traits, antibiotic resistance mechanisms, and genetic lineages in healthy humans in Spain, with detection of CC398 and CC97 strains. Int. J. Med. Microbiol. 301, 500–505.

Lu, P.L., Chin, L.C., Peng, C.F., Chiang, Y.H., Chen, T.P., Ma, L., Siu, L.K., 2005. Risk factors and molecular analysis of community methicillin-resistant *Staphylococcus aureus* carriage. J. Clin. Microbiol. 43, 132–139.

Ma, X.X., Sun, D.D., Wang, S., Wang, M.L., Li, M., Shang, H., Wang, E.H., Luo, E.J., 2011. Nasal carriage of methicillin-resistant *Staphylococcus aureus* among preclinical medical students: epidemiologic and molecular characteristics of methicillin-resistant *S. aureus* clones. Diagn. Microbiol. Infect. Dis. 70, 22–30.

Mainous 3rd, A.G., Hueston, W.J., Everett, C.J., Diaz, V.A., 2006. Nasal carriage of *Staphylococcus aureus* and methicillin-resistant S aureus in the United States, 2001–2002. Ann. Fam. Med. 4, 132–137.

Maree, C.L., Daum, R.S., Boyle-Vavra, S., Matayoshi, K., Miller, L.G., 2007. Community-associated methicillin-resistant *Staphylococcus aureus* isolates causing healthcare-associated infections. Emerg. Infect. Dis. 13, 236–242.

Mediavilla, J.R., Chen, L., Mathema, B., Kreiswirth, B.N., 2012. Global epidemiology of community-associated methicillin resistant *Staphylococcus aureus* (CA-MRSA). Curr. Opin. Microbiol. 15, 588–595.

Mehraj, J., Akmatov, M.K., Strompl, J., Gatzemeier, A., Layer, F., Werner, G., Pieper, D.H., Medina, E., Witte, W., Pessler, F., Krause, G., 2014. Methicillin-sensitive and methicillin-resistant *Staphylococcus aureus* nasal carriage in a random sample of non-hospitalized adult population in northern Germany. PLoS One 9, e107937.

Melconian, A.K., Brun, Y., Fleurette, J., 1983. Enterotoxin production, phage typing and serotyping of *Staphylococcus aureus* strains isolated from clinical materials and food. J. Hyg. (Lond.) 91, 235–242.

Melles, D.C., Tenover, F.C., Kuehnert, M.J., Witsenboer, H., Peeters, J.K., Verbrugh, H.A., van Belkum, A., 2008. Overlapping population structures of nasal isolates of *Staphylococcus aureus* from healthy Dutch and American individuals. J. Clin. Microbiol. 46, 235–241.

Miller, R.R., Walker, A.S., Godwin, H., Fung, R., Votintseva, A., Bowden, R., Mant, D., Peto, T.E., Crook, D.W., Knox, K., 2014. Dynamics of acquisition and loss of carriage of *Staphylococcus aureus* strains in the community: the effect of clonal complex. J. Infect. 68, 426–439.

Monecke, S., Coombs, G., Shore, A.C., Coleman, D.C., Akpaka, P., Borg, M., Chow, H., Ip, M., Jatzwauk, L., Jonas, D., Kadlec, K., Kearns, A., Laurent, F., O'Brien, F.G., Pearson, J., Ruppelt, A., Schwarz, S., Scicluna, E., Slickers, P., Tan, H.L., Weber, S., Ehricht, R., 2011. A field guide to pandemic, epidemic and sporadic clones of methicillin-resistant *Staphylococcus aureus*. PLoS One 6, e17936.

Monecke, S., Ehricht, R., Slickers, P., Tan, H.L., Coombs, G., 2009. The molecular epidemiology and evolution of the Panton-Valentine leukocidin-positive, methicillin-resistant *Staphylococcus aureus* strain USA300 in Western Australia. Clin. Microbiol. Infect. 15, 770–776.

Monecke, S., Ruppelt, A., Wendlandt, S., Schwarz, S., Slickers, P., Ehricht, R., Jackel, S.C., 2013. Genotyping of *Staphylococcus aureus* isolates from diseased poultry. Vet. Microbiol. 162, 806–812.

Moon, D.C., Tamang, M.D., Nam, H.M., Jeong, J.H., Jang, G.C., Jung, S.C., Park, Y.H., Lim, S.K., 2015. Identification of livestock-associated methicillin-resistant *Staphylococcus aureus* isolates in Korea and molecular comparison between isolates from animal carcasses and slaughterhouse workers. Foodborne Pathog. Dis. 12, 327–334.

Mulcahy, M.E., McLoughlin, R.M., 2016. Host-bacterial crosstalk determines *Staphylococcus aureus* nasal colonization. Trends Microbiol. 24, 872–886.

Mulders, M.N., Haenen, A.P., Geenen, P.L., Vesseur, P.C., Poldervaart, E.S., Bosch, T., Huijsdens, X.W., Hengeveld, P.D., Dam-Deisz, W.D., Graat, E.A., Mevius, D., Voss, A., Van De Giessen, A.W., 2010. Prevalence of livestock-associated MRSA in broiler flocks and risk factors for slaughterhouse personnel in The Netherlands. Epidemiol. Infect. 138, 743–755.

Munckhof, W.J., Nimmo, G.R., Carney, J., Schooneveldt, J.M., Huygens, F., Inman-Bamber, J., Tong, E., Morton, A., Giffard, P., 2008. Methicillin-susceptible, non-multiresistant methicillin-resistant and multiresistant methicillin-resistant *Staphylococcus aureus* infections: a clinical, epidemiological and microbiological comparative study. Eur. J. Clin. Microbiol. Infect. Dis. 27, 355–364.

Nashev, D., Toshkova, K., Bizeva, L., Akineden, O., Lammler, C., Zschock, M., 2007. Distribution of enterotoxin genes among carriage- and infection-associated isolates of *Staphylococcus aureus*. Lett. Appl. Microbiol. 45, 681–685.

Nasrolahei, M., Mirshafiee, S., Kholdi, S., Salehian, M., Nasrolahei, M., 2016. Bacterial assessment of food handlers in Sari city, Mazandaran Province, north of Iran. J. Infect. Public Health 10.

Nnachi, A.U., Emele, F.E., Ukae, C.O., Maduka, V.A., Udulbiam, O.E., Chukwu, O.S., Agwu, M.M., 2014. Prevalence of methicillin-resistant *Staphylococcus aureus* (MRSA) in raw meat and meat handlers in Onitsha, Nigeria. Eur. J. Prev. Med. 2, 9–15.

Normanno, G., Dambrosio, A., Lorusso, V., Samoilis, G., Di Taranto, P., Parisi, A., 2015. Methicillin-resistant *Staphylococcus aureus* (MRSA) in slaughtered pigs and abattoir workers in Italy. Food Microbiol. 51, 51–56.

Nouwen, J.L., Ott, A., Kluytmans-Vandenbergh, M.F., Boelens, H.A., Hofman, A., van Belkum, A., Verbrugh, H.A., 2004. Predicting the *Staphylococcus aureus* nasal carrier state: derivation and validation of a "culture rule". Clin. Infect. Dis. 39, 806–811.

O'Donoghue, M.M., Boost, M.V., 2004. The prevalence and source of methicillin-resistant *Staphylococcus aureus* (MRSA) in the community in Hong Kong. Epidemiol. Infect. 132, 1091–1097.

Okareh, O.T., Erhahon, O.O., 2015. Microbiological assessment of food and hand-swabs samples of school food vendors in Benin city, Nigeria. Food Public Health 5, 23–28.

Omoe, K., Hu, D.L., Takahashi-Omoe, H., Nakane, A., Shinagawa, K., 2005. Comprehensive analysis of classical and newly described staphylococcal superantigenic toxin genes in *Staphylococcus aureus* isolates. FEMS Microbiol. Lett. 246, 191–198.

Omuse, G., Kariuki, S., Revathi, G., 2012. Unexpected absence of meticillin-resistant *Staphylococcus aureus* nasal carriage by healthcare workers in a tertiary hospital in Kenya. J. Hosp. Infect. 80, 71–73.

Oteri, T., Ekanem, E.E., 1989. Food hygiene behaviour among hospital food handlers. Public Health 103, 153–159.

Ouedraogo, A.S., Dunyach-Remy, C., Kissou, A., Sanou, S., Poda, A., Kyelem, C.G., Solassol, J., Banuls, A.L., Van De Perre, P., Ouedraogo, R., Jean-Pierre, H., Lavigne, J.P., Godreuil, S., 2016. High nasal carriage rate of *Staphylococcus aureus* containing panton-valentine leukocidin- and EDIN-encoding genes in community and hospital settings in Burkina Faso. Front. Microbiol. 7, 1406.

Paling, F.P., Wolkewitz, M., Bode, L.G., Klein Klouwenberg, P.M., Ong, D.S., Depuydt, P., de Bus, L., Sifakis, F., Bonten, M.J., Kluytmans, J.A., 2016. *S. aureus* colonization at ICU admission as a risk factor for developing *S. aureus* ICU pneumonia. Clin. Microbiol. Infect. 23.

Peacock, S.J., Moore, C.E., Justice, A., Kantzanou, M., Story, L., Mackie, K., O'Neill, G., Day, N.P., 2002. Virulent combinations of adhesin and toxin genes in natural populations of *Staphylococcus aureus*. Infect. Immun. 70, 4987–4996.

Portillo, B.C., Moreno, J.E., Yomayusa, N., Alvarez, C.A., Cardozo, B.E., Perez, J.A., Diaz, P.L., Ibanez, M., Mendez-Alvarez, S., Leal, A.L., Gomez, N.V., 2013. Molecular epidemiology and characterization of virulence genes of community-acquired and hospital-acquired methicillin-resistant *Staphylococcus aureus* isolates in Colombia. Int. J. Infect. Dis. 17, e744–e749.

Price, L.B., Stegger, M., Hasman, H., Aziz, M., Larsen, J., Andersen, P.S., Pearson, T., Waters, A.E., Foster, J.T., Schupp, J., Gillece, J., Driebe, E., Liu, C.M., Springer, B., Zdovc, I., Battisti, A., Franco, A., Zmudzki, J., Schwarz, S., Butaye, P., Jouy, E., Pomba, C., Porrero, M.C., Ruimy, R., Smith, T.C., Robinson, D.A., Weese, J.S., Arriola, C.S., Yu, F., Laurent, F., Keim, P., Skov, R., Aarestrup, F.M., 2012. *Staphylococcus aureus* CC398: host adaptation and emergence of methicillin resistance in livestock. MBio 3.

Rall, V.L., Sforcin, J.M., Augustini, V.C., Watanabe, M.T., Fernandes Jr., A., Rall, R., Silva, M.G., Araujo Jr., J.P., 2010. Detection of enterotoxin genes of Staphylococcus SP isolated from nasal cavities and hands of food handlers. Braz. J. Microbiol. 41, 59–65.

Rodriguez, E.A., Correa, M.M., Ospina, S., Atehortua, S.L., Jimenez, J.N., 2014. Differences in epidemiological and molecular characteristics of nasal colonization with *Staphylococcus aureus* (MSSA-MRSA) in children from a university hospital and day care centers. PLoS One 9, e101417.

Ruimy, R., Angebault, C., Djossou, F., Dupont, C., Epelboin, L., Jarraud, S., Lefevre, L.A., Bes, M., Lixandru, B.E., Bertine, M., El Miniai, A., Renard, M., Bettinger, R.M., Lescat, M., Clermont, O., Peroz, G., Lina, G., Tavakol, M., Vandenesch, F., van Belkum, A., Rousset, F., Andremont, A., 2010. Are host genetics the predominant determinant of persistent nasal *Staphylococcus aureus* carriage in humans? J. Infect. Dis. 202, 924–934.

Ruimy, R., Armand-Lefevre, L., Barbier, F., Ruppe, E., Cocojaru, R., Mesli, Y., Maiga, A., Benkalfat, M., Benchouk, S., Hassaine, H., Dufourcq, J.B., Nareth, C., Sarthou, J.L., Andremont, A., Feil, E.J., 2009. Comparisons between geographically diverse samples of carried *Staphylococcus aureus*. J. Bacteriol. 191, 5577–5583.

Ruimy, R., Maiga, A., Armand-Lefevre, L., Maiga, I., Diallo, A., Koumare, A.K., Ouattara, K., Soumare, S., Gaillard, K., Lucet, J.C., Andremont, A., Feil, E.J., 2008. The carriage population of *Staphylococcus aureus* from Mali is composed of a combination of pandemic clones and the divergent Panton-Valentine leukocidin-positive genotype ST152. J. Bacteriol. 190, 3962–3968.

Saeed, H.A., Hamid, H.H., 2010. Bacteriological and parasitological assessment of food handlers in the Omdurman area of Sudan. J. Microbiol. Immunol. Infect. 43, 70–73.

Sakwinska, O., Kuhn, G., Balmelli, C., Francioli, P., Giddey, M., Perreten, V., Riesen, A., Zysset, F., Blanc, D.S., Moreillon, P., 2009. Genetic diversity and ecological success of *Staphylococcus aureus* strains colonizing humans. Appl. Environ. Microbiol. 75, 175–183.

Sangvik, M., Olsen, R.S., Olsen, K., Simonsen, G.S., Furberg, A.S., Sollid, J.U., 2011. Age- and gender-associated *Staphylococcus aureus* spa types found among nasal carriers in a general population: the Tromso Staph and Skin Study. J. Clin. Microbiol. 49, 4213–4218.

Saxena, A.K., Panhotra, B.R., 2003. The prevalence of nasal carriage of *Staphylococcus aureus* and associated vascular access-related septicemia among patients on hemodialysis in Al-Hasa region of Saudi Arabia. Saudi J. Kidney Dis. Transpl. 14, 30–38.

Schaumburg, F., Ngoa, U.A., Kosters, K., Kock, R., Adegnika, A.A., Kremsner, P.G., Lell, B., Peters, G., Mellmann, A., Becker, K., 2011. Virulence factors and genotypes of *Staphylococcus aureus* from infection and carriage in Gabon. Clin. Microbiol. Infect. 17, 1507–1513.

Severin, J.A., Lestari, E.S., Kuntaman, K., Melles, D.C., Pastink, M., Peeters, J.K., Snijders, S.V., Hadi, U., Duerink, D.O., van Belkum, A., Verbrugh, H.A., Prevalence Antimicrobial Resistance in Indonesia, Group Prevention Study, 2008. Unusually high prevalence of panton-valentine leukocidin genes among methicillin-sensitive *Staphylococcus aureus* strains carried in the Indonesian population. J. Clin. Microbiol. 46, 1989–1995.

Seybold, U., Kourbatova, E.V., Johnson, J.G., Halvosa, S.J., Wang, Y.F., King, M.D., Ray, S.M., Blumberg, H.M., 2006. Emergence of community-associated methicillin-resistant *Staphylococcus aureus* USA300 genotype as a major cause of health care-associated blood stream infections. Clin. Infect. Dis. 42, 647–656.

Shambat, S., Nadig, S., Prabhakara, S., Bes, M., Etienne, J., Arakere, G., 2012. Clonal complexes and virulence factors of *Staphylococcus aureus* from several cities in India. BMC Microbiol. 12, 64.

Shore, A.C., Tecklenborg, S.C., Brennan, G.I., Ehricht, R., Monecke, S., Coleman, D.C., 2014. Panton-Valentine leukocidin-positive *Staphylococcus aureus* in Ireland from 2002 to 2011: 21 clones, frequent importation of clones, temporal shifts of predominant methicillin-resistant *S. aureus* clones, and increasing multiresistance. J. Clin. Microbiol. 52, 859–870.

Shukla, S.K., Karow, M.E., Brady, J.M., Stemper, M.E., Kislow, J., Moore, N., Wroblewski, K., Chyou, P.H., Warshauer, D.M., Reed, K.D., Lynfield, R., Schwan, W.R., 2010. Virulence genes and genotypic associations in nasal carriage, community-associated methicillin-susceptible and methicillin-resistant USA400 *Staphylococcus aureus* isolates. J. Clin. Microbiol. 48, 3582–3592.

Simsek, Z., Koruk, I., Copur, A.C., Gurses, G., 2009. Prevalence of *Staphylococcus aureus* and intestinal parasites among food handlers in Sanliurfa, Southeastern Anatolia. J. Public Health Manag. Pract. 15, 518–523.

Skramm, I., Moen, A.E., Bukholm, G., 2011. Nasal carriage of *Staphylococcus aureus*: frequency and molecular diversity in a randomly sampled Norwegian community population. APMIS 119, 522–528.

Soares, M.J., Tokumaru-Miyazaki, N.H., Noleto, A.L., Figueiredo, A.M., 1997. Enterotoxin production by *Staphylococcus aureus* clones and detection of Brazilian epidemic MRSA clone (III::B: A) among isolates from food handlers. J. Med. Microbiol. 46, 214–221.

Soto, A., Saldias, M.E., Oviedo, P., Fernandez, M., 1996. Prevalence of *Staphylococcus aureus* among food handlers from a metropolitan university in Chile. Rev. Med. Chil. 124, 1142–1146.

Tan, S.L., Lee, H.Y., Mahyudin, N.A., 2014. Antimicrobial resistance of *Escherichia coli* and *Staphylococcus aureus* isolated from food handler's hands. Food Control 44, 203–207.

Tavares, A., Faria, N.A., de Lencastre, H., Miragaia, M., 2014. Population structure of methicillin-susceptible *Staphylococcus aureus* (MSSA) in Portugal over a 19-year period (1992–2011). Eur. J. Clin. Microbiol. Infect. Dis. 33, 423–432.

Tavares, A., Miragaia, M., Rolo, J., Coelho, C., de Lencastre, H., CA-MRSA MSSA working group, 2013. High prevalence of hospital-associated methicillin-resistant *Staphylococcus aureus* in the community in Portugal: evidence for the blurring of community-hospital boundaries. Eur. J. Clin. Microbiol. Infect. Dis. 32, 1269–1283.

Tenhagen, B.A., Vossenkuhl, B., Kasbohrer, A., Alt, K., Kraushaar, B., Guerra, B., Schroeter, A., Fetsch, A., 2014. Methicillin-resistant *Staphylococcus aureus* in cattle food chains – prevalence, diversity, and antimicrobial resistance in Germany. J. Anim. Sci. 92, 2741–2751.

Tenover, F.C., McDougal, L.K., Goering, R.V., Killgore, G., Projan, S.J., Patel, J.B., Dunman, P.M., 2006. Characterization of a strain of community-associated methicillin-resistant *Staphylococcus aureus* widely disseminated in the United States. J. Clin. Microbiol. 44, 108–118.

Thurlow, L.R., Joshi, G.S., Richardson, A.R., 2012. Virulence strategies of the dominant USA300 lineage of community-associated methicillin-resistant *Staphylococcus aureus* (CA-MRSA). FEMS Immunol. Med. Microbiol. 65, 5–22.

Todd, E.C., Greig, J.D., Bartleson, C.A., Michaels, B.S., 2008. Outbreaks where food workers have been implicated in the spread of foodborne disease. Part 4. Infective doses and pathogen carriage. J. Food Prot. 71, 2339–2373.

Udo, E.E., Al-Mufti, S., Albert, M.J., 2009. The prevalence of antimicrobial resistance and carriage of virulence genes in *Staphylococcus aureus* isolated from food handlers in Kuwait City restaurants. BMC Res. Notes 2, 108.

Uemura, E., Kakinohana, S., Higa, N., Toma, C., Nakasone, N., 2004. Comparative chracterization of *Staphylococcus aureus* isolates from throats and noses of healthy volunteers. Jpn. J. Infect. Dis. 57, 21–24.

Uzunovic, S., Ibrahimagi, A., Kamberovi, F., Rijnders, M.I.A., Stobberingh, E.E., 2013. Characterization of methicillin-susceptible and methicillin-resistant *Staphylococcus aureus* in food handlers in Bosnia and Herzegovina. Open Infect. Dis. J. 7, 15–20.

van Belkum, A., Melles, D.C., Snijders, S.V., van Leeuwen, W.B., Wertheim, H.F., Nouwen, J.L., Verbrugh, H.A., Etienne, J., 2006. Clonal distribution and differential occurrence of the enterotoxin gene cluster, egc, in carriage- versus bacteremia-associated isolates of *Staphylococcus aureus*. J. Clin. Microbiol. 44, 1555–1557.

van Belkum, A., Verkaik, N.J., de Vogel, C.P., Boelens, H.A., Verveer, J., Nouwen, J.L., Verbrugh, H.A., Wertheim, H.F., 2009. Reclassification of *Staphylococcus aureus* nasal carriage types. J. Infect. Dis. 199, 1820–1826.

Van Cleef, B.A., Broens, E.M., Voss, A., Huijsdens, X.W., Zuchner, L., Van Benthem, B.H., Kluytmans, J.A., Mulders, M.N., Van De Giessen, A.W., 2010. High prevalence of nasal MRSA carriage in slaughterhouse workers in contact with live pigs in The Netherlands. Epidemiol. Infect. 138, 756–763.

van Cleef, B.A., van Benthem, B.H., Verkade, E.J., van Rijen, M., Kluytmans-van den Bergh, M.F., Schouls, L.M., Duim, B., Wagenaar, J.A., Graveland, H., Bos, M.E., Heederik, D., Kluytmans, J.A., 2014. Dynamics of methicillin-resistant *Staphylococcus aureus* and methicillin-susceptible *Staphylococcus aureus* carriage in pig farmers: a prospective cohort study. Clin. Microbiol. Infect. 20, O764–O771.

van Trijp, M.J., Melles, D.C., Snijders, S.V., Wertheim, H.F., Verbrugh, H.A., van Belkum, A., van Wamel, W.J., 2010. Genotypes, superantigen gene profiles, and presence of exfoliative toxin genes in clinical methicillin-susceptible *Staphylococcus aureus* isolates. Diagn. Microbiol. Infect. Dis. 66, 222–224.

VandenBergh, M.F., Yzerman, E.P., van Belkum, A., Boelens, H.A., Sijmons, M., Verbrugh, H.A., 1999. Follow-up of *Staphylococcus aureus* nasal carriage after 8 years: redefining the persistent carrier state. J. Clin. Microbiol. 37, 3133–3140.

Vatansever, L., Sezer, C., Bilge, N., 2016. Carriage rate and methicillin resistance of *Staphylococcus aureus* in food handlers in Kars City, Turkey. Springerplus 5, 608.

Verhoeven, P.O., Gagnaire, J., Botelho-Nevers, E., Grattard, F., Carricajo, A., Lucht, F., Pozzetto, B., Berthelot, P., 2014. Detection and clinical relevance of *Staphylococcus aureus* nasal carriage: an update. Expert Rev. Anti Infect. Ther. 12, 75–89.

Vlack, S., Cox, L., Peleg, A.Y., Canuto, C., Stewart, C., Conlon, A., Stephens, A., Giffard, P., Huygens, F., Mollinger, A., Vohra, R., McCarthy, J.S., 2006. Carriage of methicillin-resistant *Staphylococcus aureus* in a Queensland Indigenous community. Med. J. Aust. 184, 556–559.

von Eiff, C., Becker, K., Machka, K., Stammer, H., Peters, G., 2001. Nasal carriage as a source of *Staphylococcus aureus* bacteremia. Study Group. N. Engl. J. Med. 344, 11–16.

Voyich, J.M., Otto, M., Mathema, B., Braughton, K.R., Whitney, A.R., Welty, D., Long, R.D., Dorward, D.W., Gardner, D.J., Lina, G., Kreiswirth, B.N., DeLeo, F.R., 2006. Is Panton-Valentine leukocidin the major virulence determinant in community-associated methicillin-resistant *Staphylococcus aureus* disease? J. Infect. Dis. 194, 1761–1770.

Wang, R., Braughton, K.R., Kretschmer, D., Bach, T.H., Queck, S.Y., Li, M., Kennedy, A.D., Dorward, D.W., Klebanoff, S.J., Peschel, A., DeLeo, F.R., Otto, M., 2007. Identification of novel cytolytic peptides as key virulence determinants for community-associated MRSA. Nat. Med. 13, 1510–1514.

Wattinger, L., Stephan, R., Layer, F., Johler, S., 2012. Comparison of *Staphylococcus aureus* isolates associated with food intoxication with isolates from human nasal carriers and human infections. Eur. J. Clin. Microbiol. Infect. Dis. 31, 455–464.

Wei, H.L., Chiou, C.S., 2002. Molecular subtyping of *Staphylococcus aureus* from an outbreak associated with a food handler. Epidemiol. Infect. 128, 15–20.

Wendlandt, S., Kadlec, K., Fessler, A.T., Monecke, S., Ehricht, R., van de Giessen, A.W., Hengeveld, P.D., Huijsdens, X., Schwarz, S., van Duijkeren, E., 2013a. Resistance phenotypes and genotypes of methicillin-resistant *Staphylococcus aureus* isolates from broiler chickens at slaughter and abattoir workers. J. Antimicrob. Chemother. 68, 2458–2463.

Wendlandt, S., Schwarz, S., Silley, P., 2013b. Methicillin-resistant *Staphylococcus aureus*: a food-borne pathogen? Annu. Rev. Food Sci. Technol. 4, 117–139.

Wertheim, H.F., Melles, D.C., Vos, M.C., van Leeuwen, W., van Belkum, A., Verbrugh, H.A., Nouwen, J.L., 2005. The role of nasal carriage in *Staphylococcus aureus* infections. Lancet Infect. Dis. 5, 751–762.

Wertheim, H.F., Verbrugh, H.A., 2006. Global prevalence of meticillin-resistant *Staphylococcus aureus*. Lancet 368, 1866 Author reply 66–7.

Wertheim, H.F., Vos, M.C., Boelens, H.A., Voss, A., Vandenbroucke-Grauls, C.M., Meester, M.H., Kluytmans, J.A., van Keulen, P.H., Verbrugh, H.A., 2004. Low prevalence of methicillin-resistant *Staphylococcus aureus* (MRSA) at hospital admission in The Netherlands: the value of search and destroy and restrictive antibiotic use. J. Hosp. Infect. 56, 321–325.

WHO, 2016. Criticall Important Antimicrobials for Human Medicine, fourth rev. .

Wieneke, A.A., Roberts, D., Gilbert, R.J., 1993. Staphylococcal food poisoning in the United Kingdom, 1969–90. Epidemiol. Infect. 110, 519–531.

Williams, R.E., 1963. Healthy carriage of *Staphylococcus aureus*: its prevalence and importance. Bacteriol. Rev. 27, 56–71.

Williamson, D.A., Bakker, S., Coombs, G.W., Tan, H., Monecke, S., Heffernan, H., 2014. Emergence and molecular characterization of clonal complex 398 (CC398) methicillin-resistant *Staphylococcus aureus* (MRSA) in New Zealand. J. Antimicrob. Chemother. 69, 1428–1430.

Witte, W., Braulke, C., Cuny, C., Heuck, D., Kresken, M., 2001. Changing pattern of antibiotic resistance in methicillin-resistant *Staphylococcus aureus* from German hospitals. Infect. Control Hosp. Epidemiol. 22, 683–686.

Wyllie, D., Paul, J., Crook, D., 2011. Waves of trouble: MRSA strain dynamics and assessment of the impact of infection control. J. Antimicrob. Chemother. 66, 2685–2688.

Zeinhom, M.M., Abdel-Latef, G.K., Jordan, K., 2015. The Use of Multiplex PCR to determine the prevalence of enterotoxigenic *Staphylococcus aureus* isolated from raw milk, feta cheese, and hand swabs. J. Food Sci. 80, M2932–M2936.

Zhang, M., O'Donoghue, M.M., Ito, T., Hiramatsu, K., Boost, M.V., 2011. Prevalence of antiseptic-resistance genes in *Staphylococcus aureus* and coagulase-negative staphylococci colonising nurses and the general population in Hong Kong. J. Hosp. Infect. 78, 113–117.

CHAPTER

MOLECULAR EPIDEMIOLOGY OF *STAPHYLOCOCCUS AUREUS* LINEAGES IN THE ANIMAL–HUMAN INTERFACE

10

Myriam Zarazaga, Paula Gómez, Sara Ceballos, Carmen Torres
Department of Food and Agriculture, University of La Rioja, Logroño, Spain

1. INTRODUCTION

Staphylococcus aureus is a commensal microorganism of the skin and nose of humans and many animals, but at the same time is an important opportunistic human and animal pathogen that can be implicated in a wide diversity of infections, from mild to severe ones. This microorganism can also be disseminated through the contaminated food and water and can be found in many different ecological niches. *S. aureus* has the ability to acquire a wide diversity of antimicrobial resistance and virulence genes, being those associated to methicillin resistance in *S. aureus* (MRSA) of special relevance for human and animal health. Moreover, this microorganism can also acquire mobile genetic elements (MGEs), which contain genes that can facilitate its adaptation to the different hosts (Lindsay, 2014).

Because of the ubiquity of this microorganism and its interest, both as commensal and as opportunistic pathogen, it is important to understand the distribution of clones and the population structure of *S. aureus* in the different environments and in the different geographical areas to track the evolution of this microorganism as well as the factors that could drive it. Different methods are applied for molecular epidemiology of *S. aureus* or to determine the population structure in different ecosystems, the most common ones are multilocus sequence typing (MLST) or *spa*-typing (Stefani et al., 2012; Vanderhaeghen et al., 2010). The MLST implicates the sequencing of internal fragments of seven housekeeping genes, which gives a combination of seven alleles that determine the sequence type (ST) of the microorganism; strains that differ in only one or two loci are called single-locus variants (SLVs) or double-locus variants (DLVs), respectively. The STs, SLVs, and DLVs are grouped into clonal complexes (CC) using the BURST (based upon related STs) analysis. In a CC, the ST that has the highest number of different SLVs and DLVs is considered as the ancestral ST, and the CC is numbered after its ancestral ST (Enright et al., 2002; Vanderhaeghen et al., 2010). The *spa*-typing implicates the sequencing of the polymorphic X region of the staphylococcal protein A gene (*spa*). There is a good correlation between the *spa*-type and the CC determined by MLST. For further details on the different typing methods it is referred to Chapter 6 of this book.

In the last years, the application of whole-genome sequencing (WGS) and microarrays for the analysis of big collections of *S. aureus* of many different origins, lineages, and characteristics is

Staphylococcus aureus. http://dx.doi.org/10.1016/B978-0-12-809671-0.00010-3

providing important information to deepen in the knowledge of the molecular epidemiology of this microorganism, identifying epidemiological shifts, transmission routes, and adaptation of major clones (Lindsay, 2014).

Along the years, different genetic lineages of *S. aureus* associated either to the hospital (HA) or to the community (CA) have been reported in the human population, mostly referred to MRSA isolates. In the last decade, new lineages have emerged in the animal population, being especially relevant the livestock-associated (LA) lineage CC398 that has been widely spread in animals, mainly in food-producing animals, and subsequently along the different farm-to-fork value chains. Moreover, CC398 MRSA has emerged as human pathogen causing severe disease, mainly among livestock professionals.

In this chapter we will review different *S. aureus* lineages that have expanded in the animal–human interface. Particular attention will be payed to the emergence, characteristics, and molecular epidemiology of the LA lineage CC398. But also *S. aureus* lineages CC9, CC130, and CC97, found among animals and in the animal–human interface, and the more human-adapted lineage CC1, are presented and discussed.

2. GENOME STRUCTURE OF *STAPHYLOCOCCUS AUREUS*

The first *S. aureus* full-genome sequence was available in 2001 (Kuroda et al., 2001). After that time, many other complete *S. aureus* genomes have been described, mostly in the last few years. The *S. aureus* genome is around 2.8 Mb in size and carries ~2800 genes (McCarthy et al., 2012a). Three different parts can be distinguished in the *S. aureus* genome: the core genome, the variable core genome (VCG), and the MGEs. The core genome is highly conserved in all isolates of this species (>97%) and represents ~75% of the genome corresponding to housekeeping genes. The VCG represents 10% of the genome and differs among isolates by presence/absence of genes or by gene polymorphisms; the isolates of different CCs present specific genes in this VCG. Finally the MGEs include bacteriophages, plasmids, *S. aureus* pathogenicity islands, staphylococcal cassette chromosome (SCCs), transposons, and insertion sequences, and could represent ~15% of the genome; MGEs can move among bacteria by horizontal gene transfer (HGT) mechanisms, can vary in the different genetic lineages and usually contain genes of clinical relevance, as is the case of virulence genes or antimicrobial and heavy metal resistance genes, or those implicated in host adaptation (McCarthy et al., 2012a; Lindsay, 2014).

The acquisition or loss of MGEs, important events in the evolution of the *S. aureus* genome, could happen by transduction (by bacteriophages, carried in many isolates), less frequently by conjugation (by conjugative plasmids, present in some isolates), and scarcely by transformation (Lindsay, 2014). HGT of MGEs is more frequent among *S. aureus* isolates of the same lineage than among those isolates of different lineages.

Point mutations are also implicated in the evolution of the *S. aureus* genome, which can generate modifications in the gene function or in gene expression, but in general the effect in the cell is low (McCarthy et al., 2012a); nevertheless, single-point mutations in the genome could alter the host tropism of this microorganism (Viana et al., 2015).

The acquisition of MGEs has been very important in the evolution of the lineage CC398, one of the most important ones located in the animal–human interface, as will be shown below.

3. THE EMERGENCE OF *STAPHYLOCOCCUS AUREUS* LINEAGE CC398

3.1 HISTORICAL ASPECTS AND CHARACTERISTICS OF *STAPHYLOCOCCUS AUREUS* CC398. A PROBLEM OF PUBLIC HEALTH

S. aureus lineage CC398 emerged, both as methicillin-sensitive *S. aureus* (MSSA) and MRSA, in pigs and pig farmers at the beginning of the 21st century, in 2004 (Armand-Lefevre et al., 2005; Voss et al., 2005). This lineage was isolated, along with several others ones, in healthy farmers in France, and the authors already suggested that pig farming was a risk factor for increased nasal *S. aureus* colonization (Armand-Lefevre et al., 2005). Parallel studies carried out in The Netherlands, a country with low human MRSA prevalence, found an unexpected high frequency of MRSA (nontypeable by pulsed-field gel electrophoresis (PFGE), mainly belonging to *spa*-type t108 and t011) among family members of pig farmers and their pigs, and alerted for the first time of the possible transmission of CC398 MRSA between animals and humans, especially in swine workers (Voss et al., 2005; Huijsdens et al., 2006). In the following years, an increasing number of CC398 MRSA isolates, called initially pig-associated MRSA and later designated as LA-MRSA, was revealed. CC398 MRSA was demonstrated to be widespread among livestock, especially pigs, in many European countries, where CC398 was the most common LA CC (EFSA, 2009a; Crombé et al., 2013). CC398 MRSA has been also detected in pigs in North America, South America, and Australia (Khanna et al., 2008; Arriola et al., 2011; Groves et al., 2014), and it has been occasionally observed in Asia, where CC9 is the major LA-MRSA (Chuang and Huang, 2015). Despite the high carriage rate, clinical LA-MRSA CC398 infections have been rarely reported in pigs (van der Wolf et al., 2012).

Certainly pigs are an important reservoir for MRSA CC398, but this lineage has also been detected as colonizer or causing infections in many other animal species such as veal calves and dairy cows (Graveland et al., 2010; Vanderhaeghen et al., 2010), poultry (Persoons et al., 2009; Argudín et al., 2013), horses (Van den Eede et al., 2012; Gómez-Sanz et al., 2014), and also pets (Floras et al., 2010). In recent years, colonization by CC398 MRSA in wildlife animals has also been reported (Gómez et al., 2016).

The risk of the presence of LA-MRSA CC398 in animals is due to the possible transfer of these strains to humans. Indeed, this lineage is very often associated with colonization of humans working in the livestock industry, such as farmers, veterinarians, slaughterhouse-personnel, among others, CC398 is also able to cause human infections, from minor diseases (the skin and soft tissue) to severe or life-threatening diseases. Numerous publications have detected and alerted of this public health threat and the implications of the emergence of this lineage in livestock and humans (see reviews Becker et al., 2015; Smith and Wardyn, 2015; Cuny et al., 2015b; Graveland et al., 2011; Smith and Pearson, 2011).

S. aureus CC398 shows the following specific genotypic and phenotypic characteristics, mostly depending on MRSA or MSSA type of isolates, or their host or geographic locations:

1. Nontypeability by standard PFGE using *SmaI* digestion (Bens et al., 2006); these isolates are typeable when PFGE protocol using Cfr9I is followed (Argudín et al., 2010).
2. Most of CC398 isolates are typed as sequence type ST398, although diversification of this CC is occurring over time and currently, it includes besides the founder ST398, 67 related STs: 52 SLVs, 8 DLVs, and 7 triple-locus variants (http://saureus.mlst.net). Recently ST291 (DLV of ST398) has been defined as a distinct genetic lineage, nonclosely related in general to ST398 or to the LA-CC398 group (Stegger et al., 2013a).

3. Many different *spa* types were described to be associated with *S. aureus* CC398, and the number is increasing day by day (Argudín et al., 2011; Vanderhaerghen et al., 2010; Lozano et al., 2012). The most prevalent CC398 associated *spa* types are t011, t034, and t108, among others.
4. CC398 do often carry SCC*mec* type IVa or V, but other SCC*mec* elements (VII, IX, X) have also been detected.
5. MRSA CC398 strains are resistant to tetracycline mediated by *tet*(M) gene, and they often also present coresistance against other antibiotics (macrolides, lincosamides, trimethoprim, among others) (Argudín et al., 2011).
6. CC398 isolates usually do not carry genes encoding for toxins, such as Panton–Valentine leukocidin (PVL) or staphylococcal enterotoxins (SEs). However, CC398 isolates, mainly MSSA carrying *luk*-PV gene, have been detected in recent years (Zhao et al., 2012; Stegger et al., 2010). Also CC398 MRSA harboring staphylococcal enterotoxin (SE) encoding genes and producing SE have been recently described among poultry isolates (Kraushaar et al., 2017).

3.2 PIG ORIGIN OF METHICILLIN-RESISTANT *STAPHYLOCOCCUS AUREUS* CC398: FIRST HYPOTHESIS

The origin of MRSA CC398 remains a subject of debate. It can be assumed that CC398 MRSA evolved from CC398 MSSA by acquisition of the SCC*mec* element. In the early years of emergence of CC398 MRSA, when CC398 MSSA had only been detected in pigs and pig farmers, the association between exposure to pigs and human carriage leads to the hypothesis of a pig origin of this lineage (Voss et al., 2005). It is important to note that, back then, the methodology for comparative analysis of genomes was not yet a common and affordable practice, as is shown below. It was believed that CC398 might originally have been a highly prevalent commensal pig strain as MSSA, which then acquired *mecA* from other staphylococci that colonize pigs or pig farmers (van Loo et al., 2007; Wulf and Voss, 2008). This hypothesis was based on the initial assumption (this opinion has subsequently changed as explained below) that CC398 MSSA was essentially pig-specific (Vanderhaeghen et al., 2010). Indeed, MSSA CC398 strains had also been detected in healthy pigs (Guardabassi et al., 2007; Hasman et al., 2010), but nevertheless, CC398 MSSA had been described more often and previously in humans than in pigs. It was also hypothesized that CC398 MRSA in pigs may had occurred earlier than 2004, and might be not really "emerging", but rather had been overlooked (Meemken et al., 2010).

CC398 MRSA from pig origin could subsequently be transmitted and disseminated from this species to other animal species and also to humans. This initial hypothesis of pig-specific and pig origin of CC398 MRSA would be in accordance with the transient observed dynamic of colonization by CC398 MRSA in humans in contact with animals. Moreover, several studies stated that MRSA CC398 was not a good colonizer in humans. van Cleef et al. (2011), for instance, found a high rate of transient contamination, without substantial persistent colonization in farm workers with short-term occupational exposure to positive animals. Furthermore, in most human colonization cases, decolonization occurs rapidly after removal of the contact with the reservoir (Graveland et al., 2011), although persistent cases have also been reported (Köck et al., 2012), as well as recolonization when exposure continued (Lozano et al., 2011a).

3.3 ORIGIN AND HOST ADAPTATION: FIRST JUMP FROM HUMANS TO PIGS

In recent years, the initial hypothesis has been dismissed given significant advances in methodology, especially in next-generation sequencing and microarray assays that have allowed bacterial WGS and comparative genome analysis contributing to a deeper understanding of the origin and evolution of *S. aureus* CC398 genetic lineage. HGT of MGEs is to be considered a major evolutionary step that accelerates genetic and phenotypic variation in *S. aureus* populations and enables adaptation to changing environments (Lindsay, 2010). Acquisition or loss of some MGEs can be associated with host adaptation in CC398 and MGE profile may help to determine the epidemiological associations among CC398 isolates (McCarthy et al., 2011). Phage $\varphi 3$ deserves special attention; it carries the genes of the immune evasion cluster (IEC), encoding human-specific immune modulatory elements, playing crucial roles in human niche adaptation and also in human invasive infections (Rooijakkers et al., 2005; McCarthy et al., 2011). The IEC encodes proteins that interact specifically with the human immune response: *chp* (chemotaxis inhibitory protein), *sak* (staphylokinase), and *scn* (staphylococcal complement inhibitor), as well as particular enterotoxin genes as *sea*, *sep*, *sek*—see Chapters 2 and 3 for further details. Generally, about 90% of all clinical and laboratory human-derived *S. aureus* isolates carry phage $\varphi 3$ in their genomes (van Wamel et al., 2006). McCarthy et al. (2011) analyzed by microarray 76 CC398 isolates, including different types of isolates: MRSA and MSSA, invasive and colonizing strains, and strains of porcine and human origin, respectively. Phage $\varphi 3$ and IEC genes were detected in 5 of 17 human invasive isolates, only. Interestingly, all isolates of porcine origin lacked these elements. In addition to phage $\varphi 3$, other MGEs have been studied and their association with CC398 host adaptation and human-to-human transmission has been analyzed (McCarthy et al., 2011, 2012b). Phage $\varphi Av\beta$ is relevant in *S. aureus* adaptation to nonmammal hosts; it carries SAAV_2008 and SAAV_2009 genes (encoding an ornithine cyclodeaminase and a putative membrane protease, respectively), involved in the avian niche-specific activities (Lowder et al., 2009). Aspects related to CC398 host adaptation have been recently reviewed by Ballhausen et al. (2017).

To study the origin and evolution of *S. aureus* CC398, Price et al. (2012) analyzed and compared genomes of a collection of 89 CC398 (40 MSSA and 49 MRSA) isolated from humans and livestock and spanning 19 countries and four continents, allowing for the generation of a highly accurate phylogenetic reconstruction of the CC398 clonal lineage. Phylogenetic results suggest that MRSA CC398 was originated in humans as MSSA CC398, and this ancestral population carried phage $\varphi 3$ with IEC cluster genes. When CC398 adapted to the livestock, i.e., pig environment, this phage was eliminated from the genome of *S. aureus*. This jump from humans to pigs was accompanied by further genomic changes, such as the acquisition of *Tn*916, which harbor the *tet*(M) gene, and the SCC*mec* with methicillin resistance *mecA* gene, probably due to the use of antimicrobials in animal production (Price et al., 2012). The tetracycline resistance gene *tet*(M) was nearly universal among LA-CC398, both in MRSA and MSSA, and was completely missing in CC398 human-associated strains. It was considered that antimicrobial treatment of food-producing animals with tetracycline may have selected for LA-CC398 (MRSA and MSSA) (Price et al., 2012).

According to Price et al. (2012), CC398 would then be formed by two major host-associated distinct populations: (1) an **LA-population**, where CC398 strains carry *tet*(M) gene and do not carry neither phage $\varphi 3$ (*scn*-negative) nor the *lukF-lukS* gene (PVL-negative), and where most of strains are MRSA; and (2) a human-associated population (**H-population**), where MSSA strains carry IEC human host-adaptation genes (*scn*-positive) and do not harbor *tet*(M) gene. Concerning LA-population, a subclade clustering a group of mainly turkey meat isolates with prophage φAvβ, was identified. Regarding H-population, a clade assembles MSSA strains from humans in North America, South America, and Europe that present mainly *spa*-type t571.

Subsequent studies have confirmed that most of the *S. aureus* CC398 found in the animal and livestock environment lack φ3 specific genes (McCarthy et al., 2012a,b; Stegger et al., 2013b; Cuny et al., 2015a). The *scn* and *tet*(M) genes are strongly associated with the H- and LA-population, respectively. Discrimination between host-associated population can be performed by simple assay based in canonical single-nucleotide polymorphism (*can*SNP) assays, that differentiate the two major host-associated *S. aureus* CC398 populations, and a duplex PCR assay for detection of *scn* and *tet*(M) genes (Stegger et al., 2013b).

On the basis of the results of further studies, we could consider that the CC398 **H-population** includes two subpopulations: the ancestral MSSA human subpopulation of LA-isolates (**H-ancestral-LA**), and the MSSA human-associated livestock-independent subpopulation (**H–LI**). The H-ancestral-LA subpopulation comprises isolates of different *spa* types (t034, t108, t011, among others), and the H–LI subpopulation comprises mainly isolates of *spa*-type t571 with highly transmissibility among humans (Uhlemann et al., 2012).

4. NEW *STAPHYLOCOCCUS AUREUS* CC398 EMERGING TOPICS ARISING IN THE LAST FEW YEARS

There is no doubt that livestock, especially pigs, is an important reservoir of MRSA CC398 (Crombé et al., 2013), and that direct contact with farm animals (pigs, cattle, and poultry) constitutes a risk for human CC398 colonization and subsequent infection (van Cleef et al., 2011; Graveland et al., 2010; Mulders et al., 2010). Moreover, high risk of colonization by this *S. aureus* lineage exists in zones with high pig-farming density (Feingold et al., 2012; van Cleef et al., 2011).

But what at first seemed relatively confined and restricted to a specific human population (farmers and veterinarians and their relatives, or zones with high-farming activities), could start to be also found in the general population, what is worrisome. The possibility of this LA-CC398 lineage to become more adapted again to humans and the potential human–human transmission in the general population (nonexposed to farm animals), although still low, is a question of concern.

The epidemiology of CC398 lineage has become more and more complex. Emerging worrisome questions regarding the increasingly frequent detection of CC398 strains (both MRSA and MSSA) in humans without livestock contact and adaptation of this lineage to humans are arisen and will be reviewed considering its health implications in the following.

4.1 LIVESTOCK-ASSOCIATED METHICILLIN-RESISTANT *STAPHYLOCOCCUS AUREUS* CC398 IN HUMANS WITHOUT ANIMAL CONTACT. HUMAN-TO-HUMAN TRANSMISSION

MRSA CC398 isolates, which have long been responsible for frequent and transient colonization in humans with direct livestock contact, belong undoubtedly to LA-population. These isolates, which lack phage φ3, are less adapted to humans; this may explain the traditional association with lower risk of human-to-human transmission of LA-MRSA CC398 isolates. CC398 seems to be less efficient to spread among humans than other human-adapted lineages. Nosocomial transmission of LA-MRSA CC398 seems to be less likely (72%) than that of non-CC398 MRSA strains (Wassenberg et al., 2011). Nevertheless, *S. aureus* CC398 is capable of adequately competing for a niche with a human-adapted strain, surviving in the human nose for at least 21 days (Slingerland et al., 2012). The absence of phage $\varphi 3$ could, at least in part, explain the transient colonization in humans without permanent contact to CC398-carrier animals. Nevertheless, a generalized increasing number of colonization or infection with MRSA CC398 (with characteristic of the LA-population, i.e., *tet*(M)-positive, *scn*-negative) in humans without animal contact are reported (Benito et al., 2014; van Rijen et al., 2014; Lekkerkerk et al., 2015; Larsen et al., 2015; Deiters et al., 2015). Therefore, LA-MRSA CC398 might also have been spread into the general human population. Fifteen percent of all MRSA CC398 human carriers detected in The Netherlands had not been in direct contact with pigs or veal calves (Lekkerkerk et al., 2015). In Denmark, contact with animals was not documented in 34% of MRSA CC398 infection cases (Larsen et al., 2015). In Spain, no livestock contact was referred in ~50% of MRSA CC398 colonization/infections (Benito et al., 2014). Other potential unknown reservoirs or transmission routes of these LA-MRSA CC398 might exist, but direct human-to-human transmission cannot be ruled out. Perhaps human host adaptation could be explained by more than phage $\varphi 3$ alone, or host adaptation might not have to be as extensive as it was assumed, to facilitate human-to-human transmission (Lekkerkerk et al., 2015). McCarthy et al. (2012b) noted that phage $\varphi 7$ was present in 5/9 CC398 isolates from Danish humans without pig contact and also in 10/11 Belgian and Dutch CC398 human-associated isolates (H-population), whereas $\varphi 7$ was absent in non-European isolates. These findings prompted the authors to suggest that the potential acquisition of phage $\varphi 7$ by LA-MRSA could increase the capacity for human-to-human LA-MRSA CC398 transmission, even in absence of phage $\varphi 3$; McCarthy et al. (2012b) did even suggest that this phage $\varphi 7$ could be a suitable marker of human-to-human transmission capability in LA-MRSA population. The emergence of a new variant of CC398 with increased capacity for human-to-human transmission is worrisome, especially in countries with intensive pig farming.

4.2 SECOND JUMP FROM ANIMALS TO HUMANS. ACQUISITION OF IMMUNE EVASION CLUSTER: READAPTATION TO HUMANS

As stated above, MRSA CC398 strains found in humans with livestock contact exhibit the same molecular patterns as MRSA of LA-population (*tet*(M)-positive, *scn*-negative) and seems to be low adapted to humans. The possibility of these strains to readapt to the human niche might enhance its invasiveness and its ability for animal-to-human or human-to-human transmissions. The acquisition of phage $\varphi 3$ is probably one of the first steps of this readaptation process. The IEC genes have been found almost exclusively in MSSA CC398 invasive isolates attributed to the H-population, but the presence of IEC genes in LA-MRSA CC398, although still infrequent, has already been described in animals (Haenni et al., 2011; Stegger et al., 2013b; Cuny et al., 2015a). LA-MRSA CC398 carrying phage $\varphi 3$ (*scn*-positive) have already been found in colonized or infected humans, including cases of severe diseases (Cuny et al., 2015a; Pérez-Moreno et al., 2016). These strains found in humans harbor the *tet*(M) gene marker of LA-population, and may have evolved from LA-MRSA population, who have acquired $\varphi 3$ phage. These strains constitute an emerging public health problem, as the reacquisition of IEC will contribute to enhance the human-to-human transmission capacity. Some other authors suggest that the acquisition of the φMR11-like helper in CC398 phage φ3-positive isolates could increase the human adaptation (van der Mee-Marquet et al., 2013). Taken together, it seems that CC398 human adaptation is a complex process, in which other elements, in addition to phage φ3, are very likely to be implicated.

4.3 EMERGING OF METHICILLIN-SUSCEPTIBLE *STAPHYLOCOCCUS AUREUS* CC398, MAINLY OF *SPA*-TYPE T571

Recently, the emergence of human infections, frequently septicemia, associated with MSSA CC398 in humans without animal contact, was noted (Valentin-Domelier et al., 2011; van der Mee-Marquet et al., 2011; Vandendriessche et al., 2011; Jiménez et al., 2011; Mediavilla et al., 2012; Verkade et al., 2012). MSSA CC398 infections have been reported worldwide including Europe (mainly France), America (mainly northeastern USA), China, and the Caribbean. These strains present *spa*-type t571, are tetracycline-susceptible and PVL-negative.

Already in 2005, MSSA CC398 strains (tetracycline-susceptible) were detected in healthy pig farmers and pigs in France (Armand-Lefevre et al., 2005). In the first studies in which MSSA CC398 bloodstream infection were reported, an animal origin was suspected, although not clearly documented in most of the cases (van der Mee-Marquet, 2011; Verkade et al., 2012). In France, MSSA CC398 accounts for 7.5% of all MSSA endocarditis cases (Chroboczek et al., 2013). CC398 *spa*-type t571 was predominant (11%) among MSSA human clinical isolates in China, where CC9 (non-CC398) was the most common LA-MRSA in pigs (Chen et al., 2010).

Subsequent studies, in which comparative genome analysis was performed, revealed particular characteristics of these strains, different from LA-MRSA CC398: presence of phage φ3 (*scn*-positive), lack of *tet*(M) and a greater virulence potential and highly transmissibility among humans (Uhlemann et al., 2012). These MSSA CC398 *spa*-t571 isolates (*scn*-positive and *tet*(M) negative) seem to have evolved from the H–LI subpopulation (Uhlemann et al., 2012). In addition, they also contain a chromosomally encoded *erm*(T) gene that can be used as an indicator of this origin, when detected in MSSA CC398 *spa*-t571 containing phage φ3 (*scn*-positive) (Vandendriessche et al., 2011; Cuny et al., 2013). MSSA *spa*-type t571 was also sporadically detected among colonized and infected pigs, but isolates involved displayed resistance to tetracycline, one of the feature characteristics of CC398 of animal origin (Hasman et al., 2010).

So far, only few studies have focused on the genetic analysis of MSSA of healthy human population, but this specific MSSA t571 variant has already been detected in human carriers in some studies (Bhat et al., 2009; Lozano et al., 2011b; Gómez-Sanz et al., 2013; Cuny et al., 2013; David et al., 2013). Human colonization of MSSA CC398 (*spa*-type t034) usually occurs for prolonged periods of time (Slingerland et al., 2012), however colonization of LA-MRSA CC398, as mentioned above, is usually more transient.

In addition, the ability of t571 isolates to acquire the *luk-PV* gene, encoding PVL, is another serious threat. Indeed, lethal necrotizing pneumonia caused by MSSA ST398 *spa*-type t571 (PVL-positive, *tet*(M)-negative) has already been reported (Rasigade et al., 2010). The presence of MSSA t571 human infections in geographically dispersed regions supports the notion that specific MSSA strains constitute the potential of a pandemic pathogen (Verkade and Kluytmans, 2014). A very recent research found that households served as major sites for CC398 MSSA and also reveals international routes of transmission for this pandemic MSSA t571 clone from France to Martinique and from New York to the Dominican Republic (Uhlemann et al., 2017).

Curiously, a recent study found that white storks are frequently tracheal carriers of MSSA CC398 *spa*-t571, positive for *scn* and *erm*(T) genes (Gómez et al., 2016). More studies concerning CC398 among wildlife should be conducted in the future to gain knowledge of this specific lineage in the human–animal interface and its epidemiology.

4.4 ACQUISITION OF PANTON–VALENTINE LEUKOCIDIN BY CC398 STRAINS (METHICILLIN-SUSCEPTIBLE *STAPHYLOCOCCUS AUREUS* AND METHICILLIN-RESISTANT *STAPHYLOCOCCUS AUREUS*)

S. aureus CC398 isolates containing *luk-PV* gene, encoding PVL toxin, are of particular interest. PVL has been mainly detected in CA-MRSA strains, but an association with CA-MSSA has also been suggested (Shallcross et al., 2010). Although CC398 MRSA seems to carry *luk-PV* gene less frequently compared with other MRSA lineages (Argudín et al., 2011; Larsen et al., 2015; Ballhausen et al., 2017), this virulence gene has sporadically been detected years ago in CC398 MRSA (mainly of *spa*-t034) in pig farmers in the Netherlands (van Belkum et al., 2008), and from patients without animal contact in Sweden, Denmark, and Finland (Welinder-Olsson et al., 2008; Lewis et al., 2008; Stegger et al., 2010; Salmenlinna et al., 2010). Curiously, three of these cases detected in Nordic countries involved adopted Chinese children. PVL-positive MRSA CC398 of *spa*-type t034 have also been reported in two patients in New Zealand, one of them with a recent travel to South-East Asia (Williamson et al., 2014). Recently, a fatal infection in a Chinese woman (without animal contact) due to an MRSA CC398 PVL-positive strain has been reported in Japan (Koyama et al., 2015). MSSA and MRSA CC398 PVL-positive seem to be highly prevalent in China in humans that do not refer any animal contact, whereas CC9 is the most prevalent lineage among LA-MRSA in animals in this region (Yu et al., 2008; Zhao et al., 2012).

Sporadic infections caused by MSSA and MRSA CC398 PVL-positive have been also detected in Germany (Cuny et al., 2015a); these authors performed genetic tests (PCR for *can*SNPs, *scn*, *tet*(M), and *tet*(K)) to discriminate between the LA-clade and the H-clade as ancestral population, and all MSSA and MRSA CC398 PVL-positive isolates might have evolved from MSSA of the H-clade (and not from the LA-clade after PVL acquisition); they harbored *scn* and *tet*(K) but not *tet*(M), and frequently corresponded to *spa*-t034 (Cuny et al., 2015a). The evolution of this type of strains should also be monitored.

4.5 A NOVEL HYBRID LIVESTOCK-ASSOCIATED METHICILLIN-RESISTANT *STAPHYLOCOCCUS AUREUS* CC9/CC398 GENOTYPE AND FOODBORNE TRANSMISSION

Special attention deserves a new genotype characteristic of strains displaying *spa*-type t899 that cluster together in an out-group that diverges in the tree from other CC398 strains. This new genotype designated as LA-MRSA CC9/CC398 consists of a hybrid of CC398 chromosomal backbone and a smaller CC9 region (Price et al., 2012).

Genome phylogenetic comparative analysis of CC9/CC398 isolates allows to cluster these strains into different clades and subclades. A recent study analyzed a collection of CC9/CC398 isolates obtained from humans, animals, and retail foods from different European countries. MRSA CC9/CC398 isolates from infected people living in urban areas in Denmark (where this genotype is not present in livestock) were also included (Larsen et al., 2016). Most of CC9/CC398 isolates from poultry and poultry meat clustered together in the subclade named poultry associated. Presence of phage *φ3* (related to human adaptation) and of avian prophage *φAvβ* (genetic marker of poultry adaptation) was also analyzed. Genomic results obtained along with epidemiologic data suggest that poultry meat may serve as a vehicle for livestock-to-human transmission and as the source of MRSA CC9/CC398 in urban-dwelling Danes (Larsen et al., 2016). Findings of a subsequent German study, that includes traceback investigations of isolates of poultry origin, support this hypothesis and alert that turkey meat may be a probable source of CC9/CC398 (Fetsch et al., 2017).

Taking into consideration all the aspects referred above in relation to CC398 lineage, we can remark its extremely high plasticity, acquiring different genetic elements related to host adaptation, antimicrobial resistance and virulence. In this sense, the epidemiology of this lineage is more complex than initially assumed. Livestock is still an important source of LA-MRSA and should be intensively monitored, but CC398 human infections should not be systematically attributed to contact with livestock, as an MSSA CC398 highly human-adapted subpopulation is in continuous evolution (with the possibility of acquiring *mecA* or PVL encoding genes).

5. OTHER *STAPHYLOCOCCUS AUREUS* LINEAGES AT THE ANIMAL–HUMAN INTERFACE

5.1 *STAPHYLOCOCCUS AUREUS* LINEAGE CC9

Although CC398 is the LA-MRSA most worldwide found, other LA-MRSA have emerged in particular geographic areas. MRSA CC9 is the prevalent LA-MRSA found in pigs and farm workers in Asian countries (Cui et al., 2009). A recent review summarizes LA-MRSA in Asia and, as in the case of CC398, MRSA CC9 prevalence observed in pigs and humans in contact with them, varies widely from country to country (Chuang and Huang, 2015).

S. aureus isolates of lineage CC9 show a great variety of *spa* types (t337, t4359, t526, and t899, among others). The *spa*-type t899 is also characteristic of CC398 isolates, as previously indicated; therefore, MLST typing of t899 isolates is necessary to discriminate between both CCs.

Although MRSA CC9 is known as "Asian" LA-MRSA, it has also sporadically been detected in Europe in animals (pigs, poultry) (EFSA, 2009a; Battisti et al., 2010; Peeters et al., 2015) and in food of animal origin (Febler et al., 2011; Dhup et al., 2015). The possible importation of these strains might explain these sporadic detections; nevertheless, some questions arise in this respect because MSSA CC9 isolates are frequently found in livestock in Europe and the United States (Hasman et al., 2010; Velasco et al., 2015). The reasons for the differences observed between Asia and the rest of the world remain unclear; although it would be possible that the SCC*mec* element was acquired only by Asian CC9 *S. aureus* strains (Butaye et al., 2016).

Cases of human infections by MRSA CC9 are less frequent than MRSA CC398; some human infections have been reported in Asia (Liu et al., 2009; Wan et al., 2013), and one case has been documented in the Netherlands (van Loo et al., 2007). The potential risk of spreading to humans of LA-MRSA is worrisome (Chuang and Huang, 2015). It is important to point out that, unlike MRSA CC398, enterotoxin genes are frequent in CC9 isolates (Febler et al., 2011; He et al., 2013).

Few reports have described the epidemiology and molecular characteristics of LA-MRSA in Asia. A recent study noted that all 19 *S. aureus* CC9 isolates from livestock workers were MRSA IEC-negative, and tetracycline resistant, indicating that these characteristics can also be considered as markers of livestock association (Ye et al., 2016).

5.2 *STAPHYLOCOCCUS AUREUS* LINEAGE CC130

Historically, *S. aureus* CC130 was considered a bovid-related lineage because MSSA-CC130 strains have been a common cause of infections in sheep (Monecke et al., 2016a; Porrero et al., 2012) and are as well present not only in healthy sheep and goats (Gharsa et al., 2012, 2015) and wild ruminants (Luzzago et al., 2014), but also in the environment (Gómez et al., 2017). Previous studies have demonstrated that this lineage could have been the result of a human-to-bovine host-jump, resulting on the bovine-associated CC130 lineage; in this sense, it has been calculated that this host-jump occurred ~5429 years ago, which is coincident historically with the spread of bovid domestication in the Old World (Weinert et al., 2012). To our knowledge, no *mecA*-positive CC130 isolates had been reported, and CC130 was initially associated only with MSSA animal isolates, detected in Europe and Africa (Smith et al., 2014). Nevertheless, during the last 6 years, MRSA CC130 is gaining more and more interest because of the fact that this is the main lineage, which contains the new *mecC* gene (*mecA* gene homolog, located at SCC*mec*XI), conferring methicillin resistance (García-Álvarez et al., 2011). It is possible that *mecC*-positive MRSA CC130 circulated for some time ago but were not detected before 2011 because of the lack of specific diagnostic tests. In this sense, in a retrospective study, a *mecC*-positive CC130 strain was found in a human sample isolated in 1975 (García-Álvarez et al., 2011). It is interesting to highlight that the new technologies applied on genome analysis have demonstrated that the CC130 linage, considered to be unique to animals, displays a low host specificity, i.e., is able to colonize and adapt to different species. According to the *S. aureus* MLST database (saureus.beta.mlst.net), 38 different STs are assigned to CC130 of which 29 are SLVs of ST130 (ST480, ST483, ST700, ST1245, ST1526, ST1627, ST1764, ST1944, ST1945, ST2024, ST2413, ST2490, ST2676, ST2944, ST3083, ST3089, ST3094, ST3095, ST3107, ST3108, ST3114, ST3115, ST3123, ST3124, ST3131, ST3132, ST3133, ST3138, and ST3139); and 8 are DLVs (ST1614, ST1739, ST2021, ST2574, ST2620, ST2899, ST3044, and ST3061). Of these 38 STs described, 11 have been found as MRSA (ST130, ST1245, ST1526, ST1764, ST1944, ST1945, ST2573, ST2574, ST2620, ST2676, and ST3061) and 4 as MSSA (ST130, ST700, ST1739, and ST2011). The lineage CC130 has been principally detected in cattle, being currently uncommon in human community infections, suggesting a zoonotic reservoir (Aires-de-Sousa, 2017). Of particular concern is the zoonotic transmission of this lineage; Harrison et al. (2013) corroborated a zoonotic transmission of MRSA CC130 between animals and humans of the same farm, detected by the analysis of single-nucleotide polymorphisms (SNPs), suggesting the ability of these strains to colonize and infect humans.

Up until now, *mecC* gene has been found in different animal-associated MRSA lineages, in addition to CC130: CC49, ST151, ST425, CC599, and CC1943. The most frequent CCs carrying *mecC* were CC130, CC1943, and ST425. *mecC*-MRSA strains of different CCs have been reported in many different European countries (Austria, Belgium, Denmark, France, Germany, Ireland, Sweden, the Netherlands, Norway, Slovenia, Spain, Switzerland, and the United Kingdom) (Aires-de-Sousa, 2017; see as well references of Fig. 10.1); recently a single case in a domestic cat (ST425) has also been described in Australia (Worthing et al., 2016). The large diversity of hosts carrying MRSA

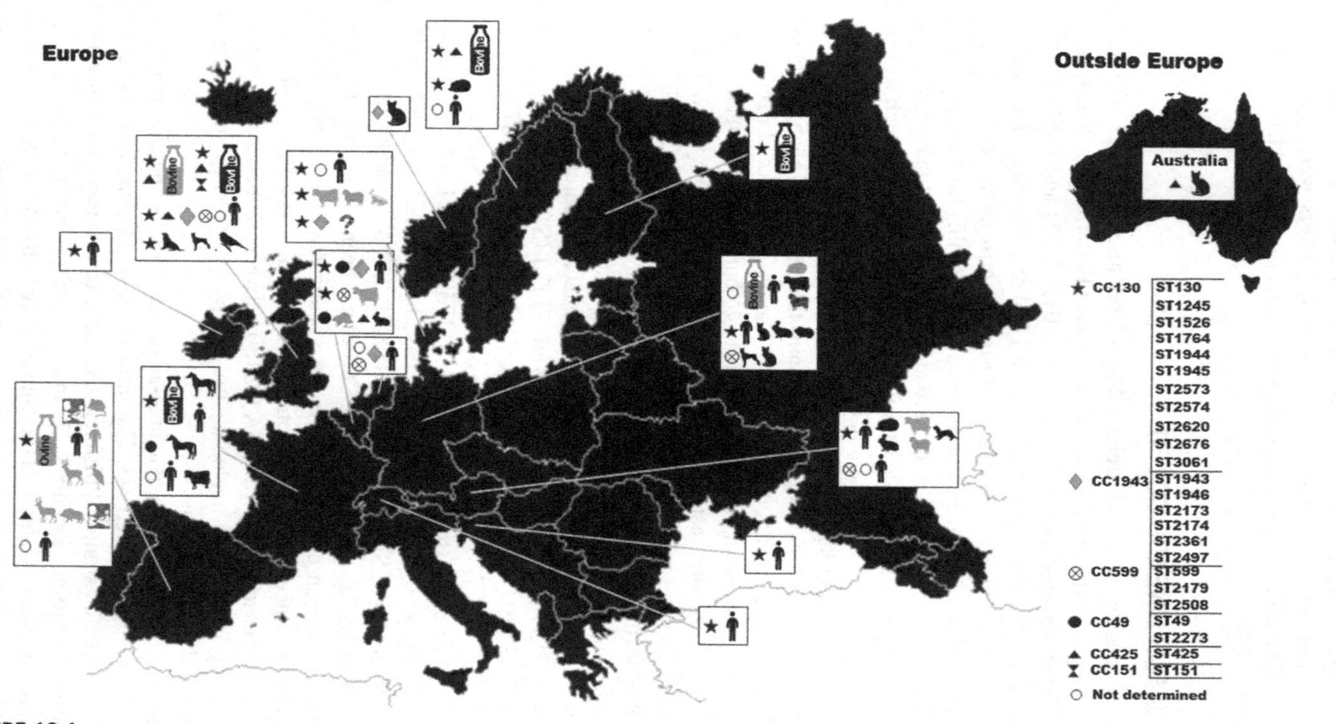

FIGURE 10.1

Genetic lineages (clonal complexes or sequence type), origin, and geographical location of MRSA isolates carrying the *mecC* gene reported in the literature. Animals, humans or milk in green (gray in print version): healthy individuals; animals, humans or milk in brown (black in print version): sick individuals; animal marbled: unknown health status; Milk bottle: milk sample; question mark: unknown origin.

Data obtained of references: Ariza-Miguel et al., 2014; Barraud et al., 2013; Basset et al., 2013; Benito et al., 2016; Cuny et al., 2011; Deplano et al., 2014; Dermota et al., 2015a,b; Eriksson et al., 2013; Espinosa-Gongora et al., 2015; García-Álvarez et al., 2011; García-Garrote et al., 2014; Gindonis et al., 2013; Gómez et al., 2014, 2015, 2016; Haenni et al., 2014, 2015; Harrison et al., 2013; Kerschner et al., 2014; Kriegeskorte et al., 2012; Laurent et al., 2012; Lindgren et al., 2016; Loncaric et al., 2013, 2014; Medhus et al., 2013; Monecke et al., 2013, 2016a,b; Paterson et al., 2012, 2013; 2014; Pichon et al., 2012; Porrero et al., 2014a,b; Romero-Gómez et al., 2013; Sabat et al., 2012; Schaumburg et al., 2012; Schlotter et al., 2014; Shore et al., 2011; Stegger et al., 2012; Unnerstad et al., 2013; Vandendriessche et al., 2013; Walther et al., 2012; Worthing et al., 2016; Zarfel et al., 2016.

mecC-positive strains, reported in livestock, wildlife and companion animals, humans, as well as in environmental samples (wastewater and river water) is shown in Fig. 10.1.

Nowadays, the detection of *mecC*-positive MRSA strains in food-producing animals was related to beef and dairy cattle (ovine and bovine), typed in most of the cases as CC130 and CC599 (Eriksson et al., 2013; Haenni et al., 2014; Laurent et al., 2012; Loncaric et al., 2014; Schlotter et al., 2014; Vandendriessche et al., 2013). On the other hand, different studies report the presence of MRSA ST130, ST425, and ST2573 in milk samples of healthy animals (bovine and ovine) and MRSA ST130, ST151, ST425, ST1245, and ST1526 in milk samples of mastitis cattle, all isolates included in lineages CC130, ST425, and ST151 (Ariza-Miguel et al., 2014; García-Álvarez et al., 2011; Gindonis et al., 2013; Paterson et al., 2013; Schlotter et al., 2014; Unnerstad et al., 2013). It is known that *S. aureus* plays an important role as a major cause of mastitis in dairy animals that potentially could enter the food chain via raw milk. Very recently, a case description of a Spanish artisan cheese producer, suffering from a cutaneous lesion from which an MRSA ST130 *mecC*-positive strain was detected has been published, suggesting a potential zoonotic transmission (Benito et al., 2016).

The genes associated to the IEC, used as a marker of human origin, have been studied in some of the *mecC*-positive MRSA isolates, rendering in the majority of the cases a negative result in isolates recovered from humans (of lineage ST130), as well as in those from livestock and wild animals (of lineages ST130, ST425, and ST3061), suggesting an animal origin of these isolates (Benito et al., 2016; Dermota et al., 2015b; Gómez et al., 2016; Haenni et al., 2014; Harrison et al., 2013; Monecke et al., 2013; Worthing et al., 2016); nevertheless, it is interesting to remark that IEC-positive strains have been recently detected in *mecC*-positive MRSA ST1945 strains recovered from wild animals, mainly red deer and small mammals (Gómez et al., 2014, 2015). It seems that *mecC*-MRSA isolates of CC130 lineage with and without the IEC system are circulating in wild animals, what rise questions about the origin and evolution of this lineage located in the animal–human interface.

Few studies report data related to enterotoxin gene detection in *mecC*-MRSA isolates; the content of *se* genes seems to be low, although the *sec*, *seg*, *sei*, *sel*, *sen*, *seo*, and *seu* genes have been detected in one ST2361 strain, and the *sel* and/or *sec* genes in two additional ST599 and ST49 strains, none of them derived from livestock or food of animal origin (Haenni et al., 2015; Kerschner et al., 2014; Sabat et al., 2012).

5.3 *STAPHYLOCOCCUS AUREUS* LINEAGE CC97

CC97 is a bovine-associated lineage that seems to have a human ancestor, which jumped from humans to bovines ~1200 years ago (Weinert et al., 2012). The lineage CC97 has been frequently implicated in bovine mastitis, both as MSSA (Rabello et al., 2007) and as MRSA (Wang et al., 2015; Luini et al., 2015). Less commonly, CC97 has also been reported to cause infections of small ruminants, pigs, and humans (Spoor et al., 2013). Variants of MSSA CC97 have also been detected in cow milk (Ben Said et al., 2016) or in nasal samples of healthy sheep (Gharsa et al., 2012). The MRSA CC97 lineage has been reported as the second most prevalent lineage in Italian pig industry (Battisti et al., 2010; Feltrin et al., 2015) and has also been detected in slaughter pigs in Spain (Gómez-Sanz et al., 2010).

CC97 is one of the most important *S. aureus* bovine lineages (Spoor et al., 2013). Nevertheless, a bovine–human host back jump of MRSA CC97 seems to have occurred about 40 years ago; later in humans, CC97 *S. aureus* has adapted and is evolving, may even become a pandemic CA-MRSA, with public health implications (Spoor et al., 2013; Feltrin et al., 2015). In this sense, livestock represent a reservoir for the emergence of new human-pathogenic *S. aureus* clones with the capacity for pandemic spread. Surveillance for early identification of emergent clones and improved transmission control measures are important to be developed for these clones that are located in the animal–human interface (Spoor et al., 2013).

S. aureus CC97 has also evolved in the ruminant niche. In this sense, a novel and widespread hybrid clone of LA *S. aureus*, named ST71, has been reported and seems to have evolved from an ancestor of the bovine lineage CC97, after multiple large-scale recombination events with other *S. aureus* lineages that shared the same ruminant ecosystem (Spoor et al., 2015; Budd et al., 2015). This novel ST71 hybrid clone acquired different pathogenic traits associated with the acquired and innate immune evasion and bovine extracellular matrix adherence (Spoor et al., 2015).

5.4 *STAPHYLOCOCCUS AUREUS* LINEAGE CC1

The MRSA CC1 lineage corresponds to a successful CA clone detected in humans (Butaye et al., 2016). This lineage, which is frequently been ascribed to the *spa*-type t127, has also been detected in livestock, i.e., in pigs or dairy cows in Italy (Alba et al., 2015; Luini et al., 2015), but has also been detected in other countries (Butaye et al., 2016). In dairy cows, MRSA CC1 have been cause of mastitis, too (Luini et al., 2015). Different toxin genes have been detected among the animal-associated CC1 isolates, but so far not the ones encoding PVL, also they can carry the IEC genes. In contrast, human-associated CC1 isolates tend to carry the PVL encoding genes (Butaye et al., 2016).

6. SUMMARY AND CONCLUSION

The epidemiology of *S. aureus* has been for a long time mostly associated to the HA and later to the CA in the human environment. Nevertheless, in the last decade, important epidemiological changes have occurred with the emergence of genetic lineages located in the animal–human interface, being CC398 the most relevant one, although other lineages are also of relevance, such as CC9, CC97, or CC130, among others. The zoonotic capacity of these lineages should be tracked, analyzing potential host-jump events that might occur in these and other lineages, to identify new emergent clones adapted to specific hosts (specially to humans). In the past, molecular epidemiology studies in *S. aureus* were mostly restricted to MRSA strains; nowadays, the importance of molecular typing of MSSA strains became also clear, to understand the evolution of specific genetic lineages in the different ecosystems, including those located in the animal–human interface. Wherever there is an interface between different host species, the opportunity exists for bacterial exchange.

Molecular epidemiology of the lineage CC398 is a complex issue. This lineage was initially assumed as LA-MRSA (mostly pig-associated), with capacity to be transmitted to in-contact humans, causing in some occasions infections, and human-to-human transmission very rarely occurred (except among relatives of in-contact humans). This lineage was later detected in other livestock animals, in addition to pigs. Later on, it was demonstrated that LA-MRSA was the result of a human–animal host-jump from an ancestral MSSA CC398 that lost the *scn* gene (related to human host adaptation) and acquired the *mecA* and *tet*(M) genes; in this way two CC398 populations were defined: **H-population** (MSSA, *scn*-positive and *tet*(M) negative) and **LA-population** (MRSA, *scn*-negative and *tet*(M)-positive, low capacity for human–human transmission). MSSA CC398 H-population includes: (1) **H-ancestral-LA**, ancestral MSSA that suffered the human–animal jump to become

LA-MRSA (mostly of *spa* types t034, t108, and t011, among others); and (2) **H–LI**, livestock independent, mostly of *spa*-type t571 (*erm*(T)-positive), with high transmissibility among humans; curiously, this MSSA t571 seems to be also frequent in some wild birds.

In the last years, emerging issues are arising in relation to CC398 *S. aureus*: (1) more and more frequently LA-MRSA CC398 (*scn*-negative) are being detected among humans with no livestock risk factors; consequently, it should be evaluated if human adaptation is increasing (even in absence of IEC system), or whether unknown CC398 reservoirs might exist; (2) Acquisition of the IEC system (human adaptation–related genes) by LA-MRSA, increasing human adaptation of this LA-population and in return, the capacity for human-to-human transmission; (3) Dissemination of the highly transmissible MSSA CC398 t571 subpopulation, with potential acquisition of *mecA* gene; (4) Acquisition of PVL encoding genes by MSSA or MRSA CC398 with increased virulence; (5) Emergence of new hybrids, LA-MRSA CC398/CC9, with higher capacity to be transmitted through the food chain. The research in this specific lineage is very intensive and is providing important information about the capacity of this lineage to evolve and to adapt to different hosts.

The lineage CC130, mostly identified and characterized in MRSA strains carrying the *mecC* gene, has also attracted the attention of researchers in the last years. It has been detected mostly in Europe in many different animals in wildlife (with low host specificity), as well as in farm animals, with some few human infections reported. This lineage seems to have an animal origin, although there are some aspects that require further study: although most of MRSA CC130 of animal origin lack the *scn* gene, there are also strains that carry this gene, as well as others of the IEC system. A more complete analysis of MRSA CC130 strains IEC-positive and -negative by WGS would give important information about the evolution of this lineage in the wildlife and in other ecosystems.

Taken together, *S. aureus* is a very dynamic microorganism that is in continuous evolution, with high capacity to acquire MGEs, which encode key proteins for host adaptation, in addition to antimicrobial resistance or virulence characteristics. The genetic plasticity of this microorganism is the primary evolutionary driving force behind the emergence of new strains with capacity to adapt to new hosts or with increased resistance or virulence, and in the interface between different host species, the opportunity for bacterial exchange exists with possible important implications for public health. Continuous surveillance should be performed to track the evolution of this species.

ACKNOWLEDGMENTS

Paula Gómez and Sara Ceballos have a predoctoral fellowship of the University of La Rioja (Logroño, Spain). The research carried out in the group has been financed by Project SAF2016-76571-R from the Agencia Estatal de Investigación (AEI) of Spain and the Fondo Europeo de Desarrollo Regional (FEDER).

REFERENCES

Aires-de-Sousa, M., 2017. Methicillin-resistant *Staphylococcus aureus* among animals: current overview. Clin. Microbiol. Infect. 23 (6), 373–380. http://dx.doi.org/10.1016/j.cmi.2016.11.002.

Alba, P., Feltrin, F., Cordaro, G., Porrero, M.C., Kraushaar, B., Argudín, M.A., Nykäsenoja, S., Monaco, M., Stegger, M., Aarestrup, F.M., Butaye, P., Franco, A., Battisti, A., 2015. Livestock-associated methicillin resistant and methicillin susceptible *Staphylococcus aureus* sequence type (CC)1 in European farmed animals: high genetic relatedness of isolates from Italian cattle herds and humans. PLoS One 10 (8), e0137143.

Argudín, M.A., Cariou, N., Salandre, O., Le Guennec, J., Nemeghaire, S., Butaye, P., 2013. Genotyping and antimicrobial resistance of *Staphylococcus aureus* isolates from diseased turkeys. Avian Pathol. 42 (6), 572–580.

Argudín, M.A., Rodicio, M.R., Guerra, B., 2010. The emerging methicillin-resistant *Staphylococcus aureus* ST398 clone can easily be typed using the Cfr9I SmaI-neoschizomer. Lett. Appl. Microbiol. 50 (1), 127–130.

Argudín, M.A., Tenhagen, B.A., Fetsch, A., Sachsenröder, J., Käsbohrer, A., Schroeter, A., Hammerl, J.A., Hertwig, S., Helmuth, R., Bräunig, J., Mendoza, M.C., Appel, B., Rodicio, M.R., Guerra, B., 2011. Virulence and resistance determinants of German *Staphylococcus aureus* ST398 isolates from nonhuman sources. Appl. Environ. Microbiol. 77, 3052–3060.

Ariza-Miguel, J., Hernández, M., Fernández-Natal, I., Rodríguez-Lázaro, D., 2014. Methicillin-resistant *Staphylococcus aureus* harboring *mecC* in livestock in Spain. J. Clin. Microbiol. 52 (11), 4067–4069.

Armand-Lefevre, L., Ruimy, R., Andremont, A., 2005. Clonal comparison of *Staphylococcus aureus* isolates from healthy pig farmers, human controls, and pigs. Emerg. Infect. Dis. 11 (5), 711–714.

Arriola, C.S., Güere, M.E., Larsen, J., Skov, R.L., Gilman, R.H., Gonzalez, A.E., Silbergeld, E.K., 2011. Presence of methicillin-resistant *Staphylococcus aureus* in pigs in Peru. PLoS One 6 (12), e28529.

Ballhausen, B., Kriegeskorte, A., van Alen, S., Jung, P., Köck, R., Peters, G., Bischoff, M., Becker, K., 2017. The pathogenicity and host adaptation of livestock-associated MRSA CC398. Vet. Microbiol. 200 (2), 39–45. http://dx.doi.org/10.1016/j.vetmic.2016.05.006.

Barraud, O., Laurent, F., François, B., Bes, M., Vignon, P., Ploy, M.C., 2013. Severe human bone infection due to methicillin-resistant *Staphylococcus aureus* carrying the novel *mecC* variant. J. Antimicrob. Chemother. 68 (12), 2949–2950.

Basset, P., Prod'hom, G., Senn, L., Greub, G., Blanc, D.S., 2013. Very low prevalence of meticillin-resistant *Staphylococcus aureus* carrying the *mecC* gene in western Switzerland. J. Hosp. Infect. 83 (3), 257–259.

Battisti, A., Franco, A., Merialdi, G., Hasman, H., Iurescia, M., Lorenzetti, R., Feltrin, F., Zini, M., Aarestrup, F.M., 2010. Heterogeneity among methicillin-resistant *Staphylococcus aureus* from Italian pig finishing holdings. Vet. Microbiol. 142 (3–4), 361–366.

Becker, K., Ballhausen, B., Kahl, B.C., Köck, R., 2015. The clinical impact of livestock-associated methicillin-resistant *Staphylococcus aureus* of the clonal complex 398 for humans. Vet. Microbiol. http://dx.doi.org/10.1016/j.vetmic.2015.11.013.

Benito, D., Gómez, P., Aspiroz, C., Zarazaga, M., Lozano, C., Torres, C., 2016. Molecular characterization of *Staphylococcus aureus* isolated from humans related to a livestock farm in Spain, with detection of MRSA-CC130 carrying *mecC* gene: a zoonotic case? Enferm. Infecc. Microbiol. Clin. 34 (5), 280–285.

Benito, D., Lozano, C., Rezusta, A., Ferrer, I., Vasquez, M.A., Ceballos, S., Zarazaga, M., Revillo, M.J., Torres, C., 2014. Characterization of tetracycline and methicillin resistant *Staphylococcus aureus* strains in a Spanish hospital: is livestock-contact a risk factor in infections caused by MRSA CC398? Int. J. Med. Microbiol. 304 (8), 1226–1232.

Ben Said, M., Abbassi, M.S., Bianchini, V., Sghaier, S., Cremonesi, P., Romanò, A., Gualdi, V., Hassen, A., Luini, M.V., 2016. Genetic characterization and antimicrobial resistance of *Staphylococcus aureus* isolated from bovine milk in Tunisia. Lett. Appl. Microbiol. 63 (6), 473–481.

Bens, C.C., Voss, A., Klaassen, C.H., 2006. Presence of a novel DNA methylation enzyme in methicillin-resistant *Staphylococcus aureus* isolates associated with pig farming leads to uninterpretable results in standard pulsed-field gel electrophoresis analysis. J. Clin. Microbiol. 44 (5), 1875–1876.

Bhat, M., Dumortier, C., Taylor, B., Miller, M., Vasquez, G., Yunen, J., Brudney, K., Sánchez-E, J., Rodriguez-Taveras, C., Rojas, R., Leon, P., Lowy, F.D., 2009. *Staphylococcus aureus* ST398, New York city and Dominican Republic. Emerg. Infect. Dis. 15, 285–287.

Budd, K.E., McCoy, F., Monecke, S., Cormican, P., Mitchell, J., Keane, O.M., 2015. Extensive genomic diversity among bovine-adapted *Staphylococcus aureus*: evidence for a genomic rearrangement within CC97. PLoS One 10 (8), e0134592.

Butaye, P., Argudin, M.A., Smith, T.C., 2016. Livestock-Associated MRSA and its current evolution. Curr. Clin. Microbiol. Rep. 3, 19–31.

Chen, H., Liu, Y., Jiang, X., Chen, M., Wang, H., 2010. Rapid change of methicillin-resistant *Staphylococcus aureus* clones in a Chinese tertiary care hospital over a 15-year period. Antimicrob. Agents Chemother. 54 (5), 1842–1847.

Chroboczek, T., Boisset, S., Rasigade, J.P., Tristan, A., Bes, M., Meugnier, H., Vandenesch, F., Etienne, J., Laurent, F., 2013. Clonal complex 398 methicillin susceptible *Staphylococcus aureus*: a frequent unspecialized human pathogen with specific phenotypic and genotypic characteristics. PLoS One 8 (11), e68462.

Chuang, Y.Y., Huang, Y.C., 2015. Livestock-associated meticillin-resistant *Staphylococcus aureus* in Asia: an emerging issue? Int. J. Antimicrob. Agents 45 (4), 334–340.

Crombé, F., Argudín, M.A., Vanderhaeghen, W., Hermans, K., Haesebrouck, F., Butaye, P., 2013. Transmission dynamics of methicillin-resistant *Staphylococcus aureus* in pigs. Front. Microbiol. 4, 57.

Cui, S., Li, J., Hu, C., Jin, S., Li, F., Guo, Y., Ran, L., Ma, Y., October 2009. Isolation and characterization of methicillin-resistant *Staphylococcus aureus* from swine and workers in China. J. Antimicrob. Chemother. 64 (4), 680–683.

Cuny, C., Abdelbary, M., Layer, F., Werner, G., Witte, W., 2015a. Prevalence of the immune evasion gene cluster in *Staphylococcus aureus* CC398. Vet. Microbiol. 177 (1–2), 219–223.

Cuny, C., Layer, F., Köck, R., Werner, G., Witte, W., 2013. Methicillin susceptible *Staphylococcus aureus* (MSSA) of clonal complex CC398, t571 from infections in humans are still rare in Germany. PLoS One 8 (12), e83165.

Cuny, C., Layer, F., Strommenger, B., Witte, W., 2011. Rare occurrence of methicillin-resistant *Staphylococcus aureus* CC130 with a novel *mecA* homologue in humans in Germany. PLoS One 6 (9), e24360.

Cuny, C., Wieler, L.H., Witte, W., 2015b. Livestock-associated MRSA: the impact on humans,. Antibiotics (Basel) 4 (4), 521–543.

David, M.Z., Siegel, J., Lowy, F.D., Zychowski, D., Taylor, A., Lee, C.J., Boyle-Vavra, S., Daum, R.S., 2013. Asymptomatic carriage of sequence type 398, *spa* type t571 methicillin-susceptible *Staphylococcus aureus* in an urban jail: a newly emerging, transmissible pathogenic strain. J. Clin. Microbiol. 51 (7), 2443–2447.

Deiters, C., Günnewig, V., Friedrich, A.W., Mellmann, A., Köck, R., 2015. Are cases of methicillin-resistant *Staphylococcus aureus* clonal complex (CC) 398 among humans still livestock-associated? Int. J. Med. Microbiol. 305 (1), 110–113.

Deplano, A., Vandendriessche, S., Nonhoff, C., Denis, O., 2014. Genetic diversity among methicillin-resistant *Staphylococcus aureus* isolates carrying the *mecC* gene in Belgium. J. Antimicrob. Chemother. 69 (6), 1457–1460.

Dermota, U., Mueller-Premru, M., Švent-Kučina, N., Petrovič, Ž., Ribič, H., Rupnik, M., Janežič, S., Zdovc, I., Grmek-Košnik, I., 2015a. Survey of community-associated-methicillin-Resistant *Staphylococcus aureus* in Slovenia: identification of community-associated and livestock-associated clones. Int. J. Med. Microbiol. 305 (6), 505–510.

Dermota, U., Zdovc, I., Strumbelj, I., Grmek-Kosnik, I., Ribic, H., Rupnik, M., Golob, M., Zajc, U., Bes, M., Laurent, F., Mueller-Premru, M., 2015b. Detection of methicillin-resistant *Staphylococcus aureus* carrying the *mecC* gene in human samples in Slovenia. Epidemiol. Infect. 143 (5), 1105–1108.

Dhup, V., Kearns, A.M., Pichon, B., Foster, H.A., 2015. First report of identification of livestock-associated MRSA ST9 in retail meat in England. Epidemiol. Infect. 143, 2989–2992.

Enright, M.C., Robinson, D.A., Randle, G., Feil, E.J., Grundmann, H., Spratt, B.G., 2002. The evolutionary history of methicillin-resistant *Staphylococcus aureus* (MRSA). Proc. Natl. Acad. Sci. U.S.A. 99 (11), 7687–7692.

Eriksson, J., Espinosa-Gongora, C., Stamphøj, I., Larsen, A.R., Guardabassi, L., 2013. Carriage frequency, diversity and methicillin resistance of *Staphylococcus aureus* in Danish small ruminants. Vet. Microbiol. 163 (1–2), 110–115.

Espinosa-Gongora, C., Harrison, E.M., Moodley, A., Guardabassi, L., Holmes, M.A., 2015. MRSA carrying *mecC* in captive mara. J. Antimicrob. Chemother. 70 (6), 1622–1624.

European Food Safety Authority (EFSA), 2009a. Analysis of the baseline survey on the prevalence of methicillin-resistant *Staphylococcus aureus* (MRSA) in holdings with breeding pigs, in the EU, 2008. Part A: MRSA prevalence estimates. EFSA J. 7 (11), 1376.

European Food Safety Authority (EFSA), 2009b. Scientific opinion of the panel on biological hazards on a request from the European Commission on assessment of the public health significance of meticillin resistant *Staphylococcus aureus* (MRSA) in animals and foods. EFSA J. 993, 1–73.

Feingold, B.J., Silbergeld, E.K., Curriero, F.C., van Cleef, B.A., Heck, M.E., Kluytmans, J.A., 2012. Livestock density as risk factor for livestock-associated methicillin-resistant *Staphylococcus aureus*, The Netherlands. Emerg. Infect. Dis. 18 (11), 1841–1849.

Feltrin, F., Alba, P., Kraushaar, B., Ianzano, A., Argudín, M.A., Di Matteo, P., Porrero, M.C., Aarestrup, F.M., Butaye, P., Franco, A., Battisti, A., 2015. A livestock-associated, multidrug-resistant, methicillin-resistant *Staphylococcus aureus* clonal complex 97 lineage spreading in dairy cattle and pigs in Italy. Appl. Environ. Microbiol. 82 (3), 816–821.

Fetsch, A., Kraushaar, B., Käsbohrer, A., Hammerl, J.A., 2017. Turkey meat as source of CC9/CC398 methicillin-resistant *Staphylococcus aureus* in humans? Clin. Infect. Dis. 64 (1), 102–103.

Febler, A.T., Kadlec, K., Hassel, M., Hauschild, T., Eidam, C., Ehricht, R., Monecke, S., Schwarz, S., 2011. Characterization of methicillin-resistant *Staphylococcus aureus* isolates from food and food products of poultry origin in Germany. Appl. Environ. Microbiol. 77 (20), 7151–7157.

Floras, A., Lawn, K., Slavic, D., Golding, G.R., Mulvey, M.R., Weese, J.S., 2010. Sequence type 398 meticillin-resistant *Staphylococcus aureus* infection and colonisation in dogs. Vet. Rec. 166 (26), 826–827.

García-Álvarez, L., Holden, M.T., Lindsay, H., Webb, C.R., Brown, D.F., Curran, M.D., Walpole, E., Brooks, K., Pickard, D.J., Teale, C., Parkhill, J., Bentley, S.D., Edwards, G.F., Girvan, E.K., Kearns, A.M., Pichon, B., Hill, R.L., Larsen, A.R., Skov, R.L., Peacock, S.J., Maskell, D.J., Holmes, M.A., 2011. Methicillin-resistant *Staphylococcus aureus* with a novel *mecA* homologue in human and bovine populations in the UK and Denmark: a descriptive study. Lancet Infect. Dis. 11 (8), 595–603.

García-Garrote, F., Cercenado, E., Marín, M., Bal, M., Trincado, P., Corredoira, J., Ballesteros, C., Pita, J., Alonso, P., Vindel, A., 2014. Methicillin-resistant *Staphylococcus aureus* carrying the *mecC* gene: emergence in Spain and report of a fatal case of bacteraemia. J. Antimicrob. Chemother. 69 (1), 45–50.

Gharsa, H., Ben Slama, K., Gómez-Sanz, E., Lozano, C., Zarazaga, M., Messadi, L., Boudabous, A., Torres, C., 2015. Molecular characterization of *Staphylococcus aureus* from nasal samples of healthy farm animals and pets in Tunisia. Vector Borne Zoonotic Dis. 15 (2), 109–115.

Gharsa, H., Ben Slama, K., Lozano, C., Gómez-Sanz, E., Klibi, N., Ben Sallem, R., Gómez, P., Zarazaga, M., Boudabous, A., Torres, C., 2012. Prevalence, antibiotic resistance, virulence traits and genetic lineages of *Staphylococcus aureus* in healthy sheep in Tunisia. Vet. Microbiol. 156, 367–373.

Gindonis, V., Taponen, S., Myllyniemi, A.L., Pyörälä, S., Nykäsenoja, S., Salmenlinna, S., Lindholm, L., Rantala, M., 2013. Occurrence and characterization of methicillin-resistant staphylococci from bovine mastitis milk samples in Finland. Acta Vet. Scand. 55, 61.

Gómez, P., Casado, C., Sáenz, Y., Ruiz-Ripa, L., Estepa, V., Zarazaga, M., Torres, C., 2017. Diversity of species and antimicrobial resistance determinants of staphylococci in superficial waters in Spain. FEMS Microbiol. Ecol. 93 (1). http://dx.doi.org/10.1093/femsec/fiw208.

Gómez, P., González-Barrio, D., Benito, D., García, J.T., Viñuela, J., Zarazaga, M., Ruiz-Fons, F., Torres, C., 2014. Detection of methicillin-resistant *Staphylococcus aureus* (MRSA) carrying the *mecC* gene in wild small mammals in Spain. J. Antimicrob. Chemother. 69 (8), 2061–2064.

Gómez, P., Lozano, C., Camacho, M.C., Lima-Barbero, J.F., Hernández, J.M., Zarazaga, M., Höfle, Ú., Torres, C., 2016. Detection of MRSA ST3061-t843-*mecC* and ST398-t011-*mecA* in white stork nestlings exposed to human residues. J. Antimicrob. Chemother. 71 (1), 53–57.

Gómez, P., Lozano, C., González-Barrio, D., Zarazaga, M., Ruiz-Fons, F., Torres, C., 2015. High prevalence of methicillin-resistant *Staphylococcus aureus* (MRSA) carrying the *mecC* gene in a semi-extensive red deer (*Cervus elaphus hispanicus*) farm in Southern Spain. Vet. Microbiol. 177 (3–4), 326–331.

Gómez-Sanz, E., Simón, C., Ortega, C., Gómez, P., Lozano, C., Zarazaga, M., Torres, C., 2014. First detection of methicillin-resistant *Staphylococcus aureus* ST398 and *Staphylococcus pseudintermedius* ST68 from hospitalized equines in Spain. Zoonoses Public Health 61 (3), 192–201.

Gómez-Sanz, E., Torres, C., Lozano, C., Zarazaga, M., 2013. High diversity of *Staphylococcus aureus* and *Staphylococcus pseudintermedius* lineages and toxigenic traits in healthy pet-owning household members. Underestimating normal household contact? Comp. Immunol. Microbiol. Infect. Dis. 36 (1), 83–94.

Gómez-Sanz, E., Torres, C., Lozano, C., Fernandez-Pérez, E., Aspiroz, C., Ruiz-Larrea, F., Zarazaga, M., 2010. Detection, molecular characterization, and clonal diversity of methicillin-resistant *Staphylococcus aureus* CC398 and CC97 in Spanish slaughter pigs of different age groups. Foodborne Pathog. Dis. 7 (10), 1269–1277.

Graveland, H., Duim, B., van Duijkeren, E., Heederik, D., Wagenaar, J.A., 2011. Livestock-associated methicillin-resistant *Staphylococcus aureus* in animals and humans. Int. J. Med. Microbiol. 301 (8), 630–634.

Graveland, H., Wagenaar, J.A., Heesterbeek, H., Mevius, D., van Duijkeren, E., Heederik, D., 2010. Methicillin resistant *Staphylococcus aureus* ST398 in veal calf farming: human MRSA carriage related with animal antimicrobial usage and farm hygiene. PLoS One 5, e10990.

Groves, M.D., O'Sullivan, M.V., Brouwers, H.J., Chapman, T.A., Abraham, S., Trott, D.J., Al Jassim, R., Coombs, G.W., Skov, R.L., Jordan, D., 2014. *Staphylococcus aureus* ST398 detected in pigs in Australia. J. Antimicrob. Chemother. 69 (5), 1426–1428.

Guardabassi, L., Stegger, M., Skov, R., 2007. Retrospective detection of methicillin resistant and susceptible *Staphylococcus aureus* ST398 in Danish slaughter pigs. Vet. Microbiol. 122 (3–4), 384–386.

Haenni, M., Châtre, P., Boisset, S., Carricajo, A., Bes, M., Laurent, F., Madec, J.Y., 2011. Staphylococcal nasal carriage in calves: multiresistant *Staphylococcus sciuri* and immune evasion cluster (IEC) genes in methicillin-resistant *Staphylococcus aureus* ST398. J. Antimicrob. Chemother. 66 (8), 1927–1928.

Haenni, M., Châtre, P., Dupieux, C., Métayer, V., Maillard, K., Bes, M., Madec, J.Y., Laurent, F., 2015. *mecC*-positive MRSA in horses. J. Antimicrob. Chemother. 70 (12), 3401–3402.

Haenni, M., Châtre, P., Tasse, J., Nowak, N., Bes, M., Madec, J.Y., Laurent, F., 2014. Geographical clustering of *mecC*-positive *Staphylococcus aureus* from bovine mastitis in France. J. Antimicrob. Chemother. 69 (8), 2292–2293.

Harrison, E.M., Paterson, G.K., Holden, M.T., Larsen, J., Stegger, M., Larsen, A.R., Petersen, A., Skov, R.L., Christensen, J.M., Bak Zeuthen, A., Heltberg, O., Harris, S.R., Zadoks, R.N., Parkhill, J., Peacock, S.J., Holmes, M.A., 2013. Whole genome sequencing identifies zoonotic transmission of MRSA isolates with the novel *mecA* homologue *mecC*. EMBO Mol. Med. 5 (4), 509–515.

Hasman, H., Moodley, A., Guardabassi, L., Stegger, M., Skov, R.L., Aarestrup, F.M., 2010. *Spa* type distribution in *Staphylococcus aureus* originating from pigs, cattle and poultry. Vet. Microbiol. 141 (3–4), 326–331.

He, W., Liu, Y., Qi, J., Chen, H., Zhao, C., Zhang, F., Li, H., Wang, H., 2013. Food-animal related *Staphylococcus aureus* multidrug-resistant ST9 strains with toxin genes. Foodborne Pathog. Dis. 10, 782–788.

Huijsdens, X.W., van Dijke, B.J., Spalburg, E., van Santen-Verheuvel, M.G., Heck, M.E., Pluister, G.N., Voss, A., Wannet, W.J., de Neeling, A.J., 2006. Community-acquired MRSA and pig-farming. Ann. Clin. Microbiol. Antimicrob. 5, 26.

Jiménez, J.N., Vélez, L.A., Mediavilla, J.R., Ocampo, A.M., Vanegas, J.M., Rodríguez, E.A., Correa, M.M., 2011. Livestock-associated methicillin-susceptible *Staphylococcus aureus* ST398 infection in woman, Colombia. Emerg. Infect. Dis. 17 (10), 1970–1971.

Kerschner, H., Harrison, E.M., Hartl, R., Holmes, M.A., Apfalter, P., 2014. First report of *mecC* MRSA in human samples from Austria: molecular characteristics and clinical data. New Microbes New Infect. 3, 4–9.

Khanna, T., Friendship, R., Dewey, C., Weese, J.S., 2008. Methicillin resistant *Staphylococcus aureus* colonization in pigs and pig farmers. Vet. Microbiol. 128 (3–4), 298–303.

Köck, R., Loth, B., Köksal, M., Schulte-Wülwer, J., Harlizius, J., Friedrich, A.W., 2012. Persistence of nasal colonization with livestock-associated methicillin-resistant *Staphylococcus aureus* in pig farmers after holidays from pig exposure. Appl. Environ. Microbiol. 78 (11), 4046–4047.

Koyama, H., Sanui, M., Saga, T., Harada, S., Ishii, Y., Tateda, K., Lefor, A.K., 2015. A fatal infection caused by sequence type 398 methicillin-resistant *Staphylococcus aureus* carrying the Panton-Valentine leukocidin gene: a case report in Japan,. J. Infect. Chemother. 21 (7), 541–543.

Kraushaar, B., Ballhausen, B., Leeser, D., Tenhagen, B.A., Käsbohrer, A., Fetsch, A., February 2017. Antimicrobial resistances and virulence markers in methicillin-resistant *Staphylococcus aureus* from broiler and Turkey: a molecular view from farm to fork. Vet. Microbiol. 200, 25–32.

Kriegeskorte, A., Ballhausen, B., Idelevich, E.A., Köck, R., Friedrich, A.W., Karch, H., Peters, G., Becker, K., 2012. Human MRSA isolates with novel genetic homolog, Germany. Emerg. Infect. Dis. 18 (6), 1016–1018.

Kuroda, M., Ohta, T., Uchiyama, I., Baba, T., Yuzawa, H., Kobayashi, I., Cui, L., Oguchi, A., Aoki, K., Nagai, Y., Lian, J., Ito, T., Kanamori, M., Matsumaru, H., Maruyama, A., Murakami, H., Hosoyama, A., Mizutani-Ui, Y., Takahashi, N.K., Sawano, T., Inoue, R., Kaito, C., Sekimizu, K., Hirakawa, H., Kuhara, S., Goto, S., Yabuzaki, J., Kanehisa, M., Yamashita, A., Oshima, K., Furuya, K., Yoshino, C., Shiba, T., Hattori, M., Ogasawara, N., Hayashi, H., Hiramatsu, K., 2001. Whole genome sequencing of meticillin-resistant *Staphylococcus aureus*. Lancet 357 (9264), 1225–1240.

Larsen, J., Petersen, A., Sørum, M., Stegger, M., van Alphen, L., Valentiner-Branth, P., Knudsen, L.K., Larsen, L.S., Feingold, B., Price, L.B., Andersen, P.S., Larsen, A.R., Skov, R.L., 2015. Meticillin-resistant *Staphylococcus aureus* CC398 is an increasing cause of disease in people with no livestock contact in Denmark, 1999 to 2011. Euro Surveill. 20 (37). http://dx.doi.org/10.2807/1560-7917.ES.2015.20.37.30021.

Larsen, J., Stegger, M., Andersen, P.S., Petersen, A., Larsen, A.R., Westh, H., Agersø, Y., Fetsch, A., Kraushaar, B., Käsbohrer, A., Feβler, A.T., Schwarz, S., Cuny, C., Witte, W., Butaye, P., Denis, O., Haenni, M., Madec, J.Y., Jouy, E., Laurent, F., Battisti, A., Franco, A., Alba, P., Mammina, C., Pantosti, A., Monaco, M., Wagenaar, J.A., de Boer, E., van Duijkeren, E., Heck, M., Domínguez, L., Torres, C., Zarazaga, M., Price, L.B., Skov, R.L., 2016. Evidence for human adaptation and foodborne transmission of livestock-associated methicillin-resistant *Staphylococcus aureus*. Clin. Infect. Dis. 63 (10), 1349–1352.

Laurent, F., Chardon, H., Haenni, M., Bes, M., Reverdy, M.E., Madec, J.Y., Lagier, E., Vandenesch, F., Tristan, A., 2012. MRSA harboring *mecA* variant gene *mecC*, France. Emerg. Infect. Dis. 18 (9), 1465–1467.

Lekkerkerk, W.S., van Wamel, W.J., Snijders, S.V., Willems, R.J., van Duijkeren, E., Broens, E.M., Wagenaar, J.A., Lindsay, J.A., Vos, M.C., 2015. What is the origin of livestock-associated methicillin-resistant *Staphylococcus aureus* clonal complex 398 isolates from humans without livestock contact? An epidemiological and genetic analysis. J. Clin. Microbiol. 53 (6), 1836–1841.

Lewis, H.C., Mølbak, K., Reese, C., Aarestrup, F.M., Selchau, M., Sørum, M., Skov, R.L., 2008. Pigs as source of methicillin-resistant *Staphylococcus aureus* CC398 infections in humans, Denmark. Emerg. Infect. Dis. 14 (9), 1383–1389.

Lindgren, A.K., Gustafsson, E., Petersson, A.C., Melander, E., 2016. Methicillin-resistant *Staphylococcus aureus* with *mecC*: a description of 45 human cases in southern Sweden. Eur. J. Clin. Microbiol. Infect. Dis. 35 (6), 971–975.

Lindsay, J.A., 2010. Genomic variation and evolution of *Staphylococcus aureus*. Int. J. Med. Microbiol. 300, 98–103.

Lindsay, J.A., 2014. *Staphylococcus aureus* genomics and the impact of horizontal gene transfer. Int. J. Med. Microbiol. 304, 103–109.

Liu, Y., Wang, H., Du, N., Shen, E., Chen, H., Niu, J., Ye, H., Chen, M., 2009. Molecular evidence for spread of two major methicillin-resistant *Staphylococcus aureus* clones with a unique geographic distribution in Chinese hospitals. Antimicrob. Agents Chemother. 53, 512–518.

Loncaric, I., Kübber-Heiss, A., Posautz, A., Stalder, G.L., Hoffmann, D., Rosengarten, R., Walzer, C., 2013. Characterization of methicillin-resistant *Staphylococcus* spp. carrying the *mecC* gene, isolated from wildlife. J. Antimicrob. Chemother. 68 (10), 2222–2225.

Loncaric, I., Kübber-Heiss, A., Posautz, A., Stalder, G.L., Hoffmann, D., Rosengarten, R., Walzer, C., 2014. *mecC*- and *mecA*-positive meticillin-resistant *Staphylococcus aureus* (MRSA) isolated from livestock sharing habitat with wildlife previously tested positive for *mecC*-positive MRSA. Vet. Dermatol. 25 (2), 147–148.

Lowder, B.V., Guinane, C.M., Ben Zakour, N.L., Weinert, L.A., Conway-Morris, A., Cartwright, R.A., Simpson, A.J., Rambaut, A., Nübel, U., Fitzgerald, J.R., 2009. Recent human-to-poultry host jump, adaptation, and pandemic spread of *Staphylococcus aureus*. Proc. Natl. Acad. Sci. U.S.A. 106 (46), 19545–19550.

Lozano, C., Aspiroz, C., Lasarte, J.J., Gómez-Sanz, E., Zarazaga, M., Torres, C., 2011a. Dynamic of nasal colonization by methicillin-resistant *Staphylococcus aureus* ST398 and ST1 after mupirocin treatment in a family in close contact with pigs. Comp. Immunol. Microbiol. Infect. Dis. 34 (1), e1–7.

Lozano, C., Gómez-Sanz, E., Benito, D., Aspiroz, C., Zarazaga, M., Torres, C., 2011b. *Staphylococcus aureus* nasal carriage, virulence traits, antibiotic resistance mechanisms, and genetic lineages in healthy humans in Spain, with detection of CC398 and CC97 strains. Int. J. Med. Microbiol. 301, 500–505.

Lozano, C., Rezusta, A., Gómez, P., Gómez-Sanz, E., Báez, N., Martin-Saco, G., Zarazaga, M., Torres, C., 2012. High prevalence of spa types associated with the clonal lineage CC398 among tetracycline-resistant methicillin-resistant *Staphylococcus aureus* strains in a Spanish hospital. J. Antimicrob. Chemother. 67 (2), 330–334.

Luini, M., Cremonesi, P., Magro, G., Bianchini, V., Minozzi, G., Castiglioni, B., Piccinini, R., 2015. Methicillin-resistant *Staphylococcus aureus* (MRSA) is associated with low within-herd prevalence of intra-mammary infections in dairy cows: genotyping of isolates. Vet. Microbiol. 178 (3–4), 270–274.

Luzzago, C., Locatelli, C., Franco, A., Scaccabarozzi, L., Gualdi, V., Viganò, R., Sironi, G., Besozzi, M., Castiglioni, B., Lanfranchi, P., Cremonesi, P., Battisti, A., 2014. Clonal diversity, virulence-associated genes and antimicrobial resistance profile of *Staphylococcus aureus* isolates from nasal cavities and soft tissue infections in wild ruminants in Italian Alps. Vet. Microbiol. 170 (1–2), 157–161.

McCarthy, A.J., Lindsay, J.A., Loeffler, A., 2012a. Are all meticillin-resistant *Staphylococcus aureus* (MRSA) equal in all hosts? Epidemiological and genetic comparison between animal and human MRSA. Vet. Dermatol. 23 (4), 267–275.

McCarthy, A.J., van Wamel, W., Vandendriessche, S., Larsen, J., Denis, O., Garcia-Graells, C., Uhlemann, A.C., Lowy, F.D., Skov, R., Lindsay, J.A., 2012b. *Staphylococcus aureus* CC398 clade associated with human-to-human transmission. Appl. Environ. Microbiol. 78 (24), 8845–8848.

McCarthy, A.J., Witney, A.A., Gould, K.A., Moodley, A., Guardabassi, L., Voss, A., Denis, O., Broens, E.M., Hinds, J., Lindsay, J.A., 2011. The distribution of mobile genetic elements (MGEs) in MRSA CC398 is associated with both host and country. Genome Biol. Evol. 3, 1164–1174.

Medhus, A., Slettemeås, J.S., Marstein, L., Larssen, K.W., Sunde, M., 2013. *Staphylococcus aureus* with the novel *mecC* gene variant isolated from a cat suffering from chronic conjunctivitis. J. Antimicrob. Chemother. 68 (4), 968–969.

Mediavilla, J.R., Chen, L., Uhlemann, A.C., Hanson, B.M., Rosenthal, M., Stanak, K., Koll, B., Fries, B.C., Armellino, D., Schilling, M.E., Weiss, D., Smith, T.C., Lowy, F.D., Kreiswirth, B.N., 2012. Methicillin-susceptible *Staphylococcus aureus* ST398, New York and New Jersey, USA. Emerg. Infect. Dis. 18, 700–702.

Meemken, D., Blaha, T., Tegeler, R., Tenhagen, B.A., Guerra, B., Hammerl, J.A., Hertwig, S., Käsbohrer, A., Appel, B., Fetsch, A., December 2010. Livestock associated methicillin-resistant *Staphylococcus aureus* (LaMRSA) isolated from lesions of pigs at necropsy in northwest Germany between 2004 and 2007. Zoonoses Public Health 57 (7–8), e143–e148.

Monecke, S., Gavier-Widén, D., Hotzel, H., Peters, M., Guenther, S., Lazaris, A., Loncaric, I., Müller, E., Reissig, A., Ruppelt-Lorz, A., Shore, A.C., Walter, B., Coleman, D.C., Ehricht, R., 2016a. Diversity of *Staphylococcus aureus* isolates in European wildlife. PLoS One 11 (12), e0168433.

Monecke, S., Gavier-Widen, D., Mattsson, R., Rangstrup-Christensen, L., Lazaris, A., Coleman, D.C., Shore, A.C., Ehricht, R., 2013. Detection of *mecC*-positive *Staphylococcus aureus* (CC130-MRSA-XI) in diseased European hedgehogs (*Erinaceus europaeus*) in Sweden. PLoS One 8 (6), e66166.

Monecke, S., Jatzwauk, L., Müller, E., Nitschke, H., Pfohl, K., Slickers, P., Reissig, A., Ruppelt-Lorz, A., Ehricht, R., 2016b. Diversity of SCC*mec* elements in *Staphylococcus aureus* as observed in South-Eastern Germany. PLoS One 11 (9), e0162654.

Mulders, M.N., Haenen, A.P., Geenen, P.L., Vesseur, P.C., Poldervaart, E.S., Bosch, T., Huijsdens, X.W., Hengeveld, P.D., Dam-Deisz, W.D., Graat, E.A., Mevius, D., Voss, A., Van De Giessen, A.W., 2010. Prevalence of livestock-associated MRSA in broiler flocks and risk factors for slaughterhouse personnel in The Netherlands. Epidemiol. Infect. 138 (5), 743–755.

Paterson, G.K., Larsen, A.R., Robb, A., Edwards, G.E., Pennycott, T.W., Foster, G., Mot, D., Hermans, K., Baert, K., Peacock, S., Parkhill, J., Zadoks, R.N., Holmes, M.A., 2012. The newly described *mecA* homologue, $mecA_{LGA251}$, is present in methicillin-resistant *Staphylococcus aureus* isolates from a diverse range of host species,. J. Antimicrob. Chemother. 67 (12), 2809–2813.

Paterson, G.K., Morgan, F., Harrison, E.M., Peacock, S.J., Parkhill, J., Zadoks, R.N., Holmes, M.A., 2013. Prevalence and properties of *mecC* methicillin-resistant *Staphylococcus aureus* (MRSA) in bovine bulk tank milk in Great Britain. J. Antimicrob. Chemother. 69 (3), 598–602.

Paterson, G.K., Morgan, F.J., Harrison, E.M., Cartwright, E.J., Török, M.E., Zadoks, R.N., Parkhill, J., Peacock, S.J., Holmes, M.A., 2014. Prevalence and characterization of human *mecC* methicillin-resistant *Staphylococcus aureus* isolates in England. J. Antimicrob. Chemother. 69 (4), 907–910.

Peeters, L.E., Argudín, M.A., Azadikhah, S., Butaye, P., 2015. Antimicrobial resistance and population structure of *Staphylococcus aureus* recovered from pigs farms. Vet. Microbiol. 180, 151–156.

Pérez-Moreno, M.O., Centelles-Serrano, M.J., Nogales-López, J., Domenech-Spanedda, M.F., Lozano, C., Torres, C., 2016. Unusual presence of the immune evasion gene cluster in livestock-associated MRSA of lineage CC398 causing peridural and psoas abscesses in a poultry farmer. Enferm. Infecc. Microbiol. Clín. pii:S0213–005X(16)30238-5.

Persoons, D., Van Hoorebeke, S., Hermans, K., Butaye, P., de Kruif, A., Haesebrouck, F., Dewulf, J., 2009. Methicillin-resistant *Staphylococcus aureus* in poultry. Emerg. Infect. Dis. 15 (3), 452–453.

Pichon, B., Hill, R., Laurent, F., Larsen, A.R., Skov, R.L., Holmes, M., Edwards, G.F., Teale, C., Kearns, A.M., 2012. Development of a real-time quadruplex PCR assay for simultaneous detection of *nuc*, Panton-Valentine leucocidin (PVL), *mecA* and homologue $mecA_{LGA251}$. J. Antimicrob. Chemother. 67 (10), 2338–2341.

Porrero, M.C., Harrison, E.M., Fernández-Garayzábal, J.F., Paterson, G.K., Díez-Guerrier, A., Holmes, M.A., Domínguez, L., 2014a. Detection of *mecC*-Methicillin-resistant *Staphylococcus aureus* isolates in river water: a potential role for water in the environmental dissemination. Environ. Microbiol. Rep. 6 (6), 705–708.

Porrero, M.C., Hasman, H., Vela, A.I., Fernández-Garayzábal, J.F., Domínguez, L., Aarestrup, F.M., 2012. Clonal diversity of *Staphylococcus aureus* originating from the small ruminants goats and sheep. Vet. Microbiol. 156 (1–2), 157–161.

Porrero, M.C., Valverde, A., Fernández-Llario, P., Díez-Guerrier, A., Mateos, A., Lavín, S., Cantón, R., Fernández-Garayzabal, J.F., Domínguez, L., 2014b. *Staphylococcus aureus* carrying *mecC* gene in animals and urban wastewater, Spain. Emerg. Infect. Dis. 20 (5), 899–901.

Price, L.B., Stegger, M., Hasman, H., Aziz, M., Larsen, J., Andersen, P.S., Pearson, T., Waters, A., Foster, J.T., Schupp, J., Gillece, J., Driebe, E., Liu, C.M., Springer, B., Zdovc, I., Battisti, A., Franco, A., Zmudzki, J., Schwarz, S., Butaye, P., Jouy, E., Pomba, C., Porrero, M.C., Ruimy, R., Smith, T.C., Robinson, D.A., Weese, J.S., Arriola, C.S., Yu, F., Laurent, F., Keim, P., Skov, R., Aarestrup, F.M., 2012. *Staphylococcus aureus* CC398: host adaptation and emergence of methicillin resistance in livestock. MBio 3 (1), 58–59.

Rabello, R.F., Moreira, B.M., Lopes, R.M., Teixeira, L.M., Riley, L.W., Castro, A.C., 2007. Multilocus sequence typing of *Staphylococcus aureus* isolates recovered from cows with mastitis in Brazilian dairy herds. J. Med. Microbiol. 56, 1505–1511.

Rasigade, J.P., Laurent, F., Hubert, P., Vandenesch, F., Etienne, J., 2010. Lethal necrotizing pneumonia caused by an ST398 *Staphylococcus aureus* strain. Emerg. Infect. Dis. 16 (8), 1330.

Romero-Gómez, M.P., Mora-Rillo, M., Lázaro-Perona, F., Gómez-Gil, M.R., Mingorance, J., 2013. Bacteremia due to methicillin-resistant *Staphylococcus aureus* carrying the *mecC* gene in a patient with urothelial carcinoma. J. Med. Microbiol. 62, 1914–1916.

Rooijakkers, S.H., Ruyken, M., Roos, A., Daha, M.R., Presanis, J.S., Sim, R.B., van Wamel, W.J., van Kessel, K.P., van Strijp, J.A., 2005. Immune evasion by a staphylococcal complement inhibitor that acts on C3 convertases. Nat. Immunol. 6 (9), 920–927.

Sabat, A.J., Koksal, M., Akkerboom, V., Monecke, S., Kriegeskorte, A., Hendrix, R., Ehricht, R., Köck, R., Becker, K., Friedrich, A.W., 2012. Detection of new methicillin-resistant *Staphylococcus aureus* strains that carry a novel genetic homologue and important virulence determinants. J. Clin. Microbiol. 50 (10), 3374–3377.

Salmenlinna, S., Lyytikäinen, O., Vainio, A., Myllyniemi, A.L., Raulo, S., Kanerva, M., Rantala, M., Thomson, K., Seppänen, J., Vuopio, J., 2010. Human cases of methicillin-resistant *Staphylococcus aureus* CC398, Finland. Emerg. Infect. Dis. 16 (10), 1626–1629.

Schaumburg, F., Köck, R., Mellmann, A., Richter, L., Hasenberg, F., Kriegeskorte, A., Friedrich, A.W., Gatermann, S., Peters, G., von Eiff, C., Becker, K., 2012. Population dynamics among methicillin-resistant *Staphylococcus aureus* isolates in Germany during a 6-year period. J. Clin. Microbiol. 50 (10), 3186–3192.

Schlotter, K., Huber-Schlenstedt, R., Gangl, A., Hotzel, H., Monecke, S., Müller, E., Reißig, A., Proft, S., Ehricht, R., 2014. Multiple cases of methicillin-resistant CC130 *Staphylococcus aureus* harboring *mecC* in milk and swab samples from a Bavarian dairy herd. J. Dairy Sci. 97 (5), 2782–2788.

Shallcross, L.J., Williams, K., Hopkins, S., Aldridge, R.W., Johnson, A.M., Hayward, A.C., 2010. Panton–Valentine leukocidin associated staphylococcal disease: a cross-sectional study at a London Hospital, England. Clin. Microbiol. Infect. 16, 1644–1648.

Shore, A.C., Deasy, E.C., Slickers, P., Brennan, G., O'Connell, B., Monecke, S., Ehricht, R., Coleman, D.C., 2011. Detection of staphylococcal cassette chromosome *mec* type XI carrying highly divergent *mecA, mecI, mecR1, blaZ,* and ccr genes in human clinical isolates of clonal complex 130 methicillin-resistant *Staphylococcus aureus*. Antimicrob. Agents Chemother. 55 (8), 3765–3773.

Slingerland, B.C., Tavakol, M., McCarthy, A.J., Lindsay, J.A., Snijders, S.V., Wagenaar, J.A., van Belkum, A., Vos, M.C., Verbrugh, H.A., van Wamel, W.J., 2012. Survival of *Staphylococcus aureus* ST398 in the human nose after artificial inoculation. PLoS One 7 (11), e48896.

Smith, E.M., Needs, P.F., Manley, G., Green, L.E., 2014. Global distribution and diversity of ovine-associated *Staphylococcus aureus*. Infect. Genet. Evol. 22, 208–215.

Smith, T.C., Pearson, N., 2011. The emergence of *Staphylococcus aureus* ST398. Vector-Borne Zoonotic Dis. 11 (4), 327–339.

Smith, T.C., Wardyn, S.E., 2015. Human infections with *Staphylococcus aureus* CC398. Curr. Environ. Health Rep. 2 (1), 41–51.

Spoor, L.E., McAdam, P.R., Weinert, L.A., Rambaut, A., Hasman, H., Aarestrup, F.M., Kearns, A.M., Larsen, A.R., Skov, R.L., Fitzgerald, J.R., 2013. Livestock origin for a human pandemic clone of community-associated methicillin-resistant *Staphylococcus aureus*. MBio 4 (4), e00356–e00413.

Spoor, L.E., Richardson, E., Richards, A.M., Wilson, G.J., Mendonca, C., Gupta, R.K., McAdam, P.R., Nutbeam-Tuffs, S., Black, N.S., O'Gara, J.P., Lee, C.Y., Corander, J., Fitzgerald, J.R., 2015. Recombination-mediated remodeling of host-pathogen interactions during *Staphylococcus aureus* niche adaptation. Microb. Genomics. 1 (4), e000036. http://mgen.microbiologyresearch.org.

Stefani, S., Chung, D.R., Lindsay, J.A., Friedrich, A.W., Kearns, A.M., Westh, H., Mackenzie, F.M., 2012. Meticillin-resistant *Staphylococcus aureus* (MRSA): global epidemiology and harmonisation of typing methods. Int. J. Antimicrob. Agents 39 (4), 273–282.

Stegger, M., Andersen, P.S., Kearns, A., Pichon, B., Holmes, M.A., Edwards, G., Laurent, F., Teale, C., Skov, R., Larsen, A.R., 2012. Rapid detection, differentiation and typing of methicillin-resistant *Staphylococcus aureus* harbouring either *mecA* or the new *mecA* homologue *mecA*(LGA251). Clin. Microbiol. Infect. 18 (4), 395–400.

Stegger, M., Aziz, M., Chroboczek, T., Price, L.B., Ronco, T., Kiil, K., Skov, R.L., Laurent, F., Andersen, P.S., 2013a. Genome analysis of *Staphylococcus aureus* ST291, a double locus variant of ST398, reveals a distinct genetic lineage. PLoS One 8 (5), e63008.

Stegger, M., Lindsay, J.A., Sørum, M., Gould, K.A., Skov, R., 2010. Genetic diversity in CC398 methicillin-resistant *Staphylococcus aureus* isolates of different geographical origin. Clin. Microbiol. Infect. 16 (7), 1017–1019.

Stegger, M., Liu, C.M., Larsen, J., Soldanova, K., Aziz, M., Contente-Cuomo, T., Petersen, A., Vandendriessche, S., Jiménez, J.N., Mammina, C., van Belkum, A., Salmenlinna, S., Laurent, F., Skov, R.L., Larsen, A.R., Andersen, P.S., Price, L.B., 2013b. Rapid differentiation between livestock-associated and livestock-independent *Staphylococcus aureus* CC398 clades. PLoS One 8 (11), e79645.

Uhlemann, A.C., McAdam, P.R., Sullivan, S.B., Knox, J.R., Khiabanian, H., Rabadan, R., Davies, P.R., Fitzgerald, J.R., Lowy, F.D., 2017. Evolutionary dynamics of pandemic methicillin-sensitive *Staphylococcus aureus* ST398 and its international spread via routes of human migration. MBio 8 (1) pii:e01375-16.

Uhlemann, A.C., Porcella, S.F., Trivedi, S., Sullivan, S.B., Hafer, C., Kennedy, A.D., Barbian, K.D., McCarthy, A.J., Street, C., Hirschberg, D.L., Lipkin, W.I., Lindsay, J.A., DeLeo, F.R., Lowy, F.D., 2012. Identification of a highly transmissible animal-independent *Staphylococcus aureus* ST398 clone with distinct genomic and cell adhesion properties. MBio 3 (2), FMW137. http://dx.doi.org/10.1128/mBio.00027-12.

Unnerstad, H.E., Bengtsson, B., Horn af Rantzien, M., Börjesson, S., 2013. Methicillin-resistant *Staphylococcus aureus* containing *mecC* in Swedish dairy cows. Acta Vet. Scand. 55, 6.

Valentin-Domelier, A.S., Girard, M., Bertrand, X., Violette, J., François, P., Donnio, P.Y., Talon, D., Quentin, R., Schrenzel, J., van der Mee-Marquet, N., Bloodstream Infection Study Group of the Réseau des Hygiénistes du Centre (RHC), 2011. Methicillin-susceptible ST398 *Staphylococcus aureus* responsible for bloodstream infections: an emerging human-adapted subclone? PLoS One 6 (12), e28369.

van Belkum, A., Melles, D.C., Peeters, J.K., van Leeuwen, W.B., van Duijkeren, E., Huijsdens, X.W., Spalburg, E., de Neeling, A.J., Verbrugh, H.A., 2008. Methicillin-resistant and -susceptible *Staphylococcus aureus* sequence type 398 in pigs and humans. Emerg. Infect. Dis. 14, 479–483.

van Cleef, B.A., Monnet, D.L., Voss, A., Krziwanek, K., Allerberger, F., Struelens, M., Zemlickova, H., Skov, R.L., Vuopio-Varkila, J., Cuny, C., Friedrich, A.W., Spiliopoulou, I., Pászti, J., Hardardottir, H., Rossney, A., Pan, A., Pantosti, A., Borg, M., Grundmann, H., Mueller-Premru, M., Olsson-Liljequist, B., Widmer, A., Harbarth, S., Schweiger, A., Unal, S., Kluytmans, J.A., 2011. Livestock-associated methicillin-resistant *Staphylococcus aureus* in humans, Europe. Emerg. Infect. Dis. 17 (3), 502–505.

Van den Eede, A., Hermans, K., Van den Abeele, A., Floré, K., Dewulf, J., Vanderhaeghen, W., Crombé, F., Butaye, P., Gasthuys, F., Haesebrouck, F., Martens, A., 2012. Methicillin-resistant *Staphylococcus aureus* (MRSA) on the skin of long-term hospitalised horses. Vet. J. 193 (2), 408–411.

van der Mee-Marquet, N., Corvaglia, A.R., Valentin, A.S., Hernandez, D., Bertrand, X., Girard, M., Kluytmans, J., Donnio, P.Y., Quentin, R., François, P., 2013. Analysis of prophages harbored by the human-adapted subpopulation of *Staphylococcus aureus* CC398. Infect. Genet. Evol. 18, 299–308.

van der Mee-Marquet, N., François, P., Domelier-Valentin, A.S., Coulomb, F., Decreux, C., Hombrock-Allet, C., Lehiani, O., Neveu, C., Ratovohery, D., Schrenzel, J., Quentin, R., 2011. Emergence of unusual bloodstream infections associated with pig-borne-like *Staphylococcus aureus* ST398 in France. Clin. Infect. Dis. 52, 152–153.

van der Wolf, P.J., Rothkamp, A., Junker, K., de Neeling, A.J., 2012. *Staphylococcus aureus* (MSSA) and MRSA (CC398) isolated from post-mortem samples from pigs. Vet. Microbiol. 158 (1–2), 136–141.

van Loo, I., Huijsdens, X., Tiemersma, E., de Neeling, A., van de Sande-Bruinsma, N., Beaujean, D., 2007. Emergence of methicillin-resistant *Staphylococcus aureus* of animal origin in humans. Emerg. Infect. Dis. 13, 1834–1839.

van Rijen, M.M., Bosch, T., Verkade, E.J., Schouls, L., Kluytmans, J.A., CAM Study Group, 2014. Livestock-associated MRSA carriage in patients without direct contact with livestock. PLoS One 9 (6), e100294.

van Wamel, W.J., Rooijakkers, S.H., Ruyken, M., van Kessel, K.P., van Strijp, J.A., 2006. The innate immune modulators staphylococcal complement inhibitor and chemotaxis inhibitory protein of *Staphylococcus aureus* are located on beta-hemolysin-converting bacteriophages. J. Bacteriol. 188 (4), 1310–1315.

Vandendriessche, S., Kadlec, K., Schwarz, S., Denis, O., 2011. Methicillin susceptible *Staphylococcus aureus* ST398-t571 harbouring the macrolide-lincosamide-streptogramin B resistance gene *erm*(T) in Belgian hospitals. J. Antimicrob. Chemother. 66, 2455–2459.

Vandendriessche, S., Vanderhaeghen, W., Soares, F.V., Hallin, M., Catry, B., Hermans, K., Butaye, P., Haesebrouck, F., Struelens, M.J., Denis, O., 2013. Prevalence, risk factors and genetic diversity of methicillin-resistant *Staphylococcus aureus* carried by humans and animals across livestock production sectors. J. Antimicrob. Chemother. 68 (7), 1510–1516.

Vanderhaeghen, W., Hermans, K., Haesebrouck, F., Butaye, P., 2010. Methicillin-resistant *Staphylococcus aureus* (MRSA) in food production animals. Epidemiol. Infect. 138, 606–625.

Velasco, V., Buyukcangaz, E., Sherwood, J.S., Stepan, R.M., Koslofsky, R.J., Logue, C.M., 2015. Characterization of *Staphylococcus aureus* from humans and a comparison with İsolates of animal origin, in North Dakota, United States. PLoS One 10 (10), e0140497.

Verkade, E., Kluytmans, J., 2014. Livestock-associated *Staphylococcus aureus* CC398: animal reservoirs and human infections. Infect. Genet. Evol. 21, 523–530.

Verkade, E., Bergmans, A.M., Budding, A.E., van Belkum, A., Savelkoul, P., Buiting, A.G., Kluytmans, J., 2012. Recent emergence of *Staphylococcus aureus* clonal complex 398 in human blood cultures. PLoS One 7 (10), e41855.

Viana, D., Comos, M., McAdam, P.R., Ward, M.J., Selva, L., Guinane, C.M., González-Muñoz, B.M., Tristan, A., Foster, S.J., Fitzgerald, J.R., Penadés, J.R., 2015. A single natural nucleotide mutation alters bacterial pathogen host tropism. Nat. Genet. 47 (4), 361–366.

Voss, A., Loeffen, F., Bakker, J., Klaassen, C., Wulf, M., 2005. Methicillin-resistant *Staphylococcus aureus* in pig farming. Emerg. Infect. Dis. 11 (12), 1965–1966.

Walther, B., Wieler, L.H., Vincze, S., Antão, E.M., Brandenburg, A., Stamm, I., Kopp, P.A., Kohn, B., Semmler, T., Lübke-Becker, A., 2012. MRSA variant in companion animals. Emerg. Infect. Dis. 18 (12), 2017–2020.

Wang, D., Wang, Z., Yan, Z., Wu, J., Ali, T., Li, J., Lv, Y., Han, B., 2015. Bovine mastitis *Staphylococcus aureus*: antibiotic susceptibility profile, resistance genes and molecular typing of methicillin-resistant and methicillin-sensitive strains in China. Infect. Genet. Evol. 31, 9–16.

Wan, M.T., Lauderdale, T.L., Chou, C.C., 2013. Characteristics and virulence factors of livestock associated ST9 methicillin-resistant *Staphylococcus aureus* with a novel recombinant staphylocoagulase type. Vet. Microbiol. 162, 779–784.

Wassenberg, M.W., Bootsma, M.C., Troelstra, A., Kluytmans, J.A., Bonten, M.J., 2011. Transmissibility of livestock-associated methicillin-resistant *Staphylococcus aureus* (ST398) in Dutch hospitals. Clin. Microbiol. Infect. 17 (2), 316–319.

Weinert, L.A., Welch, J.J., Suchard, M.A., Lemey, P., Rambaut, A., Fitzgerald, J.R., 2012. Molecular dating of human-to-bovid host jumps by *Staphylococcus aureus* reveals an association with the spread of domestication. Biol. Lett. 8 (5), 829–832.

Welinder-Olsson, C., Florén-Johansson, K., Larsson, L., Oberg, S., Karlsson, L., Ahrén, C., 2008. Infection with Panton-Valentine leukocidin-positive methicillin-resistant *Staphylococcus aureus* t034. Emerg. Infect. Dis. 14 (8), 1271–1272.

Williamson, D.A., Bakker, S., Coombs, G.W., Tan, H., Monecke, S., Heffernan, H., 2014. Emergence and molecular characterization of clonal complex 398 (CC398) methicillin-resistant *Staphylococcus aureus* (MRSA) in New Zealand. J. Antimicrob. Chemother. 69 (5), 1428–1430.

Worthing, K.A., Coombs, G.W., Pang, S., Abraham, S., Saputra, S., Trott, D.J., Jordan, D., Wong, H.S., Abraham, R.J., Norris, J.M., 2016. Isolation of *mecC* MRSA in Australia. J. Antimicrob. Chemother. 71 (8), 2348–2349.

Wulf, M., Voss, A., 2008. MRSA in livestock animals-an epidemic waiting to happen. Clin. Microbiol. Infect. 14 (6), 519–521.

Ye, X., Wang, X., Fan, Y., Peng, Y., Li, L., Li, S., Huang, J., Yao, Z., Chen, S., 2016. Genotypic and phenotypic markers of livestock-associated methicillin-resistant *Staphylococcus aureus* CC9 in humans. Appl. Environ. Microbiol. 82, 3892–3899.

Yu, F., Chen, Z., Liu, C., Zhang, X., Lin, X., Chi, S., Zhou, T., Chen, Z., Chen, X., 2008. Prevalence of *Staphylococcus aureus* carrying Panton-Valentine leukocidin genes among isolates from hospitalised patients in China. Clin. Microbiol. Infect. 14 (4), 381–384.

Zarfel, G., Luxner, J., Folli, B., Leitner, E., Feierl, G., Kittinger, C., Grisold, A., 2016. Increase of genetic diversity and clonal replacement of epidemic methicillin-resistant *Staphylococcus aureus* strains in South-East Austria. FEMS Microbiol. Lett. 363 (14), fnw137. http://dx.doi.org/10.1093/femsle/fnw137.

Zhao, C., Liu, Y., Zhao, M., Liu, Y., Yu, Y., Chen, H., Sun, Q., Chen, H., Jiang, W., Liu, Y., Han, S., Xu, Y., Chen, M., Cao, B., Wang, H., 2012. Characterization of community acquired *Staphylococcus aureus* associated with skin and soft tissue infection in Beijing: high prevalence of PVL+ ST398. PLoS One 7 (6), e38577.

PART

IV

PREVENTION AND CONTROL OF *STAPHYLOCOCCUS AUREUS* IN THE FOOD CHAIN

CHAPTER

11 HYGIENE PRINCIPLES TO AVOID CONTAMINATION/CROSS-CONTAMINATION IN THE KITCHEN AND DURING FOOD PROCESSING

Marco Ebert

University of Applied Sciences, Neubrandenburg, Germany

1. INTRODUCTION

Staphylococcus aureus is a frequent commensal on the skin and respiratory mucosal of mammals and birds, with prevalence in humans of up to 30%–50% (Le Loir et al., 2003). The frontal nasal cavities of around 20%–30% of the human adults are permanently colonized by *S. aureus* and around 60% are intermittent carriers (Mainous et al., 2006; Kluytmans and Wertheim, 2005). But the bacteria are also found in the pharynx, perineum, on the skin (predominantly on the hands, face, chest, and abdomen), and axillae of humans. Genes encoding for enterotoxins, which can cause food poisoning, are present in about 25% of the coagulase-positive strains found in humans (Bergdoll, 1989). Even though *S. aureus* can also be found in the air, soil, water, sewage, plant surfaces, and animals and therefore in products of animal origin (e.g., poultry, meat, and milk and products thereof), the food handler is the most frequent source of contamination of food with *S. aureus* (Bergdoll and Wong, 1993).

The growth of *S. aureus* and subsequent toxin production depends on several intrinsic factors of the food, such as pH-value, salt content, nutritional factors, water activity (a_W-value), concurring microorganisms, and extrinsic factors, e.g., temperature and atmospheric conditions. Some of these factors are listed in Table 11.1 with the respective optimal and range values. Staphylococcal food poisoning is most often associated with processed red meats, poultry products (especially chicken salad), sauces, dairy products (especially cheeses), and custard- or cream-filled bakery products. Ham and associated products are involved in about 30% of outbreaks of staphylococcal food poisoning (Halpin-Dohnalek and Marth, 1989). In 2014, 31 strong-evidence outbreaks caused by staphylococcal toxins were reported in the European Union (EU). Vehicles which caused more than one of these outbreaks were mixed foods (29.9%), broiler meat (*Gallus gallus*) and products thereof (9.7%), pig meat and products thereof (9.7%), cheese (6.5%), dairy products other than cheese (6.5%), fish and fish products (6.5%), vegetables and juices and other products thereof (6.5%), and bakery products (6.5%) (EFSA, 2015). Because *S. aureus* is osmotolerant, it can multiply rapidly when a high salt content in the food inhibits the growth of many other concurring bacteria. Starter cultures, such as lactic acid bacteria, can lower the

Staphylococcus aureus. http://dx.doi.org/10.1016/B978-0-12-809671-0.00011-5

Table 11.1 Factors Affecting Growth and Enterotoxin Production by *Staphylococcus aureus* (Tatini, 1973)

	***S. aureus* Growth**		**Enterotoxin Production**	
Factor	**Optimum**	**Range**	**Optimum**	**Range**
Temperature	37°C	7–48°C	40–45°C	10–48°C
pH	6–7	4–10	7–8	4–9.6
Water activity (a_w)	0.98	0.83–0.99	0.98	0.86≥0.99
NaCl (%)	0	0–20	0	0–10
Redox potential (E_h)	>+200 mV	(−200)–(+200) mV	>+200 mV	≥−100–(+200) mV
Oxygen	Aerobic	Anaerobic–aerobic	Aerobic	Anaerobic–aerobic
pH	6–7	4–10	7–8	4–9.6
Water activity (a_w)	0.98	0.83–0.99	0.98	0.86≥0.99
NaCl (%)	0	0–20	0	0–10
Redox potential (E_h)	>+200 mV	(−200)–(+200) mV	>+200 mV	≥−100–(+200) mV
Oxygen	Aerobic	Anaerobic–aerobic	Aerobic	Anaerobic–aerobic

rate of *S. aureus* growth and toxin production, e.g., in cheese (Otero et al., 1988; Vernozy-Rozand et al., 1998) and fermented raw sausages (Sameshima et al., 1998).

Because even very small amounts of toxin as low as 0.1 μg/kg body weight can lead to intoxication in sensitive consumers (Genigeorgis, 1989; Baird-Parker, 1990), the prevention of the replication of *S. aureus* is a very important issue for food safety. Furthermore, the toxins are highly resistant to thermal treatment and proteolytic degradation and therefore are not deactivated in the human digestive tract (Le Loir et al., 2003; Baird-Parker, 1990). That is why good hygiene praxis (GHP) is one of the most important measures to avoid contamination with and replication of *S. aureus* and prevent staphylococcal toxin production and food poisoning. Many different measures have to be taken to establish and maintain GHP- and hazard analysis critical control point (HACCP)-procedures, which are particularly important, when food is produced manually.

Concerning *S. aureus* food can be considered safe if there is an a_W-value of lower than 0.83 or a pH-value under 4.0, as the bacterium cannot multiply in such foods. When food is manufactured in a way that they have an a_W-value of less than 0.83, measures have to be taken to prevent rehydration during storage and distribution. These measures can, for example, be a control of the relative air humidity and the ambient temperature in the packaging areas and storage rooms to avoid condensation of water on the food, reduction in the water vapor permeability of the packaging, or the use of water absorbers in the packaging.

2. GOOD HYGIENE PRAXIS AND HAZARD ANALYSIS CRITICAL CONTROL POINT PRINCIPLES

S. aureus is present in a variety of foods and can cause serious intoxications in consumers, so it is a biological hazard in terms of the HACCP principles. That is why predisposed sites of contamination and when applicable critical control points (CCPs) have to be determined in food businesses and the following steps have to be applied to prevent or eliminate this hazard or to reduce it to an acceptable level (European Parliament and Council, 2004a):

1. identifying the hazards that must be prevented, eliminated, or reduced to acceptable levels;
2. identifying the CCPs at the step or steps at which control is essential to prevent or eliminate the hazard or to reduce it to an acceptable level;
3. establishing critical limits at CCPs, which separate acceptability from unacceptability for the prevention, elimination, or reduction of the hazard;
4. establishing and implementing effective monitoring procedures at CCPs;
5. establishing corrective actions when monitoring indicates that a CCP is not under control;
6. establishing procedures, which shall be carried out regularly, to verify that the measures outlined in subparagraphs (1)–(5) are working effectively; and
7. establishing documents and records commensurate with the nature and size of the food business to demonstrate the effective application of the measures outlined in subparagraphs (1)–(7).

Some of the most important prerequisite for a GHP are as follows (European Commission, 2016):

1. Infrastructure (building, equipment)
2. Cleaning and disinfection
3. Pest control: focus on prevention
4. Technical maintenance and calibration
5. Prevention of physical and chemical contaminations from production environment (e.g., oils, inks, use of (damaged) wooden equipment, etc.)
6. Allergen management
7. Waste management
8. Water and air control
9. Personnel (hygiene, health status)
10. Raw materials (supplier selection, specifications)
11. Temperature control of storage environment
12. Working methodology

Not all of these prerequisites are crucial to control *S. aureus* intoxications. Table 11.2 summarizes the most relevant factors contributing to foodborne outbreaks due to *S. aureus* intoxications in the United States, Canada, and England and Wales. Failure in the temperature and time management, infected food handlers, and improper personal hygiene are the most common reasons for staphylococcal intoxications (above 30%). Therefore, mainly these issues will be discussed in this chapter. However, at least all of the mentioned prerequisites have to be carried out, to produce food which is safe for the consumer. Of note, most measures described in the following will not only have an effect on the safety of foodstuffs as regard *S. aureus* but also on most other yet known mesophilic (facultative) aerobic foodborne pathogenic bacteria, such as *Salmonella* spp.

3. TEMPERATURE-RELATED FOOD HYGIENE PRINCIPLES

As shown in Table 11.1 *S. aureus* is a mesophilic bacteria multiplying at temperatures between 7.0 and 48.0°C with the temperature optimum of 37.0°C (Tatini, 1973; Wong and Bergdoll, 1993). The minimal replication temperature also depends on the food matrix. The toxin production is regulated by quorum

Table 11.2 Identified Factors That Have a Significant Impact on Staphylococcal Food Poisoning

	Percentage (%) of Outbreaks Were the Factor Was Relevant (Multiple References Were Possible)		
Factor	**United States**	**Canada**	**England/Wales**
Improper cooling	**N/A**	**60.6**	**7.2**
Improper holding temperature	**95.6**	**N/A**	**N/A**
Storage at room temperature	**N/A**	**N/A**	**45.2**
Use of remnants	N/A	11.5	6.6
Improper handling of food	N/A	13.1	N/A
Infected food-handler	**N/A**	**18.0**	**30.1**
Improper cooking	10.1	6.5	1.2
Food preparation a long time in advance	**N/A**	**N/A**	**48.2**
Contaminated equipment	21.4	3.3	N/A
Improper personal hygiene	**46.6**	**N/A**	**N/A**
Food contaminated before packaging	N/A	N/A	25.3

N/A, *not applicable.*
According to Genigeorgis, C.A., 1989. Present state of knowledge on staphylococcal intoxication. Int. J. Food Microbiol. 9, 327–360.

sensing and can begin at a concentration of 10^5 cfu/g foodstuff (Le Loir et al., 2003; Götz et al., 2006) and food temperatures from 10.0 to 48.0°C with the temperature optimum between 40.0 and 45.0°C (Tatini, 1973). For that reason, not only the bacterial replication but also the toxin production can be prevented by temperature control. Most meat products are stored at 7°C. A very slow replication of *S. aureus* might be possible at this temperature. But because of competing bacteria present in the product, it is quite unlikely that *S. aureus* is able to replicate exceeding 10^5 cfu/g. Vice versa, *S. aureus* can grow quickly at higher temperatures, especially when there are no or hardly any competing bacteria, e.g., in cases of postprocessing contamination (e.g., recontamination of thermally treated food) or in raw milk obtained from cows suffering from *S. aureus* mastitis (Kümmel et al., 2016). That is why it is very important to maintain the cold chain, chill food after heat treatment, keep it warm at temperatures above 65°C, or eat it within 30 min.

To insure food safety, food business operators have to put in place, implement, and maintain permanent temperature control procedures based on the HACCP principles. But it also has to be taken into account that the production of some foodstuffs, such as cheese or most raw sausages, requires a growth-permissive temperature during processing. In these products other hurdles should be established, e.g., starter cultures (Nussinovitch et al., 1987), reduction of the oxygen partial pressure, or preservatives (Tavakoli et al., 2015).

3.1 MAINTENANCE OF THE COLD CHAIN

One of the most important measures to avoid growth of *S. aureus* in food is the maintenance of the cold chain in food businesses, during transport and thawing processes, and in consumer's households.

3.1.1 Maintenance of the Cold Chain in Food Businesses

Some food business operators store fresh perishable food for many days, weeks, and even month in their companies. During this time frame, and without appropriate chilling facilities, *S. aureus* may grow to an unacceptable level. Therefore, the cold chain has to be maintained along the whole food chain, which requires longitudinal integrated quality assurance systems. This can be achieved by monitoring, verification, and recording of the refrigeration equipment. However, limited periods outside temperature control are permitted to take into account the practices of handling during preparation, transport, storage, display, and service of food, as long as it does not result in a risk to human health (European Parliament and Council, 2004a). During preparation at room temperatures, food should never be taken out of the cold chain for more than 30min. Especially in cutting plants where carcasses are trimmed and certain parts of the slaughtered animals are gathered from the processing of larger cuts, it may take some hours until containers are filled and moved to chilled storage. Therefore, it can be necessary to chill down processing rooms. For example, during cutting, boning, trimming, slicing, dicing, wrapping, and packaging, the ambient temperature should not exceed 12°C or an alternative system should be in place, ensuring that the product temperature will not exceed the critical limits (European Parliament and Council, 2004b).

3.1.1.1 Critical Limits in the Cold Chain

The critical limits in the cold chain depend on the stored products. Some of the critical limits for chilling temperatures and the critical limits for freezing temperatures are regulated in the food legislation. For example, a lot of the critical limits for foods of animal origin are laid down in Regulation No 853/2004 within the EU. Most food of animal origin should be stored at a temperature of 7°C or below to avoid an unacceptable replication of *S. aureus*. Depending on the type of food, different chilling temperatures are recommended or required. Fresh meat products, meat from cattle, swine, goat, sheep, and other meat from taller mammals (including game such as boar, dear, etc.) have to be chilled at 7°C. Meat preparations, fresh meat from hare, poultry, and wild birds such as pheasant need to be chilled at even lower temperature (4°C) to prevent the replication of Salmonella. Offal is usually stored at 3°C and minced meat and fish at 2°C (European Parliament and Council, 2004b). Deep-frozen products have to be stored and transported at −18°C (European Council, 1989).

When several foods with different temperature requirements are stored in the same facility the product with the lowest required temperature is crucial for the determination of the critical limit. The food business operators should always set critical limits at or below legal limits because a lower temperature usually results in a longer shelf life.

3.1.1.2 Monitoring Procedures in the Cold Chain

Systems to monitor the determined critical temperature limits have to be put in place. The temperature of any refrigerated unit, which contains perishable food, has to be monitored. The monitoring is often more complicated than it seems and many food businesses, especially small and microcatering and retail establishments, do not have proper HACCP systems in place (European Commission, 2009). The temperature in a chilling facility usually varies within the device. The cooling unit chills the air and blows it into the cooling device. Because cold air has a higher density than warm air it falls down, warms up, and ascends. This is how the air circulates in the cooling facility leading to different temperature zones, especially when the cooler is filled to overflowing, which inhibits the air circulation (Byrd-Bredbenner et al., 2007). Also the spacing between products can affect the proper air circulation.

Therefore, the different temperature zones within a cooling device should be determined. Once the different temperature zones are found, the placing of the temperature sensor can be planed. The sensor should be placed either in the warmest zone in the chilling facility or where the product with the lowest required temperature are placed. It is also of importance how far the sensor is from the door of the chilling facility. When doors are not closed during the work in a chilling room, the temperature can increase dramatically. For that reason, doors of chilling facilities should not be left opened and the opening and closing should be reduced to a minimum.

Quite often chilling devices have internal thermometers. Food business operators should find out where the temperature sensors of these thermometers are placed. Often the sensor is located close to the cold air flow, and for that reason it does not measure the temperature where the products are placed. So a separate external thermometer should always be placed where the products are located to avoid undetected exceedances of the critical temperature limits.

The temperature should be controlled at least once a day (in the EU the temperature controls of deep freezing rooms have to be carried out more frequently (European Council, 1989)). If an automatic temperature control system (e.g., temperature data logger) is used, it is crucial that audible and/or visible alarms are activated and/or an email or a message is send to the responsible person when temperature limits are close to being violated.

3.1.1.3 Corrective Actions in the Cold Chain

When failures of the temperature control procedures occur, food may no longer be fit for human consumption because of the growth of bacteria such as *S. aureus*. Therefore, actions need to be taken when interruptions of the cold chain are identified by the temperature monitoring. The actions depend on the temperature deviation and should include actions regarding any product that has been outside the cold chain (e.g., immediate heat treatment, reduction of the expiring date, disposal), the failing equipment (e.g., commissioning of a refrigeration technician), and the staff (e.g., training). All corrective actions should bring the underlying cause to an end and prevent similar incidents in the future.

When the cold chain was interrupted for several hours (at least 4 h (Bergdoll and Wong, 1993)) and it seems possible that the *S. aureus* load has exceeded unacceptable levels of 10^5 cfu/g food before consumption, toxin production may have been occurred in the foodstuff. Food business operators should then carefully consider which corrective actions are appropriate. In case of any doubts about food safety, adequate measurements should be taken, e.g., microbiological sampling or discharge of the food.

3.1.1.4 Verification of the Cold Chain Maintenance

To verify the in-house monitoring of the maintenance of the cold chain, various measures should be implemented. Also staff should be monitored to ensure that all of the definite surveillance measures are performed adequately.

Each measuring instruments for the temperature monitoring should be calibrated in a predetermined frequency to ensure that the devices are working accurately. The calibration has to be performed against a recognized national or international standard (United Kingdom. British Retail Consortium, 2015). Furthermore, a calibration has to be performed when there is reason to assume a defect of the thermometer. When an automatic temperature control system is in place, it is urgent to test regularly if the alert is functioning.

3.1.1.5 Records of the Cold Chain

The temperature-control procedures based on the HACCP principles have to be laid down in documents such as standard operating procedures. Furthermore, records have to include the critical limits, the daily documentation of the monitoring results, the corrective actions, the calibration records (including thermometer number, date, and outcome of the check), and so on and so forth.

3.1.2 Maintenance of the Cold Chain During Transport

The cold chain also has to be maintained during transport. Therefore, conveyances and/or containers used for transportation of perishable foodstuffs have to be capable to maintain products at appropriate temperatures and allow those temperatures to be monitored. Thus, even during transport a monitoring system has to be in place to measure the food temperature or the ambient temperature. The cooling system used for transport can either be active (cooling unit) or passive (use of isolating boxes and, e.g., thermal packs or crushed ice). Usually the temperature indicators of active cooling systems are in the driver's cab, so that the temperature can be continuously monitored during the transport. To keep the temperatures under the critical limits, dispatch areas should be refrigerated and vehicle doors or containers should be opened as seldom as possible, especially during the summer season when the outside temperature is high.

The probability of an interruption of the cold chain during transportation is particularly high, especially when the transport is performed by smaller and microfood businesses or private individuals. Therefore, the inspection of incoming goods is of great importance.

3.1.2.1 Incoming Goods Inspection

The incoming goods inspection also has to be performed in accordance with the HACCP-principles. There are a variety of parameters that need to be monitored, e.g., temperature, shelf life, integrity of packaging, cleanness (e.g., of the driver and the loading space of the means of transport), and quality parameters of the food itself (e.g., pH-value, color, and odor).

The critical limits for the temperature control are the same as mentioned before, but the temperature monitoring system differs from the described procedures in food businesses. As a first step, the temperature recorded by the temperature monitoring system in place during transport needs to be checked at reception of the goods. Furthermore, the temperature of a representative number of products should be measured. Because there are usually different temperature zones in the storage area of cooled food transporters, the measurements should be carried out at different locations within the means of transport. These measurements would take quite a long time when they are done with a puncture thermometer. That is why most food businesses use infrared thermometers, which can indicate a value within 1 s, as a screening method for the inspection of incoming goods. One disadvantage of these thermometers is that these measure the superficial temperature of an object only (e.g., temperature of the surface of packing in case of packaged food or the surface temperature of food), which does not always correlate with the temperature inside the product. If the screening temperature values are suspicious, the product temperature of a representative number of foodstuffs should be measured using a puncture thermometer. To carry out the measurement with the infrared thermometer correctly, staff has to watch certain criteria such as emissivity of the target, measuring angle, size of the target area, and distance to the target. Therefore, the person carrying out the measurement needs to be well trained before use of the thermometer.

In case of unsatisfactory results during inspection of the incoming goods, the rejection of the food is often the only possible corrective action. To avoid the delivery of unsatisfactory foodstuffs on a prospective basis, the results of the incoming goods inspection should be evaluated systematically for each supplier and a rating system of the supplier should be established. Suppliers who often deliver unsatisfactory products should be discontinued. Furthermore, staff being responsible for the incoming goods inspection needs to be monitored; in particular, the correct use of the infrared thermometer and the correct documentation should be controlled.

Usually there are forms for the documentation of the inspection of incoming goods, which can be filled out manually or at a computer. Furthermore, it is important for any food business to describe the incoming goods inspection procedure and the critical points very precisely in standard operating procedures.

3.1.3 Maintenance of the Cold Chain During Thawing Processes

Many foodstuffs are either distributed frozen or are frozen at retail, or in consumer's households. Freezing of food has little effect on the content of microorganisms in the food. In particular, the content of *S. aureus* is not significantly altered by freezing (Horlacher et al., 2011). Therefore, it is to be assumed that a high content of *S. aureus* may survive the freezing process and subsequently will be present when the product is thawed.

Even during thawing the cold chain must be maintained. There are different thawing procedures for food, e.g., at room temperature, in warm water, on heated surfaces, in cold water, in chilling facilities, or by microwaves. When food is thawed at high temperatures, the surface temperature of the goods will rapidly rise to ambient temperature. At these high temperatures, pathogens, e.g., *S. aureus*, which is often located on the surface of foodstuff (Schelin et al., 2011), can multiply quickly (Benli, 2016). To minimize the risk of growth of *S. aureus* or the formation of toxins in the foods, the thawing process should comply with certain hygiene rules, mainly that the ambient temperature should also correspond to the storage temperature specified before. As the thawing water quickly reaches ambient temperatures, which is often highly watery and rich of nutrients, favoring bacterial growth adequate draining need to be realized, e.g., by the use of draining racks or sieves. If possible the frozen products should be processed without thawing.

3.1.4 Maintenance of the Cold Chain at Consumer's Households

The cold chain does not end in retail shops or supermarkets. Also in consumer's households, foodstuff has to be kept at temperatures according to its labeling. Even in private households refrigerators should be equipped with an external thermometer to check the appropriate chilling temperature. As soon as the thermometer indicates an increased temperature, the cooling system should be adjusted. For example, during summer, refrigerators require slightly higher cooling capacities than during winter. Consumer should consider that the date of minimum durability or the "use by" date retains its validity only if the chilling temperature complies with the labeled temperature. Because, for example, minced meat has to be stored at 2°C, this requirement can only be met in the case of particularly suitable refrigerators.

In household kitchens deep-frozen foodstuff should be thawed in the refrigerator, in cold water, or in the microwave. When the thawing is done in a microwave the food should be cooked immediately after thawing, as some areas of the food may reach temperatures where the replication of pathogens, such as *S. aureus*, is possible (USA. United States Department of Agriculture, 2003).

3.2 HOLDING TEMPERATURES

To keep the risk of staphylococcal food intoxication as low as possible, the holding temperatures of ready-to-eat food must be kept above 48.0°C. Because other foodborne pathogens (e.g., *Bacillus cereus*) can grow at even higher temperatures, the general recommendation to keep ready-to-eat food warm is a holding temperature of >65°C.

There are some restaurants and other food service operators, such as bakeries and snack bars, where takeaway food is kept lukewarm, so that customers can eat the food right after purchase. Referring to Table 11.2 this is one of the main reasons for staphylococcal replication and toxin production. Therefore, some food business operators still have to be convinced to store their foodstuffs either at a temperature above 65°C or refrigerated and reheat it before purchase. The required holding temperatures must also be adhered and monitored in accordance with the HACCP principles as described before.

A monitoring of the holding temperature system put in place should be carried out daily using thermometers. Alternatively, a standard operating procedure can be established. For example, it can be defined at which setting the temperature regulator of a holding device must be set to maintain the defined critical limits. If the required holding temperature has fallen below the germination temperature and a toxin-production cannot be excluded, the food should be disposed because staphylococcal enterotoxins are thermostable. To verify the checks of the holding temperature, the staff should be supervised regularly and the thermometers used should be calibrated against a standard. The documentation of the holding temperature is usually done manually on certain forms. In case a standard operating procedure is established, it should be checked at least weekly whether the critical limit is exceeded at the given setting.

Especially food businesses involved in catering operations carefully have to monitor the holding temperature. Some caterers cook the food and deliver it to different kitchens where the food is portioned and given to the consumer. This way of catering is called "cook and serve". Such transport of warm food takes place in thermophores, i.e., mostly polystyrene boxes or containers where the food is kept warm. Likewise, the holding temperature in these thermophores should not fall below 65°C (Germany. Bundesinstitut für Risikobewertung, 2008). To verify this, it is important to check the temperature of the food before the food is served to the consumer. Depending on the matrix, the temperature of the food, which has been put in the thermophores, the thermal isolation properties of the thermophores, and the degree of filling the temperature of the food might fall below 65°C after 3h. In case the food that has been kept in thermophores is not been consumed within a period of 3h, thermophores should not be used. Then, alternative procedures have to be implemented, such as "cook and chill" or "cook and freeze," where the ready-to-eat food is chilled or frozen, respectively, and reheated immediately before consumption.

Generally, independent of the foodstuff the temperature range between 48.0 and 7°C should be bridged as quickly as possible to minimize the risk of staphylococcal intoxications.

4. PERSONAL HYGIENE AND TRAINING OF STAFF

S. aureus is a frequent commensal on the human skin and respiratory mucosal. Consequently, it must be assumed that in almost every food company, human carriers of *S. aureus* are found among the employees handling the food. *S. aureus* colonization in humans is usually asymptomatic and quite often intermittent (Mainous et al., 2006; Kluytmans and Wertheim, 2005). Because some carriers of *S. aureus* harbor the same strain for months in their nasal cavity, while other strains persist for only 1 week or even shorter (Genigeorgis, 1989; Kluytmans et al., 1997), food business operators would have to carry out further investigations on a regular basis to find out which food handler is colonized with pathogenic *S. aureus*. Because such verifications of the health status are not feasible, food business operators should consider that pathogenic *S. aureus* is ubiquitous in the human environment, and therefore it is also present in their food businesses.

Taken together, nasal carriers cannot be excluded from food handling. For this reason, personal hygiene is essential to avoid or at least reduce the risk of staphylococcal contamination by food handlers.

4.1 PERSONAL HYGIENE

Humans are one of the main sources of *S. aureus* found in food; thus, key element to combat staphylococcal food intoxication, in addition to temperature management, is personal hygiene. Because *S. aureus* is mainly found in the upper respiratory tract and on the human skin, hygiene measures should focus on the hygiene of the skin (including hygiene clothing) and prevention of oral and nasal contamination of food, food contact materials (including packaging, machines, cutting boards, etc.), and other possible sides of cross-contamination (e.g., handles, switches, etc.). To fulfill the defined hygiene measures, the staff has to be regularly trained. In addition, quantitative microbiological testing of *S. aureus* in ready-to-eat foods may be used as indicator parameter to check for personal hygiene.

There are some factors that increase the risk of staphylococcal contamination, such as sores, cuts, and discharge from the nose, mouth, ears, and eyes. Other factors such as the age could influence this risk. Among some patients on long-term hospital-based hemodialysis, a significant correlation between age and nasal *S. aureus* carriage has been found (Saxena et al., 2004) but not for others (Tashakori et al., 2014). In any case, it is unclear whether the results obtained in hospitals can be transferred to food companies, but it seems logical that *S. aureus* could occur more frequently in older employees.

S. aureus has a strong ability to colonize the skin of patients with eczema and atopic dermatitis and is regularly found in eczematous skin lesions (Gong et al., 2006). Therefore, any infection, sore, or cut on exposed skin (i.e., on hands, arms, face, neck, or scalp) has to be totally covered with a distinctively colored waterproof plaster during the handling of food and food contact materials. The distinctive color of the plaster is necessary to find it in the food business or even in food, if it is detached from the skin. If a used plaster has come into foodstuff, these foods must be discharged immediately. To reduce the risk of such contaminations, the plasters can be covered with gloves or fingerlings.

Food can also be contaminated with *S. aureus* when it is handled with uninjured, completely healthy hands. That is why food should be handled preferably with suitable and clean tools rather than hands. Washing and disinfecting hands significantly reduces the germ content of the hands (Chen et al., 2011; Pittet et al., 2000) and is therefore crucial in personal hygiene.

4.1.1 Washing and Disinfection of Hands

There is a direct correlation between hand disinfection and the reduction of nosocomial MRSA infections in hospitals (Pittet et al., 2000). This fact can be directly transferred into food hygiene procedures.

Hands are one of the most important vehicles for transmitting staphylococcal infections (Ireland. National Disease Surveillance Centre, 2004). Proper washing and disinfection of hands can significantly reduce the risk of microbial food contamination. It is not as easy as it seems to wash off microorganisms that are located in the skins furrows, cracks, creases and pores or under fingernails, watches, finger rings, and other jewelry. In general, it is very important to remove all jewelry before washing and disinfecting the hands. For this reason, it is forbidden to wear jewelry in most food businesses. In some companies the wedding ring is an exception to this prohibition, ignoring that also a wedding ring is a physical hazard when it is worn during the handling of food and the ring also increases the risk of bacterial contamination. In any case fingernails have to be clean, cut short, unpolished, and especially not treated with gel or acrylic. Hands have to be thoroughly washed and should be disinfected before entering the hygienic area or touching foodstuff and after several hygienically suspicious activities, such as handling raw animal-derived food (e.g., poultry, red meat, eggs, fish, etc.), cleaning work, smoking, visiting the toilet, and touching dirty objects, the nose or the face. Food business operators should define in the internal hygiene plans before and after which activities the hands have to be washed and disinfected.

Hand washing efficacy on *S. aureus* is totally dependent on how well it is executed; this is particular true as it has been recently shown that *S. aureus* was not successfully removed from a plate and cutting board, resp. by rinsing with water (Fetsch et al., 2015). Firstly, the hands have to be wetted with water. Enough soap has to be used to cover all hand surfaces. Then it is important to rub hands palm to palm, palm to dorsum, palm to palm with fingers interlaced, backs of fingers to opposing palms, thumbs with the palm, and rotational with clasped fingers in palm (World Health Organization, 2006). In addition to the use of soap, friction during the rinsing stage is also crucial to dislodge *S. aureus*. After drying hands with a single-use towel, the towel should be used to turn off the faucet to not recontaminate the hands, if no contactless water supply is available. Once the hands are dry, a germicide should be applied to the hands in the same manner as described above. Except for that alcoholic-disinfectants should not be washed off, to maximize the contact time. Because of a higher efficiency, it is recommended to use soap and disinfection solution separately. It takes around 30 s to wash the hands properly and another 30 s for the disinfection.

4.1.2 Hygiene Clothing

Besides of the mucosal of the nose and the mouth *S. aureus* is frequently found on the skin of the hands, face, hairline, chest, abdomen, and axillae of humans (Bridges-Webb et al., 1971; Hatakka et al., 2000; Knox et al., 2015). Therefore, hygiene clothing should cover these areas and ensure that perspiration does not drip from the body. Depending on the type of work to be carried out, the hygiene clothing may include outerwear (recommended are long sleeves, but short sleeves are not excluded), pants, headgear, shoes or boots, aprons, and special protective elements such as gloves and beard-, mouth-, and nose protection. Work clothes should be changed weekly in areas with low hygienic risk (e.g., handling of packaged, not perishable foodstuffs) and otherwise at least every day, in case of contamination even more often (United Kingdom. British Retail Consortium, 2015; Germany. Deutsches Institut für Normung, 2012). To be able to better identify contamination, hygiene clothing should be colored lightly or in white.

Especially in approved food businesses, clean dressing rooms with personal compartments have to be provided. A cross-contamination of personal clothing with the hygiene clothing has to be avoided. Therefore, clean textile hygiene clothing should be stored in a designated storage compartment, for example, in cabinets. The compartment should be dry and protect the clothing from contamination. Private clothing, jewelry, hygiene items such as deodorant, hairbrush, mascara, and other makeup, and other private items such as mobile phones, cigarettes, earplugs, and bags have to be separated from freshly laundered hygiene clothing, e.g., by placing them in separate compartments within the same cabinet, to avoid cross-contamination.

Other hygiene clothing, which is normally not changed over a long period, such as boots, clogs, aprons, or chain mail gloves, should be stored at their intended places after they have been cleaned and, if necessary, disinfected.

Clean hygiene clothing must be worn before entering the production area and should be taken off at specified places after leaving these areas. Special internal rules of conduct can be defined for breaks (Germany. Deutsches Institut für Normung, 2012). Contaminated hygiene clothing has to be placed in collecting devices so that it cannot adversely affect the hygiene in the company. In particular, wet hygiene clothing must be placed in liquid-tight collecting containers and be cleaned as soon as possible.

The laundry should be done by professional laundry services, which put the hygiene clothing in an optically clean and for food businesses sufficiently hygienic state. To ensure a sufficiently hygienic state, the laundry services should implement regular microbiological examinations, i.e., mesophilic bacterial count. The microbiological tests are carried out by agar contact plate methods, e.g., using replicate organism direct agar contact (RODAC) plates. The plates should contain a casein-soy-peptone-agar with an inhibitor (inactivating agent) and be incubated for 48 h at 30°C or 72 h at 25°C. The laundry is considered hygienically clean if no more than 10 colonies per plate (20 cm^2) are detectable in 9 of 10 samples (Germany. Deutsches Institut für Normung, 2012). Laundries that can meet these conditions are usually externally audited and receive a seal of quality.

Laundry of work clothes used in food businesses in private households is not recommended. To reduce the content of pathogens such as *S. aureus* to an acceptable level, the laundry must be washed at temperatures of 60°C or higher and with detergents containing bleaching agents. After washing, the laundry has to be hung up, stored, and transported in a way that no recontamination is possible. One of the primary sources of recontamination of freshly laundered hygiene clothes in private households may be young children: not only because of their higher prevalence rates of *S. aureus* compared to adults (Bridges-Webb et al., 1971; Jackson, 2000; Miyake et al., 1991) but also because children most often touch their noses and mouths without washing their hands. If nasal and oral secretions are on the hands of children, they could easily contaminate freshly laundered work clothes when they touch them. Therefore, children and also pets should have no access to the laundry. If laundry of hygiene clothes is unavoidable in private households, the employees should be trained on how laundry and transport to the food business is done under hygienic conditions.

To avoid staphylococcal contamination of perishable foodstuff especially clean long sleeves, headgears, sufficient to cover hair extensively, and face masks should be worn in addition to hygienic pants and shoes. Gloves, which are suitable for foodstuffs, can also reduce the risk of contamination in certain situations. But wearing gloves does not necessarily increase the hygiene level in the production of food. The hands often feel clean, even if the worn gloves are already strongly contaminated. Therefore gloves have to be changed frequently, depending on the work carried out (United Kingdom. British Retail Consortium, 2015). Furthermore, wearing gloves does not release the employee from cleaning and disinfecting the hands before putting the gloves on.

Likewise to own staff, all visitors and employees of external companies also have to wear appropriate hygiene clothing when they enter the production areas.

4.1.2.1 Face Masks

As *S. aureus* is often located in the upper respiratory tract of food handlers, face masks can be used to retain *S. aureus* from the expiratory air and droplets dispersed by sneezing or coughing. Therefore, many food companies use face masks to avoid oral–nasal contamination of foodstuffs. This is particularly important in packaging departments of manufacturers placing thermally treated food on the market because these foods have a low amount of competing bacteria. Furthermore, face masks are often used in the meat industry, e.g., in cutting plants and during production of minced meat. In a field study conducted in two minced meat plants, face masks were microbiological examined. In one factory 43% and in the other 64% of the face masks of the persons processing minced meat were tested positive for coagulase-positive staphylococci (CPS) including *S. aureus*. Although remarkably high CPS counts have been detected in some face masks, no adverse effects on the microbial quality of the minced meat were observed. This suggests that wearing face masks in minced meat production is a simple way to reduce the risk of microbiological contamination (Schmitt, 2006).

However, staphylococci are frequently isolated from air samples obtained in operating rooms. Despite wearing surgical masks, nasopharyngeal shedding from the operating staff has been identified as a source of the staphylococcal airborne contaminants, especially in surgeries lasting longer than 90 min (Edmiston et al., 2005). Face masks usually unfold their effect only in the dry state and should be exchanged as soon as they get wet. Coughing, sneezing, laughing and speaking, or nasal and oral discharge increases the proportion of bacteria that can pass through the face mask (Muxfeldt, 1968). Therefore, food handlers who have a cold should be banned from the production areas, and face masks should be regularly replaced by unused masks, e.g., after breaks, change of the workplace, or after the use of toilets. Used beard-, mouth-, and nose protection should be disposed or thermally treated.

4.2 TRAINING

Only well-trained and motivated staff can fulfill all requirements that are necessary to produce safe food. Quite often food business operators have GMP- and HACCP systems established, but because the motivation or knowledge of the staff is too weak the specified hygiene rules are misapplied. It is assumed that the motivation of employees to follow the hygiene rules has the highest impact on food safety. Therefore good and motivating training of the complete staff (including temporary workers) on food hygiene principles is essential and should be done at least once a year or more often in case of contempt of the rules (Germany. Deutsches Institut für Normung, 2009).

Because prerequisites of such trainings on food hygiene principles at least information on the hygienic situation, the risk analysis of the company and possible effects of any misconduct of the staff on the product should be taken into account. The training materials must be written in a language that is easy to understand for the participants (United Kingdom. British Retail Consortium, 2015). The use of self-explanatory illustrations and symbolic representations is also recommendable.

The success of the training should be verified in a suitable way, e.g., by an examination, a subject-specific discussion, or a practical success check at the place of work. The training and the results of the verification have to be accurately documented for each participant to show what instructions and trainings have been given. If further training need are identified by the success check, appropriate follow-up trainings must be carried out (Germany. Deutsches Institut für Normung, 2009).

Trainings must be carried out before employee start working for the first time (United Kingdom. British Retail Consortium, 2015). In the case of seasonal and back staff, a special instruction in hygiene, in relation to the work, should be given.

The most important training contents regarding the prevention of staphylococcal intoxications are, for example, the temperature management, the need to follow instructions concerning diseases and injuries, hygienic behavior in regard of coughing, sneezing, and toilet use as well as body hygiene, and the necessity to report failing controls promptly. To determine training needs the staff should be appropriately supervised. If particular hygienic behaviors are observed in this supervision, rewards should be provided and issue reminders in case of contempt of the rules.

5. PREVENTIVE MEASURES AGAINST *STAPHYLOCOCCUS AUREUS* IN CONSUMER'S HOUSEHOLDS

In a study of *S. aureus* strains isolated from 359 foodborne staphylococcal outbreaks in England and Wales between 1969 and 1990, most cases proved to be due to external contamination and, in particular, to a lack of hygiene in the kitchens of the final consumer (Wieneke et al., 1993). In the EU 31 strong-evidence outbreaks caused by staphylococcal toxins were reported for the year 2014. Although most of these outbreaks were general outbreaks (58.1%), a considerable number of kitchen outbreaks were reported (38.7%). The setting reported was "household" in 10 outbreaks, "restaurant, café, pub, bar, hotel" in 7 outbreaks, and "school or kindergarten" in 3 outbreaks (EFSA, 2015).

Basically, similar preventive measures as in food businesses have to be taken in consumer's households to avoid the proliferation and toxin production of *S. aureus*. The need to maintain the cold chain also at consumer's households was already described before (see Section 3.1.4).

To prevent contamination of any foodstuff with *S. aureus*, direct and indirect contact of products with the mouth, nose, and hair of humans and pets should be avoided. For example, food should not be tasted with the cooking spoon, but be given from the cooking spoon to a so-called tasting spoon and only this spoon should be taken into the mouth. Even when taking a small amount of food out of a bigger packaging (e.g., jam, olives, ham, cheese, yoghurt), the removal tool should not have been licked before. Jewelry or bracelets should be discarded from hands and hands have to be washed before food is touched (Byrd-Bredbenner et al., 2013); it may even be better to handle food with clean cutlery than with clean hands. Even during food preparation, hands should be washed regularly, especially after contact with hygienically suspicious foodstuff, such as food of animal origin, which has not been thermally treated (e.g., raw eggs, meat, poultry, raw milk cheese, cured ham, raw sausages, raw fish).

Separating "clean" from "unclean" procedures and hygienically suspect food (e.g., raw eggs, poultry, meat, fish) from ready-to-eat food (e.g., salad, cooked eggs, cooked poultry, meat, or fish) is of most importance to avoid cross-contamination. For example, thermally treated food should not be handled on a chopping board, which has been used to handle raw food before, without prior proper cleaning of the board (Byrd-Bredbenner et al., 2013). The cleaning of the board should be done with hot water and dishwashing detergent using dryly stored brushes and pot sponges, which are replaced weekly. Wooden chopping boards with very rough surfaces should be replaced.

S. aureus can multiply quickly at room temperature, e.g., in kitchen cloths, when they are wet and contaminated with food residues (Byrd-Bredbenner et al., 2013). Therefore, one-use towels such as those from a kitchen roll should be used to wipe spilled food and the dripping water coming from the thawing of raw animal products. *S. aureus* can also grow quickly in unwashed crockery, which is stored at room temperature. A dishwasher cleans objects at temperatures around 60–70°C. During the rinsing phase, the water is even heated to ~80°C, which eliminates or at least reduces remaining bacteria. When dishes are cleaned manually the water temperature is too low to eliminate *S. aureus*. Therefore, measures should be undertaken to reduce initial bacteria replication to a minimum. That is why the crockery should be washed as soon as possible after the meal. Furthermore, dishes should not be soaked in the water for a long time because warm water is a good medium for *S. aureus* replication. Also food remaining after the meal should be put in the refrigerator as soon as the food has adapted to room temperatures.

Even when food is not eaten indoors (e.g., at a barbeque) or eaten quite a while after its preparation (e.g., at a picnic or as food in lunch boxes), the cold chain should be maintained wherever possible, for instance, by using cooling boxes.

6. CONCLUSION

S. aureus is one of the leading causes of foodborne disease worldwide and therefore has to be considered as a hazard in terms of the HACCP principles. Usually, the severity of the disease is mild, but very occasionally also fatal cases are reported (EFSA, 2015); in consequence, the risk of staphylococcal intoxications has to be considered as relevant. Given the high number of foodborne outbreaks caused by *S. aureus* (compare EFSA, 2015), the risk of staphylococcal intoxication is so high that any food business operator has to consider *S. aureus* when establishing its HACCP concept.

The main reasons for staphylococcal food poisoning are improper cooling and holding temperature, inadequate personal hygiene, and food handlers as carriers. Therefore, food business operators have to put in place, implement, and maintain controls based on HACCP principles. Furthermore, as food handlers are the most frequent source of contamination of food with *S. aureus*, all food has to be handled in accordance with the GHP principles. All staff handling food has to be healthy and well trained on hygienic principles, too.

To prevent foodborne disease because of *S. aureus*, the following three measures are keys to prevent or eliminate the hazard *S. aureus* from the food production process or at least to reduce it to an acceptable level:

Measurement 1: Inactivation of S. aureus applying thermal and nonthermal processing technologies (see Chapter 13 for further details)

Measurement 2: Avoid any replication of S. aureus in food, e.g., by proper temperature management (such as maintaining a chilling temperature of <7°C also during transport and holding temperatures of at least >48.0°C), by targeting an aw-value of <0.83, a pH-value of <4.0, or a NaCl concentration of >20% in the food. Moreover, multiple hurdles, such as starter culture, modified atmosphere packaging, and preservatives, should be implemented to avoid growth of S. aureus in a given foodstuff.

Measurement 3: Avoid any (re)contamination of food with *S. aureus*, mainly be strictly following hand washing and disinfecting protocols, wearing of adequate hygiene clothing (if necessary including face masks), implementing measures for staff with wounds or discharge (e.g., from the nose, mouth, ears, and eyes), training of the staff to follow the hygiene- and HACCP procedures, and cleaning and disinfection of food contact materials.

REFERENCES

Baird-Parker, A.C., 1990. The staphylococci: an introduction. Soc. Appl. Bacteriol. Symp. Ser. 19, 1S–8S.

Benli, H., 2016. Consumer attitudes toward storing and thawing chicken and effects of the common thawing practices on some quality characteristics of frozen chicken. Asian-Australas. J. Anim. Sci. 29 (1), 100–108.

Bergdoll, M.S., Wong, A.L., 1993. *Staphylococcus*/food poisoning. In: Macrae, R., Robinson, R.K., Sadler, M.J. (Eds.), Encyclopedia of Food Science, Food Technology and Nutrition. Academic Press, USA, pp. 5556–5561.

Bergdoll, M.S., 1989. Staphylococcus aureus. In: Doyle, M.P. (Ed.), Foodborne Bacterial Pathogens. Marcel Dekker, Inc., New York, NY, USA, pp. 463–523.

Byrd-Bredbenner, C., Berning, J., Martin-Biggers, J., Quick, V., 2013. Food safety in home kitchens: a synthesis of the literature. Int. J. Environ. Res. Public Health 10 (9), 4060–4085.

Byrd-Bredbenner, C., Maurer, J., Wheatley, V., Cottone, E., Clancy, M., 2007. Food safety hazards lurk in the kitchens of young adults. J. Food Prot. 2007 (70), 991–996.

Bridges-Webb, C., Gulasekharam, J., Graydon, J.J., April 3, 1971. A bacteriological study of the upper respiratory tract in normal families. Med. J. Aust. 1 (14), 735–758.

Chen, Y.-C., Sheng, W.-H., Wang, J.-T., Chang, S.-C., Lin, H.-C., Tien, K.-L., Hsu, L.-Y., Tsai, K.-S., 2011. Effectiveness and limitations of hand hygiene promotion on decreasing healthcare–associated infections. PLoS One 6 (11), e27163.

Edmiston Jr., C.E., Seabrook, G.R., Cambria, R.A., Brown, K.R., Lewis, B.D., Sommers, J.R., Krepel, C.J., Wilson, P.J., Sinski, S., Towne, J.B., 2005. Molecular epidemiology of microbial contamination in the operating room environment: is there a risk for infection? Surgery 138 (4), 573–582.

EFSA – European Food Safety Authority, European Centre for Disease Prevention, control, 2015. The European Union Summary Report on Trends and Sources of Zoonoses, Zoonotic Agents and Food-Borne Outbreaks in 2014. [Online] Available from: http://ecdc.europa.eu/en/publications/publications/zoonoses-trends-sources-eu-summary-report-2014.pdf#page=1&zoom=auto,-274,848.

European Commission, 2016. Notice on the Implementation of Food Safety Management Systems Covering Prerequisite Programs (PRPs) and Procedures Based on the HACCP-Principles, Including the Facilitation/Flexibility of the Implementation in Certain Food Businesses. C 278/01.

European Commission, 2009. Report from the Commission to the Council and the European Parliament on the Experience Gained from the Application of the Hygiene Regulations (EC) No 852/2004, (EC) No 853/2004 and (EC) No 854/2004 of the European Parliament and of the Council of 29 April 2004. [Online] Available from:https://ec.europa.eu/food/sites/food/files/safety/docs/biosafety_hygiene_report_act_part1_en.pdf.

European Council, 1989. Directive 89/108/EEC on the Approximation of the Laws of the Member States Relating to Quick-Frozen Foodstuffs for Human Consumption. OJ L 40/34.

European Parliament and Council, 2004a. Regulation (EC) No 852/2004 on the Hygiene of Foodstuffs. L 139/01.

European Parliament and Council, 2004b. Regulation (EC) No 853/2004 Laying Down Specific Hygiene Rules for Food of Animal Origin. L 226/22.

Fetsch, A., Tenhagen, B., Leeser, D., Steege, K., Schabanowski, A., Kraushaar, B., Thoens, C., Kaesbohrer, A., Kelner-Burgos, Y., 2015. High risk of cross-contamination with ESBL *E. coli* and MRSA during handling with contaminated fresh chicken meat in household kitchens. In: 4th ASM Conference on Antimicrobial Resistance, May 5, 2015–May 8, 2015, Washington, DC [Online] Available from: http://conferences.asm.org/images/2015_amr_program.pdf.

Genigeorgis, C.A., 1989. Present state of knowledge on staphylococcal intoxication. Int. J. Food Microbiol. 9, 327–360.

Germany. Bundesinstitut für Risikobewertung (BfR), 2008. Warmhaltetemperatur von Speisen sollte über 65°C betragen. [Online] Available from: http://www.bfr.bund.de/cm/343/warmhaltetemperatur_von_speisen_sollte_ueber_65_grad_betragen.pdf.

Germany. Deutsches Institut für Normung, 2012. DIN 10524:2012-04 Food Hygiene – Work Wear in Food Business.

Germany. Deutsches Institut für Normung, 2009. DIN 10514:2009-05 Food Hygiene – Hygiene Training.

Götz, F., Bannerman, T., Schleifer, K.H., 2006. The genera *Staphylococcus* and *Macrococcus*. In: Dworkin, M., Falkow, S., Rosenberg, E., Schleifer, K.H., Stackebrandt, E. (Eds.), The Prokaryotes, Bacteria: Firmicutes, Cyanobacteria, vol. 4. third ed. Springer Sciences Business Media, USA, New York, pp. 5–75.

Gong, J.Q., Lin, L., Lin, T., Hao, F., Zeng, F.Q., Bi, Z.G., Yi, D., Zhao, B., 2006. Skin colonization by *Staphylococcus aureus* in patients with eczema and atopic dermatitis and relevant combined topical therapy: a double-blind multicentre randomized controlled trial. Br. J. Dermatol. 155 (4), 680–687.

Halpin-Dohnalek, M.I., Marth, E.H., 1989. *Staphylococcus aureus*: production of extracellular compounds and behavior in foods: a review. J. Food Prot. 52, 267–282.

Hatakka, M., Björkroth, K.J., Asplund, K., Mäki-Petäys, N., Korkeala, H.J., 2000. Genotypes and enterotoxicity of *Staphylococcus aureus* isolated from the hands and nasal cavities of flight-catering employees. J. Food Prot. 63 (11), 1487–1491.

Horlacher, S., Tichaczek-Dischinger, P.S., Hummel, A.K., 2011. Zur Überlebensfähigkeit pathogener Keime in Lebensmittelproben. ATD. 1/2011, pp. 19–23.

Ireland. National Disease Surveillance Centre (NDSC), April 2004. Preventing Foodborne Disease: A Focus on the Infected Food Handler. Report of the Food Handlers with Potentially Foodborne Diseases Subcommittee of the NDSC's Scientific Advisory Committee [Online] Available from: https://www.hpsc.ie/A-Z/Gastroenteric/FoodborneIllness/Publications/File,871,en.pdf.

Jackson, M.S., 2000. Staphylococci in the oral flora of healthy children and those receiving treatment for malignant disease. Microb. Ecol. Health D. 12, 60–64.

Kluytmans, J.A.J.W., Wertheim, H.F.L., 2005. Nasal carriage of *Staphylococcus aureus* and prevention of nosocomial infections. Infection 33, 3–8.

Kluytmans, J., van Belkum, A., Verbrugh, H., 1997. Nasal carriage of *Staphylococcus aureus*: epidemiology, underlying mechanisms, and associated risks. Clin. Microbiol. Rev. 10, 505–520.

Knox, J., Uhlemann, A.C., Lowy, F.D., 2015. *Staphylococcus aureus* infections: transmission within households and the community. Trends Microbiol. 23 (7), 437–444.

Kümmel, J., Stessl, B., Gonano, M., Walcher, G., Bereuter, O., Fricker, M., Grunert, T., Wagner, M., Ehling-Schulz, M., 2016. *Staphylococcus aureus* entrance into the dairy chain: tracking *S. aureus* from dairy cow to cheese. Front. Microbiol. 7, 1603.

Le Loir, Y., Baron, F., Gautier, M., 2003. *Staphylococcus aureus* and food poisoning. Genet. Mol. Res. 2 (1), 63–76.

Mainous 3rd, A.G., Hueston, W.J., Everett, C.J., Diaz, V.A., 2006. Nasal carriage of *Staphylococcus aureus* and methicillin-resistant *S aureus* in the United States, 2001–2002. Ann. Fam. Med. 4, 132–137. 16569716.

Miyake, Y., Iwai, T., Sugai, M., Miura, K., Suginaka, H., Nagasaka, N., 1991. Incidence and characterization of *Staphylococcus aureus* from the tongues of children. J. Dent. Res. 70 (7), 1045–1047.

Muxfeldt, H., 1968. Mouth protection in the operating room. Agnes Karll Schwest. Krankenpfl. 22 (6), 256.

Nussinovitch, A., Rosen, B., Firstenberg–Eden, R., 1987. Effects of yeasts on the survival of *S. aureus* in pickled cheese brine. J. Food Prot. 50 (12), 1023–1024.

Otero, A., García, M.C., Moreno, B., 1988. Effect of growth of a commercial starter culture on growth of *Staphylococcus aureus* and thermonuclease and enterotoxins (C1 and C2) production in broth cultures. Int. J. Food Microbiol. 6 (2), 107–114.

Pittet, D., Hugonnet, S., Harbarth, S., Mourouga, P., Sauvan, V., Touveneau, S., Perneger, T.V., October 14, 2000. Effectiveness of a hospital-wide programme to improve compliance with hand hygiene. Infection Control Programme. Lancet 356 (9238), b1307–1312.

Sameshima, T., Magome, C., Takeshita, K., Arihara, K., Itoh, M., Kondo, Y., May 5, 1998. Effect of intestinal *Lactobacillus* starter cultures on the behaviour of *Staphylococcus aureus* in fermented sausage. Int. J. Food Microbiol. 41 (1), 1–7.

Saxena, A.K., Panhotra, B.R., Chopra, R., 2004. Advancing age and the risk of nasal carriage of *Staphylococcus aureus* among patients on long-term hospital-based hemodialysis. Ann. Saudi Med. 24 (5), 337–342.

Schelin, J., Wallin-Carlquist, N., Thorup Cohn, M., Lindqvist, R., Barker, G.C., Rådström, P., 2011. The formation of *Staphylococcus aureus* enterotoxin in food environments and advances in risk assessment. Virulence 2 (6), 580–592.

Schmitt, R., 2006. Staphylokokken in Mund- und Nasenmasken von Mitarbeitern in der Hackfleischproduktion am Beispiel von zwei süddeutschen Schlacht- und Fleischverarbeitungsbetrieben (Vet med dissertation). Ludwig-Maximilians-Universität München, Germany, Munich.

Tavakoli, H.R., Mashak, Z., Moradi, B., Sodagari, H.R., 2015. Antimicrobial activities of the combined use of *Cuminum Cyminum* L. essential oil, nisin and storage temperature against *Salmonella typhimurium* and *Staphylococcus aureus* in vitro. Jundishapur J. Microbiol. 8 (4), e24838.

Tashakori, M., Mohseni Moghadam, F., Ziasheikholeslami, N., Jafarpour, P., Behsoun, M., Hadavi, M., Gomreei, M., 2014. *Staphylococcus aureus* nasal carriage and patterns of antibiotic resistance in bacterial isolates from patients and staff in a dialysis center of southeast Iran. Iran. J. Microbiol. 6 (2), 79–83.

Tatini, S.R., 1973. Influence of food environments on growth of *Staphylococcus aureus* and production of various enterotoxins. J. Milk Food Technol. 39, 432–438.

United Kingdom. British Retail Consortium, 2015. Global Standard Food Safety. Issue 7.

USA. United States Department of Agriculture (USDA), 2003. The Big Thaw — Safe Defrosting Methods for Consumers. [Online] Available from:https://www.fsis.usda.gov/wps/portal/fsis/topics/food-safety-education/get-answers/food-safety-fact-sheets/safe-food-handling/the-big-thaw-safe-defrosting-methods-for-consumers/CT_Index.

Vernozy-Rozand, C., Meyrand, A., Mazuy, C., Delignette-Muller, M.L., Jaubert, G., Perrin, G., Lapeyre, C., Richard, Y., 1998. Behaviour and enterotoxin production by *Staphylococcus aureus* during the manufacture and ripening of raw goats' milk lactic cheeses. J. Dairy Res. 65 (2), 273–281.

Wieneke, A.A., Roberts, D., Gilbert, R.J., 1993. Staphylococcal food poisoning in the United Kingdom, 1969–1990. Epidemiol. Infect. 110, 519–531.

Wong, A.L., Bergdoll, M.S., 1993. *Staphylococcus*/properties and occurence. In: Macrae, R., Robinson, R.K., Sadler, M.J. (Eds.), Encyclopedia of Food Science, Food Technology and Nutrition, USA. Academic Press, pp. 5547–5551.

World Health Organization, 2006. Practical Guidance for Hand Hygiene (When and How). [Online] Available from: http://www.who.int/injection_safety/1card_handwash_web.pdf?ua=1.

CHAPTER

INACTIVATION OF *STAPHYLOCOCCUS AUREUS* IN FOODS BY THERMAL AND NONTHERMAL CONTROL STRATEGIES

12

Dana Ziuzina, Agata Los, Paula Bourke

School of Food Science and Environmental Health, Dublin Institute of Technology, Dublin, Ireland

1. INTRODUCTION

The microbiological safety of foods persists as a major challenge facing the food industry globally, which in turn drives the development of technological interventions and efforts from food processors to improve food quality and safety. In 2010, the World Health Organization reported 600 million foodborne illnesses with 420,000 associated deaths with 230,000 of these occurring from diarrheal disease agents. A significant portion among the reported foodborne illnesses has been attributed to staphylococcal infection (WHO, 2015). *S. aureus* is an opportunistic human pathogen, causing a wide array of infections varying from minor skin infection to severe life-threatening diseases (Argudín et al., 2010). *Staphylococcus aureus* pathogenicity is attributed to toxin-mediated virulence, invasiveness, and antibiotic resistance (Kadariya et al., 2014).

Considering *S. aureus* abundance in nature, there is a high potential for foods to become contaminated. The major source of staphylococcal foods contamination is via contact with *S. aureus* carriers (Ortega et al., 2010). Besides poor personal hygiene, inadequate food preparation, and storage temperature conditions, the cross-contamination of foods and food processing equipment is also among the prime factors involved in staphylococcal food contamination. Emergence of *S. aureus* antimicrobial resistant strains and formation of biofilms associated with foods and food contact surfaces are another important areas of the research focused on the food-associated human staphylococcal infection. Although companion, livestock, and wildlife animals play an important role in the transmission of methicillin-resistant *S. aureus*, personal hygiene and food-processing contact surfaces serve as a main reservoir for antimicrobial resistant *S. aureus* (Doulgeraki et al., 2016; Kamal et al., 2013; Gutierrez et al., 2012).

Therefore the first part of this chapter focuses on the inactivation capacity of conventional thermal as well as novel thermal technologies and combinations of thermal and nonthermal approaches against *S. aureus*, whereas in the second part, emerging nonthermal technologies, which show high potential for *S. aureus* control, are presented.

Staphylococcus aureus. http://dx.doi.org/10.1016/B978-0-12-809671-0.00012-7

2. INACTIVATION OF *STAPHYLOCOCCUS AUREUS* BY THERMAL PROCESSING

Thermal preservation of foods retains a dominating position in food industry since the discovery of the process in 1800s and its subsequent commercialization (Li and Farid, 2016; Ahmed and Ramaswamy, 2007). Although the main advantage of thermal treatment is the production of food that is free from spoilage microorganisms and other agents harmful to humans, who together extend the shelf life, it can also result in some other desirable changes, such as protein coagulation, texture softening, and formation of aromatic components (Ahmed and Ramaswamy, 2007). Thermal processing is classified on the basis of temperatures applied to the product and forms two categories: pasteurization and sterilization. Although pasteurization destroys only vegetative bacterial cells and reduces microbiological load to a safe level, sterilization inactivates all forms of microorganisms including spores; this also termed commercial sterilization intended for application in the food industry. Pasteurization temperatures and times vary and usually carried at temperatures below 100°C depending on the product nature and the microorganisms of the safety concerns. For example, pasteurization of milk can be achieved by 30 min heating at 63°C by long time low-temperature pasteurization method or by 15 s heating at 72°C through high-temperature short treatment processing (Ahmed and Ramaswamy, 2007). Thermal sterilization is usually achieved through two well-known methods: in container processing (or retorting) and ultrahigh temperature processing, i.e., with aseptic filling, which induces the temperature of food products to higher than 100°C (Li and Farid, 2016).

Commercial sterilization processes are designed around spores of *Clostridium botulinum* (Richardson, 2001), as other pathogenic microorganisms of food safety concerns, including cells of *S. aureus*, are not capable of surviving at sterilization temperatures during food processing. For instance, 99.6% of *S. aureus* cells are destroyed by the pasteurization of milk at 72°C for 15 s, and at 72°C for 35 s all cells are killed. However, D values at 121 and 100°C for staphylococcal enterotoxins (SEs) in milk can range from 9.9–11.4 to 70.0 min, respectively (Medveďová and Valík, 2012). Nevertheless, the thermal resistance of *S. aureus* has been evaluated by Kennedy et al. (2005), which demonstrated that pasteurization at a time–temperatures combination of anything less than 70°C for 20 min (a minimum core pasteurization regime) might result in the survival of *S. aureus*, thus increasing the potential for the subsequent growth and toxin formation if no chilling temperatures were introduced with limited effects of further reheating on heat-stable SEs. The authors recommended the use of combination of at least 75°C for at least 1 min instead. This study also revealed that *S. aureus* has a greater D-value than *Listeria monocytogenes*, the target microorganism, on which many cooking regulatory guidelines for the thermal destruction of vegetative cells are based, suggesting that this norm should be revised. In fact, thermoresistance of *S. aureus* can be greatly influenced by environmental conditions such as treatment medium composition as well as bacterial strain and bacterial growth phase. The decimal reduction time (D) value at 60°C for *S. aureus* in meat macerate with NaCl content of 0% was 6 min, which increased to 25 min when the NaCl content increased from 0% to 8.4% (Paulin et al., 2012), whereas the D value obtained at 58°C was a maximum 0.93 min for cells suspended in citrate phosphate buffer (Cebrián et al., 2007). In the same work it was demonstrated that the exponentially growing *S. aureus* cells had higher sensitivity to heat treatment at 58°C than cells in stationary phase, and the increase in D value was more pronounced for the pigmented than for the nonpigmented strains. Hassani et al. (2006) in their work demonstrated that treatment medium pH and the type of thermal processing, i.e., nonisothermal heating, associated with both heat-resistant and heat-sensitive *S. aureus* strains had also influenced inactivation rates. Under medium pH 7.4 and nonisothermal conditions, which were set at temperatures from 40 to 70°C and 0.5°C/min, the thermotolerance of heat-sensitive strain was almost at the same levels as the thermotolerance of the heat-resistant strain, making the initially called "heat-sensitive strain" as resistant as the other strain (Hassani et al., 2006). Recently, the thermal resistance of *S. aureus*, isolated from cereals and vegetables and thermally treated feed ingredients was characterized by Amado et al. (2014) using heat-treatment temperatures of 57.5, 60, 62.5, and 65°C. Moreover, to investigate if the type of the treatment media had any effects on *S. aureus* resistance, the authors heat treated the cells either in phosphate buffer solution or inoculated on cattle feed with different particle diameter. After 2 min of treatment at highest temperature (65°C) a drop in *S. aureus* viability by 2–3 log-units was observed when cells were suspended in buffer solution. In contrast, 30 min of 65°C heat treatment was necessary to obtain reductions by similar levels. Particle size also influenced the treatment inactivation activity against *S. aureus* where an increase in microbial thermal resistance was associated with the increase in particle size. In addition, due to the fact that *S. aureus* isolates showed marked tailing behavior at temperature of 57.5, 60, and 62.5°C as compared to other microorganisms tested, this study also pointed out on the presence of a highly resistant cell subpopulations, which might be responsible for the enhanced thermal resistance observed in cattle feed (Amado et al., 2014). In another study, *S. aureus* isolated from fermented sausages was subjected to the temperature of 80°C for 20 min. An inactivation of about 6 log-units in the first 2–4 min of treatment, followed by constant survivor counts (2–4 log CFU/mL) was achieved. The authors suggested that the observed thermotolerance could be due to the acquired resistance within the population as a metabolic response during stress application because the strain was isolated from food matrices pasteurized at this temperature, or the presence before the treatment of a subpopulation intrinsically predisposed to the resistance (Montanari et al., 2015). The ability of *S. aureus* to develop homologous stress resistance responses was investigated by Cebrián et al. (2009), which exposed bacterial cells to acidic (pH 2.5) and alkaline (pH 12.0) pH, hydrogen peroxide (50 mM), and heat (58°C). Under these adaptation conditions, times for the first decimal reduction to a lethal treatment at acid pH, alkaline pH, hydrogen peroxide, and heat were increased by a factor of 1.6, 2, and 6, respectively (Cebrián et al., 2009). A comparative study was conducted by Pearce et al. (2012), where the most heat-resistant strain of each of the milk-borne pathogens, including *S. aureus*, was selected to obtain the worst-case scenario in heat inactivation trials using a pilot-plant-scale pasteurizer. The inoculated milk was subjected to pasteurization under commercial-type conditions of turbulent flow for 15 s over a temperature range from 56 to 66°C and at 72°C. The mean $\log_{10}$ reductions and temperatures of inactivation of the six pathogens during a 15 s treatment were *S. aureus* >6.7 at 66.5°C, *Yersinia enterocolitica* >6.8 at 62.5°C, *Escherichia coli* >6.8 at 65°C, *Cronobacter sakazakii* >6.7 at 67.5°C, *L. monocytogenes* >6.9 at 65.5°C, and *Salmonella* ser. *Typhimurium* >6.9 at 61.5°C, confirming the reliability of current standard pasteurization conditions of 63°C for 30 min and 72°C for 15 s for inactivation of *S. aureus*.

Although *S. aureus* cells are not able to survive during standard pasteurization temperatures, SE can resist both the process of pasteurization and sterilization (Fig. 12.1). With the aim of clarifying how heat treatment used in industrial production of poultry products can affect the viability of *S. aureus* and its toxins, Pepe et al. (2006) inoculated breaded chicken cutlets with SEA-producing *S. aureus* and monitored SEA activity during manufacture at each processing step (preparation, frying, baking, and storage). As anticipated, thermal treatments were able to inactivate bacterial cells resulting in an

~7 $\log_{10}$CFU/g reduction. By contrast, SEA was detected in the cutlets after both cooking steps (frying and baking), thus indicating that SEA resisted double heat treatment at 180°C. Moreover, SEA was still present after 1 week of storage at 4°C, confirming that although thermal processes used during food manufacture can limit staphylococcal contamination, it cannot eliminate preformed staphylococcal toxins (Pepe et al., 2006). In a more recent work conducted by Necidova et al. (2016) milk inoculated with *S. aureus* strains capable of producing SE types A, B, or C was heat treated for 15 s at 72, 85, and 92°C. Before pasteurization, SEB was detected in the lowest amount compared with other SE types. However, after pasteurization at 72, 85, and 92°C, all SEs were markedly reduced and detected in 87.5%, 52.5%, and 45.0% of samples, respectively. This study clearly demonstrated that SE could still persist in milk even when the cells of *S. aureus* are inactivated through pasteurization, suggesting that any cold chain disruption during milk production and processing should be avoided to exclude increase in enterotoxins levels if initially present in milk and provide required safety.

Besides the multiple advantages of thermal inactivation of *S. aureus*, it can be associated with losses of food nutritional value, alterations in color, flavor, and texture of final products and is limited by the condition of packaging process, increasing the potential of recontamination of foods (Li and Farid, 2016). Other limitations of the traditional well-established heating methods, which rely essentially on the generation of heat outside the product to be heated, include considerable losses of heat on the surfaces of the equipment and installations, reduction of heat transfer efficiency, and thermal damage by overheating, because of the time required to conduct sufficient heat into the thermal center of foods and therefore additional financial costs (Pereira and Vicente, 2010). Together, these limitations have led to the development of new thermal technologies capable of delivering microbiologically safe foods and meeting consumer demands for high-quality food products. These are described with respect to their antistaphylococcal potential in the following section.

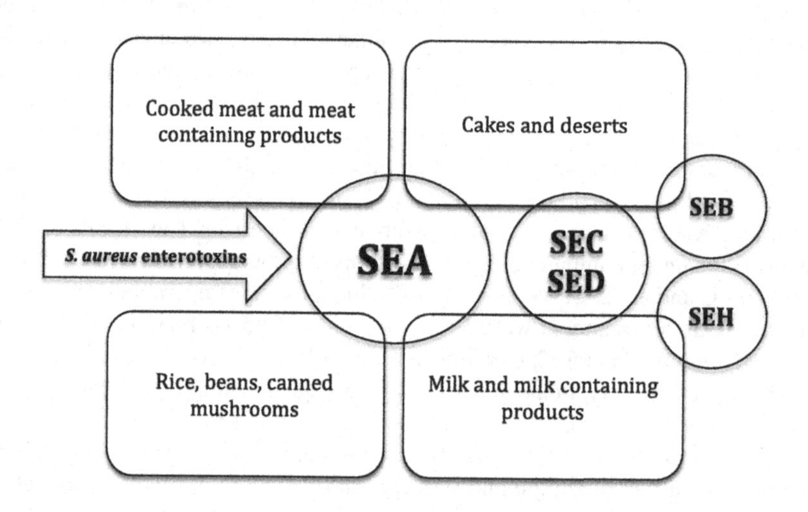

FIGURE 12.1

Staphylococcal enterotoxins (SEs) and suspected source involved in staphylococcal food poisoning outbreaks during the period from 1983 to 2013.

Adapted from Baptista, I., Rocha, S.M., Cunha, Â., Saraiva, J.A., Almeida, A., 2016. Inactivation of Staphylococcus aureus *by high pressure processing: an overview. Innov. Food Sci. Emerg. Technol. 36, 128–149.*

2.1 INACTIVATION OF *STAPHYLOCOCCUS AUREUS* BY NEW THERMAL TECHNOLOGIES

The key thermal processing parameter aligned with the reduction of the thermal damage to products is the duration of product heating. The new thermal technologies, for example, ohmic heating, microwave, radio frequency, or infrared (IR) technologies, offer shorter product exposure to heat and are referred to as a volumetric form of heating where all parts of the food is heated uniformly (Pereira and Vicente, 2010). Table 12.1 presents a summary of the research published to date on inactivation of *S. aureus* associated with different treatment food matrices using new thermal technologies, the results of which are discussed below.

2.1.1 Ohmic Heating

Ohmic heating as a novel thermal technology employs a low frequency (50–60 Hz) electric current passed through a food material, placed in direct contact with a pair of electrodes, leading to internal heat generation in food (Jojo and Mahendran, 2013; Knirsch et al., 2010; Ruan et al., 2001). Ohmic heating has found a wide range of industrial applications including blanching, evaporation, dehydration, fermentation, extraction, sterilization, and pasteurization of foods (Knirsch et al., 2010) with demonstrated advantages over conventional thermal processing in terms of nutritional and microbiological safety of different food matrices, such as vegetables and fruits (Kaur et al., 2016), fruit juices (Kim and Kang, 2015a), milk and creams (Kim and Kang, 2015b), vegetable baby foods (Mesias et al., 2016), red pepper paste (Cho et al., 2016), liquid whole egg (Bozkurt and Icier, 2012), meat (Yildiz-Turp et al., 2013), and seafood (Lascorz et al., 2016). A full review on applications and opinion on the use of ohmic heating in food industry can be found in Jaeger et al. (2016). In ohmic heating, microorganisms are inactivated by means of high temperature and mild cell wall electroporation occurring at low operational frequency (50–60 Hz) (Ruan et al., 2001), and there is no pathogenic microorganism that can resist ohmic heating (Knirsch et al., 2010). Moreover, ohmic treatment demonstrated even greater inactivation efficacy against thermophilic *Geobacillus stearothermophilus* spores by comparison with conventional heating (Somavat et al., 2012). Recently, the detrimental effect of ohmic heating on bacterial cells was confirmed by Sengun et al. (2014). In this study, the natural microflora including *S. aureus* of meat samples were at 2.54 and 2.61 log CFU/g levels for raw meat and uncooked meatball samples, respectively. Ohmic treatment at 75°C and 0 s holding time reduced *S. aureus* counts to undetectable levels. In another study, different ohmic processing parameters were examined against higher levels of artificially inoculated *S. aureus* on minced pork meat samples. A combination of higher degree temperature of 81°C and 5 min of total holding time, completely eliminated populations of *S. aureus* at 10^5 CFU/g levels (Mitelut et al., 2011).

2.1.2 Microwave and Radio Frequency Heating

The microwave technology has found wide range of applications in food manufacturing because of the rapid and more controlled heating within the food matrices leading to improved quality and consistency of food products. In microwave heating, the high frequency (300–3000 MHz) field is passed through the food, stimulating the vibrational frequencies of chemical bonds to heat the material (Richardson, 2001). This technology is currently used for thawing, drying, and baking of foods, as well as for the inactivation of microorganisms in foods (Woo et al., 2000). The current literature suggests that bacterial inactivation activity of microwave processing relies on both thermal and nonthermal effects; however,

Table 12.1 A Summary of the Research Published to Date Focusing on Inactivation of *Staphylococcus aureus* Associated With Different Treatment Food Matrices Using New Thermal Technologies

Technology	Treatment Conditions	Treatment Matrix	Effects	References
Ohmic heating	Holding time 5 min, temperature 81°C	Minced pork	Not detected	Mitelut et al. (2011)
	Voltage gradient 5.26 V/cm, holding time 0 s, temperature 75°C	Beef meatballs	Not detected	Sengun et al. (2014)
Microwave heating	Frequency 2450 MHz, power 550 W, treatment time 30 s, temperature 90°C	Liquid media	Not detected	Atmaca et al. (1996)
	Frequency 2450 MHz, power 800 W, treatment time 110 s, temperature 61.4°C	Stainless still disks, sterile distilled water	Not detected	Yeo et al. (1999)
	Frequency 2450 MHz, voltage 100 V, power 500 W, treatment time 50 s, Temperature almost 100°C	Phosphate buffer pH 7.0	Cells decreased to near 0.1%	Watanabe et al. (2000)
	Frequency 2450 MHz, power 800 W, treatment time 180 s, temperature above 90°C	Phosphate buffer pH 7.0	Reduction from $67 \pm 3 \times 10^5$ CFU/mL to $2 \pm 1 \times 105$ CFU/mL	Tahir et al. (2009)
	Frequency 60 Hz, voltage 20 V, treatment time 7 min, (temperature not available)	Surface of microwave plate	Almost complete inactivation	Boumarah et al. (2016)
Infrared heating	Power 4.36×10^3 W/m^2, treatment time 6 min, temperature ~70°C	Bacteriological agar media	Complete inactivation, i.e., 200–300 CFU/plate	Hashimoto et al. (1992)
	Power 500 W, voltage (input) 120 V, lamp temperature 619°C, treatment time 4 min	Milk	8.4 $\log_{10}$ CFU/mL	Krishnamurthy et al. (2008a,b)
High-pressure thermal treatment	Pressure 600 MPa, treatment time 8 min, temperature 45°C	Milk	8.4 $\log_{10}$ CFU/mL	Guan et al. (2006)
	Pressure 350 MPa, treatment time 5 min, temperature 40°C	Apple, orange, apricot, and sour cherry juices	Not detected (>8 $\log_{10}$ CFU/mL)	Bayindirli et al. (2006)
	Pressure 400 MPa, treatment time 7 min, temperature 55°C	Ham model system	>6 $\log_{10}$ CFU/mL	Tassou et al. (2008)
UV–C combined with mild temperature	UV treatment 27.1 mJ/L, treatment time 3.6 min, temperature 52.5°C for treatment of buffer and fruit juices or 13.6 mJ/L, treatment time 1.8 min, temperature 55.0°C for treatment of buffer and vegetable and chicken broths	Citrate buffer pH 3.0–7.0, apple juice pH 3.3, orange juice pH 3.5, vegetable broth pH 5.9, chicken broth pH 5.2	5 $\log_{10}$ CFU/mL for buffer with either pH, >4.5 $\log_{10}$ CFU/mL for fruit juices, >5.5 $\log_{10}$ CFU/mL for vegetable and chicken broths	Gayán et al. (2014)

Table 12.1 A Summary of the Research Published to Date Focusing on Inactivation of *Staphylococcus aureus* Associated With Different Treatment Food Matrices Using New Thermal Technologies—cont'd

Technology	Treatment Conditions	Treatment Matrix	Effects	References
UV, preheating and pulsed electric field	UV light bulb 30W, electric filed 40kV/cm, treatment time 100μs, temperature 58°C	Orange juice	9.5 $\log_{10}$ CFU/mL	Walkling-Ribeiro et al. (2008)
Ultrasound combined with mild temperature	Frequency kHz, power 600W, amplitude 117.27μm, treatment time 12min, temperature 60°C	Raw cow milk	1.4 $\log_{10}$ CFU/mL	Herceg and Jambrak (2012)
Ultrasound combined with pressure and mild temperature	Frequency 20kHz, power 750W, acoustic intensity 90W/cm^2, pressure 225kPa, treatment time 4min, temperature 36°C	Raw bovine milk	1.05 $\log_{10}$ CFU/mL	Cregenzán-Alberti et al. (2014)

it is often proportional to the increase in power, temperature, and treatment time. For example, Woo et al. (2000) found that microwaves had detrimental effects on various microorganisms (other than *S. aureus*) through pasteurization processing with higher treatment associated temperatures (up to 80°C). In the case of *S. aureus* suspended in liquid media, complete inactivation of cells was achieved after 30s of microwave exposure at frequency 2450MHz and power 550W in combination with 90°C temperature (Atmaca et al., 1996). Complete inactivation of *S. aureus* cells on stainless steel discs was achieved at the lower temperature in conjunction with longer treatment duration, and complete bacterial inactivation was obtained at 2450MHz and 800W after 110s and the resulting temperature of 61.4°C (Yeo et al., 1999). In the later research, the heat generated during microwave processing was considered to be primarily responsible for *S. aureus* cell inactivation (Tahir et al., 2009; Watanabe et al., 2000). The treatment for 180s generated at 800W, 2450MHz, and temperature above 95°C significantly reduced the number of *S. aureus* cells suspended in phosphate buffer (pH 7) to $2 \pm 1 \times 10^5$ CFU/mL from the initial $67 \pm 3 \times 10^5$ CFU/mL, whereas associated temperature below 40°C led to insignificant bacterial reduction with remaining numbers of $58 \pm 2 \times 10^5$ CFU/mL recorded (Tahir et al., 2009). Microwave radiation generated at 2450MHz, 100V, 500W, and temperature of almost 100°C for 50s reduced the relative numbers of the surviving *S. aureus* cells to near 0.1% of the untreated control and no reduction in the cell viability (other than *S. aureus*) was observed when 30s of treatment was accompanied with 28°C (Watanabe et al., 2000). However, a thermal effect was not always associated with microwave-induced bacterial inactivation. For instance, Dreyfuss and Chipley (1980) suggested the nonthermal nature of microwave treatment. In this study the authors characterized the effects of sublethal microwave irradiation in cells of *S. aureus* and compared resulting effects with the effect of a conventional heat treatment on the basis of bacterial enzymatic activity. This study revealed that microwave radiation affects *S. aureus* differently as compared with the effects of conventional heat treatment (Dreyfuss and Chipley, 1980). A nonthermal effect of microwave treatment was also suggested in a recent study conducted by Kushwah et al. (2013), where microwave-mediated alteration of bacterial (other than *S. aureus*) growth, enzyme activity, and extracellular polymeric substances (EPS) production was associated with ice-cold temperature conditions during the microwave treatment. In another study, aiming to observe the thermal effect of microwave treatment on the viability of *S. aureus* associated with microwave glass plate, cultures were exposed to microwave treatment at 60 Hz, 120 V, and relatively low temperatures for different treatment times. The cell population was significantly reduced as the exposure time increased, with almost complete inactivation achieved after longer microwave exposure of 7 min, similarly assuming nonthermal nature of treatment (Boumarah et al., 2016).

Radio frequency processing, also called high-frequency heating, is well-established thermal processing technology, which has found applications in many areas including food industry (Rowley et al., 2001). In contrast to the microwave heating, the frequencies required for generation of heat within food material are lower, ranging from 1 to 300 MHz (Jojo and Mahendran, 2013). In recent years, radio frequency technology was investigated for decontamination of different types of food products, such as liquid whole egg, fruit juices, soybean milk, whole milk, meat products, shell eggs, and peanut butter cracker sandwiches, contaminated with a wide range of pathogenic agents, including *Salmonella* and *E. coli* (Geveke et al., 2016; Li et al., 2016; Rincon and Singh, 2016; Schlisselberg et al., 2013; Uemura et al., 2010; Awuah et al., 2005; Geveke et al., 2002). A more detailed review on radio frequency application in food industry can be found in Zhao (2000) and Jojo and Mahendran (2013).

However, because the information on inactivation of *S. aureus* on complex food matrices is limited, further research is required to demonstrate the technological advances of electroheating, including ohmic, microwave, and radio frequency heating, over the conventional thermal sterilization approaches for control of *S. aureus* and its enterotoxins.

2.1.3 Infrared Heating

IR heating is a thermal technology, which has received increased attention in food industry because of several reported advantages such as efficiency in energy input, precise and rapid heating, high levels of time and temperature control, no heating of surrounding air, easy to use and program, and with no chemical residues after the application (Ramaswamy et al., 2012; Hamanaka et al., 2006; Skjoldebrand, 2001). Basically, IR heating is the application of electromagnetic radiation in the wavelength range of 0.75–1000 μm (near-IR: 0.75–1.4 μm, mid-IR: 1.4–3 μm, and far-IR: 3–1000 μm) to generate heat in the exposed materials, which can also be used to achieve decontamination in foods (Krishnamurthy et al., 2008b). IR heating found industrial applications in drying of low-moisture foods (e.g., breadcrumbs, cocoa, flours, grains, malt, pasta products and tea, fish, rice, potato), dehydration, blanching, thawing, pasteurization, sterilization, and other food applications such as roasting, frying, broiling, and cooking (Skjoldebrand, 2001; Krishnamurthy et al., 2008a,b). The main mechanism underlying microbial inactivation by IR heating is the effect of heat on intracellular components, such as DNA, RNA, ribosome, cell envelope, and proteins (Ramaswamy et al., 2012). The effects of IR heating have been investigated for control of *S. aureus* using a heating system comprising six IR lamps. Phosphate buffer *S. aureus* cell suspension was treated at 700°C lamp temperature for 20 min. The microscopic observation clearly indicated that there was cell wall damage, cytoplasmic membrane shrinkage, cellular content leakage, and mesosome disintegration after IR treatment (Krishnamurthy et al., 2010). However, in common with other technologies, the antimicrobial efficacy of IR heating can be influenced by many factors, including treatment processing parameters and target characteristics. The peak wavelength and surface temperature of the IR heating element, the physical state of the microorganism, penetration depth, composition, shape, and surface characteristics of food (Krishnamurthy et al., 2008a) are key influencing parameters. To date, several studies demonstrated high potential of this technology in inactivating of *S. aureus* from food safety perspectives. For example, Hashimoto et al. (1992) studied the influence of pasteurization by using far-IR heating on *S. aureus* for decontamination of wet solid foods, where agar medium was used as a food model and compared the results with conventional pasteurization by hot air heating at 50–90°C (5–30 min). The authors clearly demonstrated that far-IR pasteurization was more effective than conventional heating for *S. aureus* on the agar plate. The highest inactivation (200–300 CFU/plate) of bacterial cells was achieved after 6 min of treatment when there was no medium added on top of the inoculum retaining the distance between the treating surface and the agar plate of 0 mm, and temperature of the treated medium reached 70°C, whereas similar reduction levels could be achieved only after extended treatment (~30 min) at similar temperature by conventional approach. Additionally, when the medium was added on top of the inoculated agar, which resulted in the formation of layer with thickness of 1–2 mm, half of the reported reduction was obtained (~100–200 CFU/plate). Later, the same research group compared the effects of irradiation power, where near-IR and far-IR powers were used, on treatment inactivation effects against *S. aureus*. The authors concluded that far-IR heating is more effective than near-IR heating, with the reduction of ~4 and 1 $\log_{10}$ CFU/mL of *S. aureus*, respectively (Hashimoto et al., 1993). Krishnamurthy et al. (2008a) applied IR heating to milk processing and studied IR heating for inactivation of *S. aureus* in raw milk by varying several treatment parameters, such as lamp temperatures (536 and 619°C), volumes of treated milk samples (3, 5, and 7 mL), and treatment times (1, 2, and 4 min). The authors demonstrated that the highest inactivation of *S. aureus* could be achieved through application of lower volume of milk (3 or 5 mL), longer treatment time (4 min), and higher lamp temperature (619°C) with sample temperature below 100°C. These treatment conditions resulted in 8.4 $\log_{10}$ CFU/mL reduction of *S. aureus* cells. However, a positive growth was noted during the enrichment indicating that some of the cells were sublethally injured because of the heating. To ensure complete inactivation of cells with no further recovery of injured cells after enrichment, a combination of longer treatment duration (>4 min) and the highest lamp temperature was required. This study also demonstrated that the treatment using studied conditions exhibited poor penetration ability because of opaqueness of milk as higher reductions were achieved mainly using thinner layers of milk. Therefore, optimization of treatment and process parameters, such as temperature and treatment time as well as product type and volume, is still required to ensure sustainable inactivation of microorganisms, which will result in a commercially successful decontamination method.

2.1.4 Pressure-Assisted Thermal Processing

High hydrostatic pressure (HHP) processing is considered as a nonthermal preservation technology (~15°C at 600 MPa) and utilized by many countries for pasteurization of food products (Bermúdez-Aguirre and Barbosa-Canovas, 2011; Aymerich et al., 2008). The benefits of food processing by high pressure include immediate and uniform effect throughout different media, less deteriorated essential vitamins, phytochemicals, aroma compounds compared with the classical heat treatment, and the ability to inactivate wide range of microorganism (Gao et al., 2006). Although the exact mechanism of action of high-pressure processing (HPP) on microbial cells is still under research, cell membrane damage, cell wall rupture, denaturation of proteins and enzymes, and DNA degradation are the possible reasons responsible for bacterial inactivation (Yang et al., 2012; Aymerich et al., 2008). Antimicrobial properties of high pressure may vary with respect to the target pathogen. For instance, vegetative forms of yeasts and molds are most pressure sensitive and inactivated by pressures between 200 and 300 MPa. Pressures of about 300 MPa can inactivate gram-negative bacteria; for inactivation of gram-positive bacteria, pressures higher than 400 MPa are required (Ludikhuyze et al., 2001), whereas pressure of up to 1500 MPa at ambient temperature can fail to inactivate bacterial spores

(Wilson et al., 2008). A solution to the problem associated with pressure-resistant strains of vegetative cells is the application of pressure in combination with elevated temperatures (Patterson, 2005), also known as pressure-assisted thermal sterilization, which have been already approved by Food and Drug Administration for commercial sterilization in 2009 (Bermúdez-Aguirre and Barbosa-Canovas, 2011). In general, gram-positive bacteria are considered to be more resistant to pressure than gram-negatives, and cocci are more resistant than rod-shaped bacteria (Scheinberg et al., 2014; Patterson, 2005). *S. aureus* appeared to be the most resistant microorganism to high-pressure treatment at 600 MPa and 20°C when compared with *L. monocytogenes* and *P. aeruginosa*, and 15 min was required to achieve an inactivation of 7 log cycles in either in raw skimmed bovine milk, whey, or phosphate buffer (pH 7.0) (Ramos et al., 2015). In another work, high-pressure treatment at 550 MPa and 22°C for 60 s produced only minor reductions of gram-positive pathogens, including *S. aureus*, inoculated on beef jerky, resulting in reductions by 1.32 $\log_{10}$ CFU/strip, however, with boiling water treatments (100°C, 20–30 s) it was possible to reduce *S. aureus* cell populations by >5.0 $\log_{10}$ CFU/strip (Scheinberg et al., 2014). Diels et al. (2003) compared the effects of pressure in combination with different temperatures on inactivation of *S. aureus* and *Y. enterocolitica*. They found that *S. aureus* was more resistant to HPP than *Y. enterocolitica*, and pressure treatment at 100–300 MPa at temperatures between 5 and 40°C did not affect the viability of *S. aureus* as compared with inactivation of *Y. enterocolitica*. To date, although most research, focusing on antimicrobial effects of pressure-assisted thermal processing, utilized moderate temperatures, some studies investigated the influence of elevated temperature in combination with high pressure on inactivation rates of *S. aureus*. For example, Guan et al. (2006) exposed whole milk inoculated with *S. aureus* to 600 MPa at different temperatures (4, 21 and 45°C) for different treatment times. The authors demonstrated that the temperature had limited effect on *S. aureus* inactivation when shorter treatment was applied with log reductions of 4.3 at 4°C, 4.3 at 21°C, and 4.0 at 45°C recorded after 4 min of treatment. However, when samples were pressurized for 8 min, reductions in log values increased to 5.7 for 4°C, 7.3 for 21°C, and 8.4 for 45°C with overall conclusion that higher inactivation levels of *S. aureus* cell can be achieved when pressure treatment is combined with higher temperatures (Guan et al., 2006). Similarly, higher inactivation rates were obtained for *S. aureus* suspended in different type of fruit juices, such as apple, orange, apricot, and sour cherry, when relatively lower pressure and shorter treatment times were accompanied with higher temperature (Bayindirli et al., 2006). In this work, if a combination of 5 min at 350 MPa and 30°C resulted on average in 1 log of remaining populations of *S. aureus* cells, after treatment combining 5 min at 350 MPa and 40°C bacterial cells were not detected resulting in more than 8 log reduction cycles. Based on microbiological analysis of soymilk and total aerobic mesophilic bacteria, aerobic spores, and *Bacillus cereus*, Poliseli-Scopel et al. (2012) also demonstrated that the combination of 300 MPa and 75°C produced a commercially sterile high-quality product. According to the studies reported to date, the nature of the treatment matrix can greatly affect decontamination efficiency of treatment. Thus, Tassou et al. (2008) investigated the effect of pressure processing in combination with mild temperatures against *S. aureus* in a ham model system, where at ambient temperatures (25°C), the application of higher pressure (600 MPa) was required to achieve inactivation of microbial cells by more than 5 $\log_{10}$ CFU/mL, which was recorded after 7 min of treatment. However, the same level of inactivation of more than 6 log CFU/mL was achieved at lower pressure levels (400 MPa) when mild heating (55°C) was simultaneously applied. The authors concluded that elevated temperatures allowed application of lower pressure levels and shorter processing times to achieve efficient inactivation of bacterial cells, as compared with the ambient processing temperatures (Tassou et al., 2008).

As already stated, the most significant virulence factor related to *S. aureus* is the formation of heat-stable SEs, and the possibility of their occurrence in processed foods cannot be ignored. In fact, there is scarce information available on the effect of thermal processing on SE. Nonetheless, the effect of combined pressure (0.1–800 MPa) and temperature treatment on the activity of SE, namely SEA to SEE, after 30 min at 5 and 20°C, and after 30 and 120 min at 80°C was studied by Margosch et al. (2005). This study showed, that pressurization (0.1–800 MPa) of heat-stable monomeric staphylococcal toxins at 5 and 20°C had no effect and only after 120 min at 800 MPa and 80°C of treatment SE could be significantly reduced to a maximum of 20%. In addition, it was demonstrated that thermal stability varied strongly within the groups with the heat resistance at 80°C being as follows: SEA = SEC = SEE > SED > SEB. However, to date there is no information available on the heat resistance of SE types other than SEA–SEE.

3. INACTIVATION OF *STAPHYLOCOCCUS AUREUS* BY NOVEL NONTHERMAL TECHNOLOGIES

To avoid thermal degradation of the food components and to preserve the nutritional and sensory quality of food products, recent works have focused on developing novel nonthermal technologies. Technologies such as ultrasound (US), HPP, pulsed electric fields (PEF), ultraviolet (UV), pulsed light (PL) treatment, irradiation, and nonthermal plasma (NTP), have the ability to inactivate microorganisms at near-ambient temperatures and overcome the limitations of thermal sterilization (Li and Farid, 2016; Pereira and Vicente, 2010). Although these technologies are developed for nonthermal treatment of heat-sensitive foods, these emerging approaches may confer advantages for controlling *S. aureus*. The effects of key nonthermal technologies on *S. aureus* are outlined below, and the gaps in knowledge are highlighted.

3.1 ULTRASOUND

Ultrasonic treatment refers to sound waves beyond the audible frequency range, i.e., equal or more than 20 kHz (Chandrapala et al., 2012). When used alone, US technology is not very effective for microbial inactivation (Li and Farid, 2016), however, combining with other preservation processes, such as heat and/or pressure, increases treatment efficacy (Chemat et al., 2011). By comparison with other microorganisms, *S. aureus* was found to be very resistant to US treatment. Marchesini et al. (2015) evaluated US treatment efficacy against various microorganisms in full fat milk. The strongest treatment (100% of amplitude × 300 s) reduced the bacterial populations by 4.61, 2.75, and 2.09 for *Debaryomyces hansenii*, *Pseudomonas fluorescens*, and *E. coli*, respectively. In contrast, for *S. aureus* only 0.55 log reduction was achieved. Also Birmpa et al. (2013) found *S. aureus* to be the most resistant to USs—among tested bacteria, i.e., *E. coli*, *S. aureus*, *Salmonella enteritidis*, and *Listeria innocua*, the maximum reductions for *S. aureus* achieved on fresh produce in buffered peptone water were the lowest (1.71 and 2.41 log reduction of *S. aureus* on lettuce and strawberries, respectively).

3.2 HIGH-PRESSURE PROCESSING

HPP, also known as HHP is achieved by subjecting food products to pressure in a pressure vessel capable of sustaining the high pressure, usually between 50 and 1000 MPa. Inactivation on foodborne spoilage and pathogenic microorganisms is primarily due to disruption of bacterial cell membrane (Tao et al., 2014). Pressure treatments of around 600 MPa at ambient temperature are very effective against vegetative cells of bacteria; however, even a pressure treatment of up to 1500 MPa at ambient temperature can fail to inactivate bacterial spores (Wilson et al., 2008). Inactivation of *S. aureus* by HPP on various matrices, including foods, has been widely studied. It was found that *S. aureus* can be completely inactivated during HPP treatment, but SEs are hardly affected by this technology. However, the efficacy of the treatment is higher against strains of *S. aureus* producing SE than those without enterotoxins. A full review can be found in Baptista et al. (2016).

3.3 PULSED ELECTRIC FIELDS

Pulsed electric field (PEF) uses short pulses of electricity, which cause electroporation in the cell wall of microorganisms and leads to their inactivation. This technology can be used to process liquid, semi-liquid, and solid food by applying high electric field (20–80 kV/cm) for a very short time (1–100 μs). The primary effect of PEF on microbial cells is related to local structural changes and breakdown of the cell membrane (Amiali and Ngadi, 2012; Raso et al., 2014). It was observed that PEF antimicrobial efficacy against *S. aureus* heavily depends on the temperature at which the treatment is conducted. Sharma et al. (2014) noted that the effectiveness of PEF increased with increasing temperature and preheating prior to the treatment was necessary to reduce population of *S. aureus* in milk. Also Cregenzán-Alberti et al. (2015) increased the treatment temperature, achieving 5.2 log CFU/mL reduction of *S. aureus* in milk at 32.5°C. Monfort et al. (2010) suggested that PEF technology by itself may not be sufficient to assure food safety, therefore, combination of PEF with other nonthermal or thermal preservation technologies is recommended to control *S. aureus*.

3.4 ULTRAVIOLET

UV light consists of electromagnetic radiation in the wavelength range from 100 to 400 nm and has been used widely for inactivation of a broad range of spoilage and pathogenic microorganisms. The antimicrobial effect of UV radiation is primarily caused by DNA structure alternation, leading to mutagenesis and cell death (Keklik et al., 2012; Gayán et al., 2014). Inactivation of *S. aureus* by UV treatment was investigated using different food matrices, such as fresh produce (Birmpa et al., 2013; Ha et al., 2011), liquid foods (Walkling-Ribeiro et al., 2008; Gayán et al., 2014; Gouma et al., 2015), fish and seafood (Lee et al., 2016), and meat (Sommers et al., 2010). Because of low inactivation rates UV radiation is often combined with other technologies e.g., mild heat treatments (Gayán et al., 2014) or sanitizers (Ha et al., 2011).

3.5 PULSED LIGHT

PL involves the use of short-duration, high-power pulses of a broad spectrum of white light from the UV to the near-IR region (FDA, 2015). Successfully patented or launched in the market for other applications, the technique is not yet applied at industrial scale in food processing (Ortega-Rivas, 2012). Pulsed UV light is able to kill vegetative cells and bacterial spores, as well as fungal spores, viruses, and protozoan oocysts (Keklik et al., 2012) and is considered to be more efficient in microbial inactivation than continuous UV light (Krishnamurthy et al., 2010). Krishnamurthy et al. (2008a,b) investigated the efficacy of pulsed UV light for continuous-flow treatment for the inactivation of *S. aureus* in milk and achieved complete inactivation. Krishnamurthy et al. (2010) observed severe damage to *S. aureus* cells inoculated in phosphate buffer. The damages caused by PL include cell wall disruption and cytoplasmic membrane shrinkage. Internal cellular structure collapse was also a consequence of the treatment, which eventually led to cellular content leakage and cell death. The photochemical effect of pulsed UV light was also confirmed in the study—some of the *S. aureus* cells were inactivated without visible structural damage. A 5-s treatment with pulsed UV light resulted in complete inactivation of *S. aureus*.

3.6 IRRADIATION

Food irradiation is a process of exposing food to a certain amount of ionizing radiation. Three major types of food irradiation include gamma rays generated from the radionuclides cobalt-60 (^{60}Co) or cesium-137 (^{137}Cs), electron beams created by electron accelerators, and X-rays produced by electron accelerators with converters (Farkas et al., 2014). Microbial inactivation by irradiation can be caused by direct DNA damage and the production of reactive molecules, e.g., hydrogen peroxide, hydroxyl radicals, and hydrogen atoms, which consequently damage cellular metabolic pathways to promote intracellular oxidation and lead to cell lysis (Farkas et al., 2014; Lung et al., 2015).

Irradiation was found to be very effective for inactivation of *S. aureus* on foods—reduction of the number of viable cells below detection limits was reported in many studies, using various food matrices, including ready-to-eat and prepared foods (Lamb et al., 2002; Chawla et al., 2003; Jo et al., 2004, 2005; Vural et al., 2006; Chung et al., 2007; Kim et al., 2010), meat (Badr, 2004), seafood (Song et al., 2009), and vegetables (Shim et al., 2012).

3.7 NONTHERMAL PLASMA (COLD PLASMA)

NTP is a partially or fully ionized gas. NTP generated at atmospheric pressure consists of UV photons, neutral or excited atoms and molecules, negative and positive ions, free radicals, and free electrons. It has recently found an extensive range of applications for microbiological decontamination due to chemical and bioactive radicals generated during electrical discharge, including reactive oxygen species and reactive nitrogen species (Laroussi and Leipold, 2004; Scholtz et al., 2015). Food surface feature and nutrient composition can significantly affect the inactivation of microorganisms using NTP. Kim et al. (2014) reported that 3–4 log reduction of *S. aureus* inoculated on beef jerky required 10 min treatment, whereas to achieve similar reduction on the polystyrene and agar only 2 min treatment was necessary. Han et al. (2016) observed a protective effect of the nutritive components in a meat model media; *S. aureus* in phosphate-buffered saline was undetectable after 60 s of plasma treatment, whereas treatment of *S. aureus* in the meat model required posttreatment refrigeration to retain antimicrobial effect.

4. INACTIVATION OF *STAPHYLOCOCCUS AUREUS* BY NONTHERMAL TECHNOLOGIES COMBINED WITH HIGH TEMPERATURE

The use of combinations of thermal and nonthermal approaches has become more prevalent because of their flexibility in antimicrobial activity and ability to impact on multitarget characteristics. For example, short-wave UV radiation (UV–C light, 200–280 nm) is known to have bactericidal effects, however, it exhibits low-penetration capacity into low UV transmittance or almost opaque treatment substrates (Gomez-Lopez et al., 2012). One of the possible ways to compensate for UV light low-penetration depths is the application of UV–C in combination with mild heat treatment. The advantage of combined UV-heat treatments for inactivation of enterotoxin-producing strains of *S. aureus* inoculated in buffer solution with different pH and water activity was demonstrated by Gayán et al. (2014). Applying the maximum dose in one pass of 27.1 mJ/L for 3.6 min at room temperature (25.0°C), only 1.6 and 1.7 $\log_{10}$ inactivation cycles were achieved in buffers with absorption coefficient 22.7 cm^{-1} and of pH 7.0 and pH 3.5, respectively. However, the combination of UV light with mild heat strongly improved the UV effectiveness at both pH values. For instance, a UV treatment of 23.7 mJ/L for 3.2 min at 55.0°C increased the UV inactivation by 4.2 $\log_{10}$ cycles and more than 4.8 $\log_{10}$ cycles at pH 7.0 and pH 3.5, respectively, and inactivation by 5 $\log_{10}$ cycles of *S. aureus* by applying UV–C treatments of 27.1 mJ/L for 3.6 min at 52.5°C or 13.6 mJ/L for 1.8 min at 55.0°C was achieved. A synergistic lethal effect was also observed when UV-heat treatments were applied in food matrices. This combination treatment provided 2.7, 3.3, 2.4, and 1.8 extra $\log_{10}$ cycles reductions of *S. aureus* in apple juice, orange juice, vegetable broth, and chicken broth, respectively, in comparison with the sum of inactivation obtained through UV and heat treatments alone, demonstrating high-potential UV-heat technology for control of *S. aureus* in liquid foods (Gayán et al., 2014). The effect of hurdle sequence including UV for 30 min at 20°C, preheating from 35 to 50°C, and PEF with field strength from 28 to 40 kV/cm in comparison with conventional pasteurization (26 s at 94°C) on inactivation of *S. aureus* in orange juice was studied by Walkling-Ribeiro et al. (2008), who demonstrated that higher inactivation levels of 9.5 $\log_{10}$ CFU/mL could be achieved with a hurdle approach, i.e., a combination of UV, temperature of 58°C for pulsed electric field outlet, and pulsed electric field of 40 kV/cm with treatment time of 100 μs in comparison with conventional pasteurization (8.2 $\log_{10}$ CFU/mL). However, bacterial inactivation increased by increasing electric field strength and treatment time rather than increasing preheating temperatures. Overall this study demonstrated the potential of the hurdle approach for achieving product safety and quality requirements, however, to achieve industrial feasibility, further process optimization will be needed to increase cost efficacy and reduce processing times (Walkling-Ribeiro et al., 2008).

Results on investigation of the effect of combination of US and heat treatment on the inactivation of *S. aureus* in milk clearly indicated improved microbial inactivation (Herceg and Jambrak, 2012), although the reductions were not as high as that achieved with combining heat and UV. In this work, optimal treatment parameters for achieving maximum inactivation of *S. aureus* (1.4 $\log_{10}$ CFU/mL) were temperature of 60°C, treatment time of 12 min, and amplitude of 117.27 μm. In contrast, a combination of US, mild temperature, and pressure (225 kPa) resulted in relatively lower reduction of *S. aureus* in milk (1.05 $\log_{10}$ CFU/mL), which was associated with lower treatment temperatures of 36°C and shorter treatment duration (4 min) (Cregenzán-Alberti et al., 2014).

5. CONCLUSION

According to the work reported to date, whether relating to thermal technologies and/or combinations with nonthermal technologies, it can be concluded that, although results can be some times difficult to compare because of the differences in the equipment, experimental conditions, and the absence of standardized food models, by utilization of new thermal preservation technologies it is possible to achieve desired food produce characteristics established by the consumers and regulatory bodies without compromising food safety. However, certain issues associated with *S. aureus* require further research, these include evaluating the impact of inherent food characteristics on bacterial behavior to take novel food formulations into account, evaluating novel technologies in a comparative design to highlight where advantages can be presented, the possibility of bacterial cell recovery and stress resistance to emerging technologies, or combinations of the same as well as potential of SE formation in already processed foods.

Consumer demands for less heavily processed foods that retain high quality and microbiological safety with natural flavor and extended shelf life have led the academic community and food manufacturers to develop more energy efficient, nontoxic, and novel technological approaches. According to EU regulation, SEs are considered as one out of six human food-associated health hazards, which are defined as microbiological criteria for process hygiene and food safety control (Hennekinne et al., 2015). Nevertheless, there is a limited availability of data on inactivation of *S. aureus* and its enterotoxins associated with different food matrices and type of produce by means of novel thermal technologies. Thus, comprehensive research is still required to bring insights into the potential applicability of novel thermal and nonthermal technologies for inactivating *S. aureus* and SEs associated with different types of food models because bacteria (and even more prominent, toxins) are often observed to be more stable in real foods than in buffer or laboratory media. A thorough microbiological analysis of foods during the extended storage accompanied with temperature abuse conditions before and after treatment should also be considered to reflect more realistic environmental conditions. Another aspect to drive safety could be combinatorial approaches to inactivate the microorganism itself as well as any toxins utilizing multistage processing steps to achieve different aims.

REFERENCES

Ahmed, J., Ramaswamy, H.S., 2007. Microwave pasteurization and sterilization of foods. Handbook of Food Preservation, pp. 691–711.

Amado, I.R., Vázquez, J.A., Guerra, N.P., Pastrana, L.M., 2014. Thermal resistance of *Salmonella enterica*, *Escherichia coli* and *Staphylococcus aureus* isolated from vegetable feed ingredients. J. Sci. Food Agric. 94 (11), 2274–2281.

Amiali, M., Ngadi, M.O., 2012. Microbial decontamination of food by pulsed electric fields (PEFs). In: Microbial Decontamination in the Food Industry, Novel Methods and Applications, pp. 407–449.

Argudín, M.Á., Mendoza, M.C., Rodicio, M.R., 2010. Food poisoning and *Staphylococcus aureus* enterotoxins. Toxins 2 (7), 1751–1773.

Atmaca, S., Akdag, Z., Dasdag, S., Celik, S., 1996. Effect of microwaves on survival of some bacterial strains. Acta Microbiol. Immunol. Hung. 43 (4), 371–378.

Awuah, G.B., Ramaswamy, H.S., Economides, A., Mallikarjunan, K., 2005. Inactivation of *Escherichia coli* K-12 and *Listeria innocua* in milk using radio frequency (RF) heating. Innov. Food Sci. Emerg. Technol. 6 (4), 396–402.

Aymerich, T., Picouet, P.A., Monfort, J.M., 2008. Decontamination technologies for meat products. Meat Sci. 78 (1–2), 114–129.

Badr, H.M., 2004. Use of irradiation to control foodborne pathogens and extend the refrigerated market life of rabbit meat. Meat Sci. 67 (4), 541–548.

Baptista, I., Rocha, S.M., Cunha, Â., Saraiva, J.A., Almeida, A., 2016. Inactivation of *Staphylococcus aureus* by high pressure processing: an overview. Innov. Food Sci. Emerg. Technol. 36, 128–149.

Bayindirli, A., Alpas, H., Bozoglu, F., Hizal, M., 2006. Efficiency of high pressure treatment on inactivation of pathogenic microorganisms and enzymes in apple, orange, apricot and sour cherry juices. Food Control 17 (1), 52–58.

Bermúdez-Aguirre, D., Barbosa-Canovas, G.V., 2011. An update on high hydrostatic pressure, from the laboratory to industrial applications. Food Eng. Rev. 3 (1), 44–61.

Birmpa, A., Sfika, V., Vantarakis, A., 2013. Ultraviolet light and ultrasound as non-thermal treatments for the inactivation of microorganisms in fresh ready-to-eat foods. Int. J. Food Microbiol. 167 (1), 96–102.

Boumarah, K., Al Abdullah, A., Al Mubarak, S., Alkhawaja, S., Ali, S., Sidiq, N., 2016. The effect of microwave radiation on the growth of *Staphylococcus aureus*. Int. J. Sci. Res. 5 (5), 1350–1353.

Bozkurt, H., Icier, F., 2012. The change of apparent viscosity of liquid whole egg during ohmic and conventional heating. J. Food Process Eng. 35, 120–133.

Cebrián, G., Sagarzazu, N., Aertsen, A., Pagán, R., Condón, S., Mañas, P., 2009. Role of the alternative sigma factor σb on *Staphylococcus aureus* resistance to stresses of relevance to food preservation. J. Appl. Microbiol. 107 (1), 187–196.

Cebrián, G., Sagarzazu, N., Pagán, R., Condón, S., Mañas, P., 2007. Heat and pulsed electric field resistance of pigmented and non-pigmented enterotoxigenic strains of *Staphylococcus aureus* in exponential and stationary phase of growth. Int. J. Food Microbiol. 118 (3), 304–311.

Chandrapala, J., Oliver, C., Kentish, S., Ashokkumar, M., 2012. Ultrasonics in food processing. Ultrason. Sonochem. 19 (5), 975–983.

Chawla, S.P., Kim, D.H., Jo, C., Song, H.P., Byun, M.W., 2003. Effect of gamma irradiation on the survival of pathogens in kwamegi, a traditional Korean semidried seafood. J. Food Prot. 66, 2093–2096.

Chemat, F., Zill-E-Huma, Khan, M.K., 2011. Applications of ultrasound in food technology: processing, preservation and extraction. Ultrason. Sonochem. 18 (4).

Cho, W.I., Yi, J.Y., Chung, M.S., 2016. Pasteurization of fermented red pepper paste by ohmic heating. Innov. Food Sci. Emerg. Technol. 34, 180–186.

Chung, H.J., Lee, N.Y., Jo, C., Shin, D.H., Byun, M.W., 2007. Use of gamma irradiation for inactivation of pathogens inoculated into kimbab, steamed rice rolled by dried laver. Food Control 18 (2), 108–112.

Cregenzán-Alberti, O., Halpin, R.M., Whyte, P., Lyng, J.G., Noci, F., October 2015. Study of the suitability of the central composite design to predict the inactivation kinetics by pulsed electric fields (PEF) in *Escherichia coli*, *Staphylococcus aureus* and *Pseudomonas fluorescens* in milk. Food Bioprod. Process. 95, 313–322.

Cregenzán-Alberti, O., Halpin, R.M., Whyte, P., Lyng, J., Noci, F., 2014. Suitability of ccRSM as a tool to predict inactivation and its kinetics for *Escherichia coli*, *Staphylococcus aureus* and *Pseudomonas fluorescens* in homogenized milk treated by manothermosonication (MTS). Food Control 39 (1), 41–48.

Diels, A.M.J., Wuytack, E.Y., Michiels, C.W., 2003. Modelling inactivation of *Staphylococcus aureus* and *Yersinia enterocolitica* by high-pressure homogenisation at different temperatures. Int. J. Food Microbiol. 87 (1–2), 55–62.

Doulgeraki, A.I., Di Ciccio, P., Ianieri, A., Nychas, G.-J.E., August 2016. Methicillin-resistant food-related *Staphylococcus aureus*: a review of current knowledge and biofilm formation for future studies and applications. Res. Microbiol. 1–15.

Dreyfuss, M.S., Chipley, J.R., 1980. Comparison of effects of sublethal microwave radiation and conventional heating on the metabolic activity of *Staphylococcus aureus*. Appl. Environ. Microbiol. 39 (1), 13–16.

Farkas, J., Ehlermann, D.A.E., Mohácsi-Farkas, C., 2014. Food technologies: food irradiation. Encyclopedia of Food Safety, vol. 3, pp. 178–186.

Food and Drug Administration, 2015. In: Kinetics of Microbial Inactivation for Alternative Food Processing Technologies – Pulsed Light Technology. http://www.fda.gov/Food/FoodScienceResearch/SafePracticesforFoodProcesses/ucm103058.htm.

Gao, Y.L., Ju, X.R., Jiang, H.H., 2006. Use of response surface methodology to investigate the effect of food constituents on *Staphylococcus aureus* inactivation by high pressure and mild heat. Process Biochem. 41 (2), 362–369.

Gayán, E., García-Gonzalo, D., Álvarez, I., Condón, S., 2014. Resistance of *Staphylococcus aureus* to UV-C light and combined UV-heat treatments at mild temperatures. Int. J. Food Microbiol. 172, 30–39.

Geveke, D.J., Bigley, A.B.W., Brunkhorst, C.D., 2016. Pasteurization of shell eggs using radio frequency heating. J. Food Eng. 193, 53–57.

Geveke, D.J., Kozempel, M., Scullen, O.J., Brunkhorst, C., 2002. Radio frequency energy effects on microorganisms in foods. Innov. Food Sci. Emerg. Technol. 3 (2), 133–138.

Gomez-Lopez, V.M., Koutchma, T., Linden, K., 2012. Ultraviolet and pulsed light processing on fluid foods. In: Cullen, P.J., Tiwari, B., Valdramidis, V. (Eds.), Novel Thermal and Non-thermal Technologies for Fluid Foods. Food Science and Technology, International Series, Elsevier Inc., UK.

Gouma, M., Álvarez, I., Condón, S., Gayán, E., 2015. Modelling microbial inactivation kinetics of combined UV-H treatments in apple juice. Innov. Food Sci. Emerg. Technol. 27, 111–120.

Guan, D., Chen, H., Ting, E.Y., Hoover, D.G., 2006. Inactivation of *Staphylococcus aureus* and *Escherichia coli* O157:H7 under isothermal-endpoint pressure conditions. J. Food Eng. 77 (3), 620–627.

Gutierrez, D., Delgado, S., Vazquez-Sanchez, D., Martinez, B., Cabo, M.L., Rodriguez, A., Herrera, J.J., Garcia, P., 2012. Incidence of *Staphylococcus aureus* and analysis of associated bacterial communities on food industry surfaces. Appl. Environ. Microbiol. 78 (24), 8547–8554.

Ha, J.H., Jeong, S.H., Ha, S.D., 2011. Synergistic effects of combined disinfection using sanitizers and uv to reduce the levels of *Staphylococcus aureus* in oyster mushrooms. J. Appl. Biol. Chem. 54 (3), 447–453.

Hamanaka, D., Uchino, T., Furuse, N., Han, W., Tanaka, S-i., 2006. Effect of the wavelength of infrared heaters on the inactivation of bacterial spores at various water activities. Int. J. Food Microbiol. 108 (2), 281–285.

Han, L., Ziuzina, D., Heslin, C., Boehm, D., Patange, A., Sango, D.M., Bourke, P., June 2016. Controlling microbial safety challenges of meat using high voltage atmospheric cold plasma. Front. Microbiol. 7, 977.

Hashimoto, A., Igarashi, H., Shimizu, M., 1992. Far-infrared irradiation effect on pasteurization of bacteria on or within wet-solid medium. J. Chem. Eng. Jpn. 25, 666–671.

Hashimoto, A., Sawai, J., Igarashi, H., Shimizu, M., 1993. Irradiation power effect on ir pasteurization below lethal temperature of bacteria. J. Chem. Eng. Jpn. 26 (3), 331–333.

Hassani, M., Cebrián, G., Mañas, P., Condón, S., Pagán, R., 2006. Induced thermotolerance under nonisothermal treatments of a heat sensitive and a resistant strain of *Staphylococcus aureus* in media of different pH. Lett. Appl. Microbiol. 43 (6), 619–624.

Hennekinne, J.-A., Herbin, S., Firmesse, O., Auvray, F., 2015. European food poisoning outbreaks involving meat and meat-based products. Procedia Food Sci. 5, 93–96.

Herceg, Z., Jambrak, A.R., 2012. The effect of high intensity ultrasound treatment on the amount of *Staphylococcus aureus* and *Escherichia coli* in milk. Food Technol. Biotechnol. 50 (1), 46–52.

Jaeger, H., Roth, A., Toepfl, S., Holzhauser, T., Engel, K.-H., Knorr, D., Vogel, R.F., Bandick, N., Kulling, S., Heinz, V., Steinberg, P., 2016. Opinion on the use of ohmic heating for the treatment of foods. Trends Food Sci. Technol. 55, 84–97.

Jo, C., Lee, N.Y., Kang, H., Hong, S., Kim, Y., Kim, H.J., Byun, M.W., 2005. Radio-sensitivity of pathogens in inoculated prepared foods of animal origin. Food Microbiol. 22 (4), 329–336.

Jo, C., Lee, N.Y., Kang, H.J., Shin, D.H., Byun, M.W., 2004. Inactivation of foodborne pathogens in marinated beef rib by ionizing radiation. Food Microbiol. 21 (5), 543–548.

Jojo, S., Mahendran, R., 2013. Radio frequency heating and its application in food processin: a review. Int. J. Curr. Agric. Res. 1 (9), 042–046.

Kadariya, J., Smith, T.C., Thapaliya, D., 2014. *Staphylococcus aureus* and staphylococcal food-borne disease: an ongoing challenge in public health. BioMed Res. Int. 2014, 827965.

Kamal, R.M., Bayoumi, M.A., Abd El Aal, S.F.A., 2013. MRSA detection in raw milk, some dairy products and hands of dairy workers in Egypt, a mini-survey. Food Control 33 (1), 49–53.

Kaur, R., Gul, K., Singh, A.K., Yildiz, F., 2016. Nutritional impact of ohmic heating on fruits and vegetables – a review. Cogent Food Agric. 21159000.

Keklik, N.M., Krishnamurthy, K., Demirci, A., 2012. 12-Microbial decontamination of food by ultraviolet (UV) and pulsed UV light. Woodhead Publishing Series in Food Science, Technology and NutritionWoodhead Publishing Limited.

Kennedy, J., Blair, I.S., McDowell, D.A., Bolton, D.J., 2005. An investigation of the thermal inactivation of *Staphylococcus aureus* and the potential for increased thermotolerance as a result of chilled storage. J. Appl. Microbiol. 99 (5), 1229–1235.

Kim, H.J., Ham, J.S., Lee, J.W., Kim, K., Ha, S.D., Jo, C., 2010. Effects of gamma and electron beam irradiation on the survival of pathogens inoculated into sliced and pizza cheeses. Radiat. Phys. Chem. 79 (6), 731–734.

Kim, S.-S., Kang, D.-H., 2015a. Effect of pH for inactivation of *Escherichia coli* O157:H7, *Salmonella typhimurium* and *Listeria* monocytogenes in orange juice by ohmic heating. LWT – Food Sci. Technology 62 (1), 83–88.

Kim, S.-S., Kang, D.-H., 2015b. Comparative effects of ohmic and conventional heating for inactivation of *Escherichia coli* O157:H7, *Salmonella enterica* serovar *typhymurium*, and *Listeria* monocytogenes in skim milk and cream. J. Food Protect. 6, 1208–1214.

Kim, J.S., Lee, E.J., Choi, E.H., Kim, Y.J., 2014. Inactivation of *Staphylococcus aureus* on the beef jerky by radio-frequency atmospheric pressure plasma discharge treatment. Innov. Food Sci. Emerg. Technol. 22, 124–130.

Knirsch, M.C., Alves dos Santos, C., Martins de Oliveira Soares Vicente, A.A., Vessoni Penna, T.C., 2010. Ohmic heating – a review. Trends Food Sci. Technol. 21 (9), 436–441.

Krishnamurthy, K., Jun, S., Irudayaraj, J., Demirci, A., 2008a. Efficacy of infrared heat treatment for inactivation of *Staphylococcus aureus* in milk. J. Food Process Eng. 31 (6), 798–816.

Krishnamurthy, K., Khurana, H.K., Soojin, J., Irudayaraj, J., Demirci, A., 2008b. Infrared heating in food processing: an overview. Compr. Rev. Food Sci. Food Saf. 7 (1), 2–13.

Krishnamurthy, K., Tewari, J.C., Irudayaraj, J., Demirci, A., 2010. Microscopic and spectroscopic evaluation of inactivation of *Staphylococcus aureus* by pulsed UV light and infrared heating. Food Bioprocess Technol. 3, 93. http://dx.doi.org/10.1007/s11947-008-0084-8.

Kushwah, P., Mishra, T., Kothari, V., 2013. Effect of microwave radiation on growth, enzyme activity (amylase and pectinase), and/or exopolysaccharide production in *Bacillus subtilis*, *Streptococcus mutans*, *Xanthomonas campestris* and *Pectobacterium carotovora*. Br. Microbiol. Res. J. 3 (4), 645–653.

Lamb, J.L., Gogley, J.M., Thompson, M.J., Solis, D.R., Sen, S., 2002. Effect of low-dose gamma irradiation on *Staphylococcus aureus* and product packaging in ready-to-eat ham and cheese sandwiches. J. Food Prot. 65 (11), 1800–1805.

Laroussi, M., Leipold, F., 2004. Evaluation of the roles of reactive species, heat, and UV radiation in the inactivation of bacterial cells by air plasmas at atmospheric pressure. Int. J. Mass Spectrom. 233 (1–3), 81–86.

Lascorz, D., Torella, E., Lyng, J.G., Arroyo, C., 2016. The potential of ohmic heating as an alternative to steam for heat processing shrimps. Innov. Food Sci. Emerg. Technol. 37.

Lee, K.H., Kim, H.-J., Woo, K.S., Jo, C., Kim, J.-K., Kim, S.H., Kim, W.H., 2016. Evaluation of cold plasma treatments for improved microbial and physicochemical qualities of brown rice. LWT – Food Sci. Technol. 73, 442–447.

Li, R., Kou, X., Cheng, T., Zheng, A., Wang, S., 2016. Verification of radio frequency pasteurization process for in-shell almonds. J. Food Eng. 192, 103–110.

Li, X., Farid, M., 2016. A review on recent development in non-conventional food sterilization technologies. J. Food Eng. 182, 33–45.

Ludikhuyze, L., Van Loey, A., Hendrickx, M., 2001. Combine high pressure thermal treatment of foods. In: Richardson, P. (Ed.), Thermal Technologies in Food Processing. CRC Press, Cambridge, UK, pp. 266–278.

Lung, H.M., Cheng, Y.C., Chang, Y.H., Huang, H.W., Yang, B.B., Wang, C.Y., 2015. Microbial decontamination of food by electron beam irradiation. Trends Food Sci. Technol. 44 (1), 66–78.

Marchesini, G., Fasolato, L., Novelli, E., Balzan, S., Contiero, B., Montemurro, F., Segato, S., 2015. Ultrasonic inactivation of microorganisms: a compromise between lethal capacity and sensory quality of milk. Innov. Food Sci. Emerg. Technol. 29, 215–221.

Margosch, D., Moravek, M., Gaenzle, M.C., Maertlbauer, E., Vogel, R.F., Ehrmann, M.A., 2005. Effect of high pressure and heat on bacterial toxins. Food Technol. Biotechnol. 43 (3), 211–217.

Medveďová, A., Valík, Ľ., 2012. *Staphylococcus aureus*: characterisation and quantitative growth description in milk and artisanal raw milk cheese production. Struct. Funct. Food Eng. 4, 71–102.

Mesias, M., Wagner, M., George, S., Morales, F.J., 2016. Impact of conventional sterilization and ohmic heating on the amino acid profile in vegetable baby foods. Innov. Food Sci. Emerg. Technol. 34, 24–28.

Mitelut, A., Popa, M., Geicu, M., Niculita, P., Vatuiu, D., Vatuiu, I., Gilea, B., Balint, R., Cramariuc, R., 2011. Ohmic treatment for microbial inhibition in meat and meat products. Romanian Biotechnol. Lett. 16 (1 Suppl.), 149–152.

Monfort, S., Gayán, E., Saldaña, G., Puértolas, E., Condón, S., Raso, J., Álvarez, I., 2010. Inactivation of *Salmonella typhimurium* and *Staphylococcus aureus* by pulsed electric fields in liquid whole egg. Innov. Food Sci. Emerg. Technol. 11 (2), 306–313.

Montanari, C., Serrazanetti, D.I., Felis, G., Torriani, S., Tabanelli, G., Lanciotti, R., Gardini, F., 2015. New insights in thermal resistance of staphylococcal strains belonging to the species *Staphylococcus epidermidis*, *Staphylococcus lugdunensis* and *Staphylococcus aureus*. Food Control 50, 605–612.

Necidova, L., Bogdanovicova, K., Harustiakova, D., Bartova, K., 2016. Short communication: pasteurization as a means of inactivating staphylococcal enterotoxin A, B, and C in milk. J. Dairy Sci. 99.

Ortega, E., Abriouel, H., Lucas, R., Gálvez, A., 2010. Multiple roles of *Staphylococcus aureus* enterotoxins: pathogenicity, superantigenic activity, and correlation to antibiotic resistance. Toxins 2 (8), 2117–2131.

Ortega-Rivas, E., 2012. Pulsed light technology. In: Non-Thermal Food Engineering Operations. Food Engineering Series, Springer Science+Business Media, New York, pp. 263–273.

Patterson, M.F., 2005. Microbiology of pressure-treated foods. J. Appl. Microbiol. 98 (6), 1400–1409.

Paulin, S., Horn, B., Hudson, J.A., 2012. Factors Influencing Staphylococcal Enterotoxin Production in Dairy Products, vol. 9. Ministry for Primary Industries. New Zealand Government.

Pearce, L.E., Smythe, B.W., Crawford, R.A., Oakley, E., Hathaway, S.C., Shepherd, J.M., 2012. Pasteurization of milk: the heat inactivation kinetics of milk-borne dairy pathogens under commercial-type conditions of turbulent flow. J. Dairy Sci. 95 (1), 20–35.

Pepe, O., Blaiotta, G., Bucci, F., Anastasio, M., Aponte, M., Villani, F., 2006. *Staphylococcus aureus* and staphylococcal enterotoxin A in breaded chicken products: detection and behavior during the cooking process. Appl. Environ. Microbiol. 72 (11), 7057–7062.

Pereira, R.N., Vicente, A.A., 2010. Environmental impact of novel thermal and non-thermal technologies in food processing. Food Res. Int. 43 (7), 1936–1943.

Poliseli-Scopel, F.H., Hernandez-Herrero, M., Guamis, B., Ferragut, V., 2012. Comparison of ultra high pressure homogenization and conventional thermal treatments on the microbiological, physical and chemical quality of soymilk. LWT – Food Sci. Technol. 46 (1), 42–48.

Ramaswamy, R., Krishnamurthy, K., Jun, S., 2012. Microbial decontamination of food by infrared (IR) heating. In: Demirci, A., Ngadi, M.O. (Eds.), Microbial Decontamination in the Food Industry. Woodhead publishing Limited, UK, pp. 450–471.

Ramos, S.J., Chiquirrin, M., Garcia, S., Condon, S., Perez, M.D., 2015. Effect of high pressure treatment on inactivation of vegetative pathogens and on denaturation of whey proteins in different media. LWT – Food Sci. Technol. 63 (1), 732–738.

Raso, J., Condon, S., Alvarez, I., 2014. Non-thermal processing | pulsed electric field. In: Batt, C.A. (Ed.), Encyclopedia of Food Microbiology, second ed. Elsevier, pp. 966–973.

Richardson, P., 2001. Introduction. Thermal Technologies in Food Processing. Woodhead publishing Ltd, UK.

Rincon, A.M., Singh, R.K., 2016. Inactivation of Shiga toxin-producing and nonpathogenic *Escherichia coli* in non-intact steaks cooked in a radio frequency oven. Food Control 62, 390–396.

Rowley, E.A., Technology Ltd., Chester, 2001. Radio frequency heating. In: Richardson, P. (Ed.), Thermal Technologies in Food Processing. CRC Press, Cambridge, UK, pp. 161–177.

Ruan, R., Ye, X., Chen, P., 2001. Ohmic heating. In: Richardson, P. (Ed.), Thermal Technologies in Food Processing. CRC, Cambridge, UK, pp. 241–264.

Scheinberg, J.A., Svoboda, A.L., Cutter, C.N., 2014. High-pressure processing and boiling water treatments for reducing *Listeria* monocytogenes, *Escherichia coli* O157: H7, *Salmonella* spp., and *Staphylococcus aureus* during beef jerky processing. Food Control 39 (1), 105–110.

Schlisselberg, D.B., Kler, E., Kalily, E., Kisluk, G., Karniel, O., Yaron, S., 2013. Inactivation of foodborne pathogens in ground beef by cooking with highly controlled radio frequency energy. Int. J. Food Microbiol. 160 (3), 219–226.

Scholtz, V., Pazlarova, J., Souskova, H., Khun, J., Julak, J., 2015. Nonthermal plasma – a tool for decontamination and disinfection. Biotechnol. Adv. 33 (6), 1108–1119.

Sengun, I.Y., Yildiz Turp, G., Icier, F., Kendirci, P., Kor, G., 2014. Effects of ohmic heating for pre-cooking of meatballs on some quality and safety attributes. LWT – Food Sci. Technol. 55 (1), 232–239.

Sharma, P., Bremer, P., Oey, I., Everett, D.W., 2014. Bacterial inactivation in whole milk using pulsed electric field processing. Int. Dairy J. 35 (1), 49–56.

Shim, W.B., Je, G.S., Kim, K., Mtenga, A.B., Lee, W.G., Song, J.U., Yoon, Y., 2012. Effect of irradiation on kinetic behavior of *Salmonella typhimurium* and *Staphylococcus aureus* in lettuce and damage of bacterial cell envelope. Radiat. Phys. Chem. 81 (5), 566–571.

Skjoldebrand, C., 2001. Infrared heating. In: Richardson, P. (Ed.), Thermal Technologies in Food Processing. CRC Press, Cambridge, UK, pp. 208–228.

Somavat, R., Mohamed, H.M.H., Chung, Y.K., Yousef, A.E., Sastry, S.K., 2012. Accelerated inactivation of *Geobacillus stearothermophilus* spores by ohmic heating. J. Food Eng. 108 (1), 69–76.

Sommers, C.H., Scullen, O.J., Sites, J.E., 2010. Inactivation of foodborne pathogens on frankfurters using ultraviolet light and gras antimicrobials. J. Food Saf. 30 (3), 666–678.

Song, H.P., Kim, B., Yun, H., Kim, D.H., Kim, Y.J., Jo, C., 2009. Inactivation of 3-strain cocktail pathogens inoculated into *Bajirak jeotkal*, salted, seasoned, and fermented short-necked clam (*Tapes pilippinarum*), by gamma and electron beam irradiation. Food Control 20 (6), 580–584.

Tahir, A., Mateen, B., Univerdi, S., KaraGoban, O., Zengin, M., 2009. Simple method to study the mechanism of thermal and non thermal bactericidal action of microwave radiations on different bacterial species. J. Bacteriol. Res. 1 (5), 58–63.

Tao, Y., Sun, D.-W., Hogan, E., Kelly, A.L., 2014. High pressure processing of foods: an overview. In: Sun, Da-W. (Ed.), Emerging Technologies for Food Processing, second ed. Academic Press/Elsevier, San Diego, California, USA, pp. 3–24.

Tassou, C.C., Panagou, E.Z., Samaras, F.J., Galiatsatou, P., Mallidis, C.G., 2008. Temperature-assisted high hydrostatic pressure inactivation of *Staphylococcus aureus* in a ham model system: evaluation in selective and non-selective medium. J. Appl. Microbiol. 104 (6), 1764–1773.

Uemura, K., Takahashi, C., Kobayashi, I., 2010. Inactivation of *Bacillus subtilis* spores in soybean milk by radio-frequency flash heating. J. Food Eng. 100 (4), 622–626.

Vural, A., Aksu, H., Erkan, M.E., 2006. Low-dose irradiation as a measure to improve microbiological quality of Turkish raw meat ball (cig kofte). Int. J. Food Sci. Technol. 41 (9), 1105–1107.

Walkling-Ribeiro, M., Noci, F., Cronin, D.A., Riener, J., Lyng, J.G., Morgan, D.J., 2008. Reduction of *Staphylococcus aureus* and quality changes in apple juice processed by ultraviolet irradiation, pre-heating and pulsed electric fields. J. Food Eng. 89 (3), 267–273.

Watanabe, K., Kakita, Y., Kashige, N., Miake, F., Tsukiji, T., 2000. Effect of ionic strength on the inactivation of micro-organisms by microwave irradiation. Lett. Appl. Microbiol. 31 (1), 52–56.

Wilson, D.R., Dabrowski, L., Stringer, S., Moezelaar, R., Brocklehurst, T.F., 2008. High pressure in combination with elevated temperature as a method for the sterilisation of food. Trends Food Sci. Technol. 19 (6), 289–299.

Woo, I., Rhee, I., Park, H., 2000. Differential damage in bacterial cells by microwave radiation on the basis of cell wall structure differential damage in bacterial cells by microwave radiation on the basis of cell wall structure. Appl. Environ. Microbiol. 66 (5), 2243–2247.

World Health Organisation, 2015. WHO Estimated of the Global Burden of Foodborne Diseases: Foodborne Disease Burden Epidemiology Reference Group 2007–2015. WHO, Geneva. Available from: www.who.int.

Yang, B., Shi, Y., Xia, X., Xi, M., Wang, X., Ji, B., Meng, J., 2012. Inactivation of foodborne pathogens in raw milk using high hydrostatic pressure. Food Control 28 (2), 273–278.

Yeo, C.B.A., Watson, I.A., Stewart-Tull, D.E.S., Koh, V.H.H., 1999. Heat transfer analysis of *Staphylococcus aureus* on stainless steel with microwave radiation. J. Appl. Microbiol. 87 (3), 396–401.

Yildiz-Turp, G., Sengun, I.Y., Kendirci, P., Icier, F., 2013. Effect of ohmic treatment on quality characteristic of meat: a review. Meat Sci. 93 (3), 441–448.

Zhao, Y., 2000. Using capacitive (radio frequency) dielectric heating in food processing and preservation – a review. J. Food Process Eng. 23, 25–55.

CHAPTER

13 MITIGATION STRATEGIES TO COMBAT *STAPHYLOCOCCUS AUREUS* IN THE FOOD CHAIN: INTERNATIONAL FOOD STANDARDS, GUIDELINES, AND CODES OF PRACTICE

Sandra M. Tallent, John F. Sheehan

U.S. Food and Drug Administration (FDA), Silver Spring, MD, United States

1. INTRODUCTION

One limitation of evaluating the burden of staphylococcal food poisoning (SFP) events is the lack of information available for *Staphylococcus aureus* contamination especially because illnesses are not generally reported unless there is a large outbreak. For example, the US Centers for Disease Control and Prevention (CDC) 2011 report included estimates of 323 SFP outbreaks affecting 241,000 people. The incidence of foodborne outbreaks was applied to the number of people in the country to provide an estimate for this report (Scallan et al., 2011). French authorities reported SFP was confirmed in 16/530 outbreaks (3%) representing 25.6/6451 (0.4%) of cases in 2000 (Haeghebaert et al., 2002). Australia and New Zealand reported that between 2001–2013, there were 107 bacterial toxin-mediated outbreaks of foodborne illness linked with 2219 cases and that 16 of the outbreaks and 200 of the cases were confirmed as SFP (May and Polkinghorne, 2016).

Staphylococcal foodborne outbreaks have been associated with poor human hygiene, inadequate equipment cleaning, cross-contamination by raw ingredients, and time/temperature abuse of food products (Todd et al., 2007; Argudín et al., 2012; Kusumaningrum et al., 2003). Mitigation strategies that help control the risks are sorely needed, but a universal plan must consider factors such as variability in farming, food processing, and cooking practices (Toyofuku, 2006) as well as variability of enterotoxin production in various food products (Schelin et al., 2011). One consideration could be the development of complete risk analysis plans specific for different food products as the United States has created for milk and seafood discussed in the third paragraph below.

The risk assessment concept was initiated in the 1970s by US federal agencies to standardize regulatory decisions because of chemical exposures that had adverse effects with respect to human health. This concept has mushroomed into microbial risk analysis that incorporates risk management, risk assessment, and risk communication. Microbial risk analysis is intended to provide a context that combines policy and science-based knowledge with aims to identify and protect consumers from potential risk and ultimately not only reduce foodborne disease but also economic impact and costs to industry (Lammerding, 1997).

Staphylococcus aureus. http://dx.doi.org/10.1016/B978-0-12-809671-0.00013-9

The Codex Alimentarius Commission prepared a document entitled "Principles and Guidelines for the Application of microbial Risk Assessment" based on the General Agreement on Trade and Tariffs and Sanitary and Phytosanitary Measure agreements. By 1998, these documents became the reference standard for international trade. Codex now acts as the international body to develop and coordinate standards that safeguard the food supply and food trade (Dawson role of Codex Food Control 6:261). Numerous countries have initiated microbial risk analysis including the European Union (EU), individual countries in Europe, such as the Netherlands and Germany, and federal agencies in the United States Some countries also use risk factors to manage inspection of imported foods (Hoffmann, 2010).

Numerous programs and tools have been developed to include both qualitative and quantitative analyses, but all embrace the same basic principles designed to identify the possibility of adverse events. The tools have become more sophisticated making allowances for a variety of foods, food preparation, and food preservation techniques, but at the core the microbial risk analysis must determine both the presence of the pathogen as well as the level of a pathogen to assess the potential for an undesirable outcome (Buchanan et al., 2000; Lammerding, 1997; Schelin et al., 2011). To this end, each nation has developed policies and standards regarding acceptable levels of pathogens in a variety of food products.

This chapter briefly describes specific food safety standards for coagulase-positive staphylococci (CPS) enumeration and staphylococcal enterotoxin (SEs) applicable within a total of 55 countries, representing ~60% of the world's population (CIA World Fact Book; Population Reference Bureau). Moreover, different strategies to reduce the risk of undesirable outcomes of food contamination with CPS, *S. aureus*, and/or SEs are discussed.

2. REGULATIONS/POLICIES/STANDARDS FOR COAGULASE-POSITIVE *STAPHYLOCOCCI* ENUMERATION REGARDING FOOD SAFETY: ACCEPTABLE LEVELS OF *STAPHYLOCOCCUS AUREUS* IN DIFFERENT FOODS OR FOOD TYPES

In the following, a detailed overview of standards set by governmental agencies on CPS, *S. aureus*, and/or SEs is presented. Please refer to the Annex at the end of the chapter for detailed information regarding food types and acceptable levels per country regulation, policies, and standards. The countries are arranged by continent in the text but appear alphabetically in the Annex.

As will be seen, most countries will apply either a two-class or three-class attribute sampling plan when sampling and analyzing foodstuffs for any of the aforesaid parameters. Two-class attribute sampling plans are typically used when deciding acceptance of any given lot of food based on the presence or absence of any given parameter, e.g., a pathogen or a toxin. Such plans consist of three criteria, viz. the number of samples, which must be taken from any given lot (denoted as "n"), the maximum number of those samples, which can yield an unsatisfactory result (denoted as "c" and usually set at zero), and a microbiological limit (denoted as "m"). Three-class attribute sampling plans are generally utilized where decisions on lot acceptability are made on the basis of quantitative data as opposed to mere presence or absence of any given parameter. With three-class attribute sampling plans, four criteria are utilized. Because with two-class attribute sampling plans, the first is the number of samples to be taken from any given lot ("n"), the second is a limit ("m") that separates good product from marginal product, the third is a limit ("M") beyond which product is unacceptable, and the fourth is the number of samples ("c"), which may exceed "m" before a product is deemed unacceptable.

2.1 THE UNITED STATES OF AMERICA

2.1.1 Food and Drug Administration

2.1.1.1 General Policy/Practice

Foods regulated by the United States Food and Drug Administration (FDA) are considered adulterated and thus subject to regulatory action, if they are determined to contain any SE. The presence of any preformed SE in a food is considered to be a violation of the Federal Food, Drug, and Cosmetic Act (FFDCA), Section 402(a) (1) (21 U.S.C. Chapter 9), which recites in pertinent part that "A food shall be deemed to be adulterated… If it bears or contains any poisonous or deleterious substance which may render it injurious to health;".

Most of the foods regulated by FDA, except for low-acid canned foods, will, when sampled by FDA, be analyzed for the presence of *S. aureus*. (see, for example, FDA Compliance Programs for Domestic Food Safety; Domestic and Imported Cheese; Imported Foods General; Import Seafood Products; Medical Foods; Infant Formula). Generally, where levels of 10,000 colony-forming units (CFU)/mL (g) are obtained on analysis, the foods yielding such results will be further analyzed for the presence of preformed enterotoxins. Enterotoxin testing is also performed if time and temperature abuse of a product is suspected. Any food determined to contain *S. aureus* at levels at or above 10,000 CFU/mL (g) will be considered, by virtue of that fact alone, to be adulterated and so subject to regulatory action because such a food is considered to be in violation of the FFDCA Section 402(a) (FDA Food Code), which recites in pertinent part that "A food shall be deemed to be adulterated… if it has been prepared, packed, or held under insanitary conditions whereby it may have become contaminated with filth, or whereby it may have been rendered injurious to health;" (U.S.C. Chapter 9). Elevated numbers of *S. aureus* (at or above 10,000 CFU/mL (g)) are evidence that a food was prepared, packed, or held under insanitary conditions whereby it may have been rendered injurious to health.

While the above describes FDA's general approach to the regulation of foods relative to *S. aureus* and SE, FDA does not describe the policy itself publicly in any document apart from the Compliance Policy Guide for Dairy Products–Microbial Contaminants and Alkaline Phosphatase Activity, Section 527.300 and the Fish and Fishery Products Hazards and Controls Guidance.

2.1.2 The United States Department of Agriculture

2.1.2.1 General Policy/Practice

USDA's Food Safety and Inspection Service (FSIS) regulates meat from amenable species as identified in the Federal Meat Inspection Act (FMIA), poultry as identified in the Poultry Products Inspection Act (PPIA), and liquid, dried, and frozen egg products as per the Egg Products Inspection Act (EPIA). The definitions of adulteration under the FMIA, PPIA, and EPIA are approximately identical to that in the FFDCA.

FSIS does not have any limits for *S. aureus* in its regulations. However, it does require that establishments conduct a hazard analysis (9 CFR 417.2(a) (U.S.C. Chapter 9)), and there is an expectation (followed by verification) that with respect to certain ready-to-eat (RTE) products (e.g., those that are fermented, salt-cured, and dried have long come-up times during cooking or are shelf-stable) as well as certain raw products (e.g., those containing raw batter or brine solutions that are reused without adequate temperature control), manufacturers will consider that *S. aureus* is a hazard reasonably likely to occur, which must be controlled by one or more critical control points or if determined to be not reasonably likely to occur is addressed through a prerequisite program or other supporting program or documentation. *S. aureus* may be present on raw meat and poultry products because of cross-contamination during slaughter. After slaughter and cooking, meat and poultry products can be contaminated from handling by individuals carrying the organism. FSIS baseline testing from the 1990s indicates *S. aureus* prevalence is highest on poultry and further processed products with ground chicken having the highest prevalence (90%) followed by young turkeys (66.7%), broiler chickens (64%), ground turkey (57.3%), ground beef (31%), market hogs (16%), cows/bulls (8.4%), and steers/heifers (4.2%). For more information, see USDA FSIS Baseline Data.

S. aureus growth is a concern during fermentation, salt-curing, and drying of RTE meat and poultry products because the addition of ingredients (salt) creates what is called microbial inversion when the environment favors Gram-positive bacteria such as *S. aureus* as opposed to Gram-negative bacteria such as *E. coli* O157:H7 and *Salmonella* (which grow better on raw product without ingredients added). Controls during fermentation include use of commercial starter cultures (Raccach, 1981) and following best practices for production of fermented meats (e.g., by following the degree-hour concept described in the American Meat Institute Foundation document). For salt-cured and dried products manufacturing establishments should ensure sufficient brine concentration/low enough water activity during processing is achieved to limit *S. aureus* growth (Burnham et al., 2008; Reynolds et al., 2001). *S. aureus* growth is also a concern when there are slow come-up times during cooking because excessive dwell times within the optimum temperature range for growth can allow for high levels of *S. aureus* growth and in turn the production of its heat-stable enterotoxins. Because of these concerns, FSIS recommends limiting the come-up time for cooking to 6 h or less between 50 and 130°F (10–54.4°C) as described in FSIS Appendix A. Following this recommendation is particularly important for products such as hams, with a high salt content, because of *S. aureus*' salt tolerance. During storage of shelf-stable products, one or more hurdles should be applied to prevent *S. aureus* growth. Prevention of *S. aureus* growth is critical because of the pathogen's ability to survive and grow at a lower water activity and pH than most other bacterial pathogens. Controls during storage of shelf-stable products include low water activity limits, pH, or a combination of both. For recommended parameters to prevent *S. aureus* growth during storage, see the FSIS Jerky Compliance Guideline as well as Borneman et al. (2009) and Tilkens et al. (2015).

During the assessment of process deviations, FSIS does have guidelines that it will use, which are based on scientific literature for the levels of toxigenic *S. aureus* generally considered necessary for enterotoxin production in products. Establishments may use computer modeling to estimate the relative growth of bacteria during a heating deviation. The purpose of this modeling is to determine if conditions exist where enterotoxin formation could occur. In addition to modeling results, sanitary conditions that occurred during production that could lead to contamination with *S. aureus* may be taken into account. When evaluating modeling results, FSIS considers conditions that allow for 3-log growth or higher to be a public health concern because normal levels of *S. aureus* in raw meat and poultry are about 2 logs/g and growth of 3-logs or higher would result in total levels (5 logs/g or higher) that could result in the SEs being produced (ICMSF, 1996; NACMCF, 2015). Depending on modeling results, sampling may be used to support product safety. If an establishment was testing their RTE meat and poultry products for *S. aureus* or its enterotoxins, FSIS would normally recommend that they follow Table 8.3 from ICMSF book #7 (ICMSF, 2002) (Microbiological Testing in Food Safety Management) or other recognized statistically based sampling plan for product acceptance. For concerns with slow come-up time during cooking, FSIS recommends establishments test for *S. aureus* enterotoxins because any vegetative cells would be destroyed by cooking. This testing recommendation also applies to fermentation deviations where product undergoes a heat treatment. Further information regarding evaluating fermentation deviations may be found in the American Meat Institute Foundation (1997) Good Manufacturing Practices for Fermented Dry and Semidry Sausage Products.

2.2 CANADA

Canada, through the Canadian Food Inspection Agency, has established via regulations (Regulations under the Food and Drugs Act (The Regulations) (Standards and Guidelines for Microbiological Safety of Foods, 2008) microbiological standards for certain foods and microbiological guidelines for other foods. Both, the microbiological standards promulgated under The Regulations and the guidelines are used to adjudge compliance with the Food and Drug Act. Canada's microbiological standards and guidelines are expressed in terms of either two-class or three-class attribute sampling plans. Canada also characterizes, on a per food per microbiological parameter basis, the health risk associated with a violation of any given standard. Health Risks are ranked numerically from 1 to 3, with 1 being the highest level of concern and 3 the lowest. The Health Risks rankings essentially determine what actions should be taken and how timely they should be taken, whenever a standard or guideline is determined to have been exceeded. Canada examines foods for *S. aureus* and generally, either with respect to a microbiological standard or guideline, the three-class attribute sampling regimens for foods, which have a *S. aureus* standard, include a value for "M" at 10,000 CFU/mL (g). Exceptions to this generalization are made for instant infant cereal and powdered infant formula as well as powdered protein, meal replacements, and dietary supplements, where "M" is 100 CFU/mL (g).

2.3 THE EUROPEAN UNION

The EU has established microbiological criteria for foodstuffs through issuance of Commission Regulation (EC) No. 2073/2005. The EU has two sets of criteria, viz. "food safety" and "process hygiene". While the food safety criterion means a criterion defining the acceptability of a product or a batch of foodstuff applicable to products placed on the market, the process hygiene criterion indicates the acceptable functioning of the production process. Such a process hygiene criterion is not applicable to products placed on the market but sets an indicative contamination value above which corrective actions are required to maintain the hygiene of the process in compliance with food law. In case of *Staphylococcus*-related standards the EU has set one food safety criterion applicable to SEs in cheeses, milk powder, and whey powder. When testing a 25 g sample for SEs, these foods must be negative-by-test. This is an example of a two-class attribute sampling plan.

Relative to process hygiene criteria, the EU has *Staphylococcus*-related standards for several food categories, including milk and dairy products and fishery products (shelled and shucked products of cooked crustaceans and molluscan shellfish). Within those categories of foods, the hygiene parameter is CPS. In the milk and dairy products category, four separate standards exist, each for different subcategories of milk products (i.e., cheeses made from raw milk; cheeses made from milk that has undergone a lower heat treatment than pasteurization and ripened cheeses made from milk or whey that has undergone pasteurization or a stronger heat treatment; unripened soft cheeses made from milk or whey that has undergone pasteurization or a stronger heat treatment; milk powder and whey powder). The three-class attribute sampling plans for these four separate standards are identical in terms of the sampling plan itself (i.e., "n" and "c"). The value for "M" varies by product subcategory, with the largest "M" being allowed for raw milk cheeses (at 100,000 CFU/mL (g)) and the lowest "M" value (at 100 CFU/mL (g)) for unripened soft cheeses (fresh cheeses) made from milk or whey that has undergone pasteurization or a stronger heat treatment and for milk and whey powders.

Within the fishery products category, shelled and shucked products of cooked crustaceans and molluscan shellfish have a standard with "M" at 1000 CFU/mL (g).

2.4 TURKEY

In Turkey, the Regulation on Turkish Food Codex, microbiological criteria contain *Staphylococcus*-related standards for several specific foods. As with the EU, Turkey utilizes both food safety and process hygiene criteria. Unlike the EU, however, Turkey utilizes both CPS and SEs as food safety criteria. With the exception of melted cheeses and melted cheese products, for all milk products considered in the Turkish Regulation, the criterion used is CPS. Melted cheeses and melted cheese products have a SEs criterion. For all the foods described within the RTE meals category of the Turkish Regulation, the criterion is SEs. Foods that are not specifically identified within the food safety criteria section of the Regulation are held to standards that exist within a separate annex. There are different three-class attribute sampling plans for *Staphylococcus*-related parameters within this annex, the one for RTE foods and the other for not-RTE foods. The sampling plans ("n" and "c") are identical for both types of foods. Not-RTE foods are permitted to have a higher "M" (10,000 CFU/mL (g)) than RTE foods (1000 CFU/mL (g)). Five different process hygiene criteria are set in the Turkish Regulation on CPS: four on dairy food and the fifth for shelled and shucked crustacean and molluscan seafood. For the dairy foods, the three-class attribute sampling plans are identical with respect to "n" (at 5) and "c" (at 2), with the highest "M" being at 100,000 CFU/mL (g), which is for raw milk cheeses. As with the EU, for shucked and shelled crustacean and molluscan seafood products, "m" is set at 1000 CFU/mL (g).

2.5 RUSSIA AND THE EURASIAN CUSTOMS UNION

In 2011, Russia and its partners in the customs union (then Belarus and Kazakhstan but expanded in 2015 to include Armenia and Kyrgyzstan) adopted a Technical Regulation on Food Safety, which took effect in July of 2013 and contains staphylococcal-related standards within Annexes 1–2 of said document. Annex 1 recites the standards that apply to foods for pathogenic microorganisms and microbial toxins. The only microbial toxin for which a standard exists within Annex 1 is for SEs. The standard is a negative-by-test result on analysis of a total of 125 g of product (5 × 25 g samples composited). Testing for the presence of SEs is predicated on the finding of *S. aureus* in product, but the standard does not disclose a threshold value for the initiation of enterotoxin testing. The standard in Annex 1 applies to cheeses and cheese products, cheese pastes, dried baby food products that are milk-based, and specialized food products for dietary therapeutic nutrition.

Annex 2 comprises two tables, the first of which, entitled "microbiological food safety standards," contains microbiological standards for several broad categories of foods, within which are typically several subcategories. The second table in Annex 2 is specifically concerned with "preserved foods".

Table 1 contains standards for several parameters, which might be said to be generally considered as quality-related (e.g., "quantity of mesophilic aerobic and facultative anaerobic microorganisms"), or indicator organisms (e.g., *Escherichia coli*); however, it also contains standards for *S. aureus* for several subcategories of foods. Generally, the standards for *S. aureus* are very strict. When tested, foods are expected to be negative-by-test.

Table 2 contains a standard for "*S. aureus* and other coagulase-positive staphylococci" for "semi-preserves of Group E", which includes pasteurized preserved foods from beef and pork, minced and Liubitelskaya ham, and pasteurized semipreserved foods from fish in glassware. The standard is negative-by-test.

There are two other Technical Regulations, which contain standards for *S. aureus*, viz. the Technical Regulation on the Safety of Meat Products and the Technical Regulation on the Safety of Milk and Dairy Products, both of which were adopted in 2013 and came into effect in 2014. With both, for all the products described within the regulations that have a standard for *S. aureus*, it is negative-by-test. The contents of these two regulations are not described within the table.

2.6 THE GULF COOPERATION COUNCIL

The Gulf Cooperation Council comprises six member states (Bahrain, Kuwait, Oman, Qatar, Saudi Arabia, and the United Arab Emirates). In 2015, it published a standard entitled Microbiological Criteria for Foodstuffs, GCC (1016:2015), which replaces standard 1016/1998. That standard contains limits for *S. aureus* for many food subcategories contained within larger categories, e.g., for the category "dairy products" there is a subcategory for "whipped cream" and that subcategory has a limit for *S. aureus*.

Generally, most of the limits for *S. aureus* are expressed within three-class attribute sampling plans; however, there are several foodstuffs for which two-class attribute sampling plans for *S. aureus* exist. Where three-class attribute sampling plans are provided, the level for M is typically at or below 1000 CFU/mL (g), with only a handful of foods with a level for "M" higher than that, and wherever "M" is greater than 1000 CFU/mL (g), it does not exceed 10,000 CFU/mL (g). Foods with a level of "M" at 10,000 CFU/mL (g) include cured and/or smoked poultry meat, mortadella, frankfurters, turkey, smoked turkey breast, cooked poultry meat frozen to be reheated (e.g., chicken nuggets), frozen/chilled breaded fish (e.g., fish sticks/fingers), smoked fish, pizza, meat pies, frozen dough with or without filling, and coleslaw. Most usually, $n=5$ for all of the sampling plans, whether two-class or three-class, but there are exceptions for this general rule for "dietetic foods to be eaten by high risk category of consumers" and "cured and/or smoked poultry meat; mortadella, frankfurters, turkey, smoked turkey breast," where $n=10$ for both.

Food subcategories for which two-class attribute sampling plans are prescribed include dried and instant products requiring reconstitution, body-building foods, dehydrated egg mix, butter (salted and unsalted), margarine, Arabic sweets, drink powders, liquorice root extract (concentrates or drinks), sterilized soya drink, and, in the category of miscellaneous foods, nutritious powder. For two-class attribute sampling plans, "c" and "m" are at "0," with one exception, for "dehydrated egg mix," where "m" is listed at "10."

2.7 CHINA

China examines foods for *S. aureus*. Chinese standards for *S. aureus* in foods are expressed typically as three-class attribute sampling plans, although there are a few foods for which two-class attribute sampling plans have been established (China Standards GB 29921-2013).

Typically, n = 5 for both three-class and two-class attribute sampling plans. "M" is usually no greater than 1000 CFU/mL (g); however, with several subcategories of RTE seasonings, "M" is at 10,000 CFU/mL (g). These include products such as soy sauce, paste and paste products, seafood-based seasonings, and salad dressing. Where "M" is lower than 1000 CFU/mL (g), it is at 100 CFU/mL (g) and such standards are for milk-derived products, e.g., whey powder and whey protein powder, milk powder, cream, butter, and anhydrous milkfat as well as for foods such as infant formula, special medical purposes formulated foods for infants, special medical purposes formulated foods generally, and sports nutrition foods.

Where China utilizes two-class attribute sampling plans (e.g., for pasteurized milk, fermented milk, evaporated milk, sweetened condensed milk, and formulated condensed milk as well as fermented alcoholic beverages), both "c" and "m" are at 0.

2.8 JAPAN

In Japan, national limits for *S. aureus* have been established in Japan's Food Standards (Japan, Food Hygiene Regulation, 1959) but only for certain categories of meat products (nonheat-treated meat products, heat-treated meat products, and heat-treated meat products sterilized after packing). The standards are uniformly set at 1000/g maximum. Japan also has Hygiene Norms, which contain limits for *S. aureus* for certain types of RTE foods, such as prepared side dishes, fresh cakes and pastries, fresh pasta, boiled pasta, and toppings and other ingredients for fresh pasta, which are heat-treated. When examined, these foods are expected to be negative-by-test.

2.9 INDIA

India examines seafood, fruits, and vegetable products for *S. aureus*. Milk and milk products are examined for CPS. Where foods are examined for *S. aureus*, the standards (India Food Safety and Standards Regulation, 2010) are either negative-by-test or not greater than 100 CFU/mL (g). Milk and milk products are examined for CPS using either a three-class or two-class attribute sampling plan. "N" is uniformly set at 5 for either type of sampling plan, whereas "M" is not larger than 1000 CFU/mL (g) for any milk product category. Where two-class attribute sampling plans are utilized (e.g., for pasteurized milk, cream, flavored milk, dried powders, ice creams, and processed cheese), "c" is set at zero and the maximum permissible level is set at less than 10 CFU/mL (g).

2.10 PERU

Peru examines foods for *S. aureus*, and its food safety standards (Peru NTS No. 071, MINSA/DIGESA-V.01) contain sampling plans for foods and drinks, which are based on the risk to health posed by the microbe and the normal conditions of manufacturing, handling, and consumption of the food. Peru utilizes a categorical approach to assigning risk to pathogen/food combinations, with a range in categories from numbers 1–15, with 1 being assigned to those food/pathogen combinations expected to pose the least risk and 15 assigned to those food/pathogen combinations that are expected to pose the greatest risk. For *S. aureus*, most foods have been placed in category 8 (meaning there is moderate but direct health risk, with limited possibility of spreading the disease), with a few food categories (such as fresh, soft, unripened cheeses, butters and margarines, ground or chopped raw meat, raw fish and shellfish products, raw mollusks and crustaceans, refrigerated or frozen precooked and cooked mollusks, frozen breaded raw fish and shellfish, and prepared foods that are not heat-treated such as fresh salads) being placed in category 7. One food category, i.e., heat-smoked fish and shellfish, was classified at category 1 for *S. aureus*, meaning that it was considered that there was no direct risk to health during the shelf life of the product, even though growth during the shelf life of the product was possible. The sampling plans utilized for all foods are three-class attribute, with "n" equal to 5 uniformly. Category 8 foods have "c" set at 1, whereas category 7 foods have "c" set at 2. For the singular food category classified as 1, "c" is set at 1. For most foods, "M" is set at 100CFU/mL (g) or lower. The highest level of M permitted is 10,000CFU/mL (g) and that is for one category of food (precooked frozen fowl meat requiring heating before consumption), with the most stringent standards being applied to infant formula and special dietary foods, where "M" is 10CFU/mL (g) and "m" is set at <3CFU/mL (g).

2.11 SOUTH KOREA

South Korea standards (Korea Food Code) mandate that for edible meat (excluding raw material to be used for manufacturing and processing purposes) or food products that have been pasteurized, sterilized, or can be directly consumed without further processing or thermal treatment, "food poisoning bacteria," which includes *S. aureus*, the standard is n=5, c=0, and m=0/25g.

South Korea has additional standards for certain categories of "livestock products" and they are as follows:

1. Raw hams, fermented sausages, and natural and processed cheeses: n=5, c=2, m=10CFU/mL (g), and M=100CFU/mL (g).
2. Heated hams and sausages: n=5, c=1, m=10CFU/mL (g), and M=100CFU/mL (g) (Sterilized products have to be n=5, c=0, m=0/25g).

Moreover, South Korea has set standards for cream breads (which appears to be pastries with cream-based fillings), which must be negative-by-test for *S. aureus* and RTE foods, raw seafood, and seasoned dried fish/shellfish fillets, which all have a standard of <100CFU/mL (g).

2.12 MEXICO

Mexico standards require analysis of foods for *S. aureus*. The microbiological standards for individual food categories are located within a series of separate standards. The standards contain maxima for *S. aureus*, where such is a microorganism for which a food category or a food within a category should be analyzed. The highest permissible levels (1000CFU/mL (g)) are associated with ground beef, shellfish (fresh refrigerated and frozen) and fresh, refrigerated, or frozen fish. The most stringent standard is associated with infant formula, which must be negative-by-test when analyzed for *S. aureus*. Standards for milk and milk products vary by the type of milk product, with products such as milk-based infant formula, and pasteurized milk products having a very low standard of <10CFU/mL (g), butters, creams and fermented or acid milk products having a maximum of <100CFU/mL (g), and cheese having a range of permissible levels ranging from <100CFU/g for processed cheeses to 100CFU/mL (g) for aged cheeses to 1000CFU/mL (g) for fresh cheeses. While most cereal products do not have specifically prescribed *S. aureus* standards, sweet bread with a dairy filling (pan dulce relleno) has a limit of <100CFU/mL(g), whereas Pasteles (cakes), panqués (such as banana breads), and pays (pies) have a limit of 100CFU/mL (g).

2.13 SOUTH AFRICA

South Africa analyzes foods for either *S. aureus* or CPS. The microbiological standards for foods generally are found within one regulation (South Africa GNR 1551, 1997) with standards for milk and milk products being in another (South Africa GNR 692, 1997). For most of the categories of foods listed within the general standard, the expectation is that when tested for *S. aureus*, coagulase-positive, or otherwise, they be found negative-by-test. However, one category of foods, viz. partly cooked or uncooked seawater and freshwater foods (such as prawns, shrimps, crayfish, lobsters, crab meat, eels, or fish), does allow up to 10 CFU/mL (g) of *S. aureus* to be present. One food and one category of foods, i.e., desiccated coconut and cooked seawater or freshwater foods (such as prawns, shrimps, crayfish, lobsters, crab meat, oysters, mussels, clams), and fish, respectively, are specifically tested for coagulase-positive *S. aureus* and must be found negative-on-test when analyzed.

The regulation relating to milk and milk products does not specifically mention *S. aureus* as being an analytical parameter for any of the food categories defined within, but it instead contains a requirement that the foods defined be free of pathogenic organisms. While the regulation itself does not define the term "pathogenic organisms" and said term is not defined either in the General Regulation or in the Foodstuffs, Cosmetics and Disinfectants Act of 1972, *S. aureus* is identified as a "foodborne pathogen" within South Africa's "Guidelines for Environmental Health Officers on the Interpretation of Microbiological Analysis Data of Food".

2.14 AUSTRALIA AND NEW ZEALAND

The microbiological limits for foods sold within Australia and New Zealand are located within the Australia New Zealand Food Standards Code. They contain limits for CPS for a handful of food categories. The standards are stated as three-class attribute sampling plans, with the exception of infant formula, which has a two-class attribute sampling plan. Values for "M" vary, with the lowest "M" level (10 CFU/mL (g)) being assigned to infant formula and the highest (10,000 CFU/mL (g)) to uncooked, comminuted fermented meat.

3. SAFEGUARDING FOODS

Safeguarding the food supply is complicated by many factors including a global food supply with new and different cuisine options in distant marketplaces (Kataoka et al., 2016; Schelin et al., 2011; Toyofuku, 2006). The aspects of fair trade and competition along with cost-cutting measures to reduce risk of foodborne illness have drawn attention to different strategies to address food safety from a comprehensive viewpoint (Hoffmann, 2010; Lammerding, 1997; Toyofuku, 2006). However, the basics of food safety and risk analysis regarding staphylococcal contamination and potential outgrowth with enterotoxin production remain unchanged. These include sanitation, time, and temperature control measures as well as monitoring food handlers to reduce introduction of bacteria from human carriers or the environment (Kusumaningrum et al., 2003; Todd et al., 2007), for example, as described for milk and seafood.

3.1 RETAIL FOODS AND FOODSERVICE PRACTICES

Modernly, most foodborne outbreaks involving staphylococcal intoxication are associated with food preparation and handling (Martin and Iandolao, 2014; Todd, 2014) and so efforts should be made to control all possible routes of contamination of foods by food preparers or handlers. Governmental authorities around the globe have set various modes of safeguarding foods, mostly on a national-wide basis. For instance, the United States Food Code is a model code produced by the FDA, CDC, and USDA's FSIS, and it is used by state, city, county, and tribal agencies as a means of regulating restaurants, retail food stores, food vendors, and foodservice operations in schools, hospitals, assisted living facilities, nursing homes, and child care centers. The Food Code has a section on Employee Health, which contains text describing the responsibilities of employers when employees present with symptoms of illness, including lesions containing pus, such as boils or infected wounds on the hands or exposed portions of the arms. Under the Food Code, persons with such symptoms would be excluded or restricted from the workplace, unless a remedy could be affected, such as an impermeable covering of some sort that would cover the area of concern. The Food Code does also deal with personal cleanliness and contains specific instructions on how to clean hands and arms, when to do so, and where to do so to prevent contamination of food by employees. Moreover, the Food Code requires that all egg products be pasteurized and that milk products obtained by establishments be pasteurized and comply with Grade A standards (see above). The Food Code also deals with the destruction of microorganisms of public health concern and contains specific requirements for microwave cooking of raw animal foods and reheating of foods for hot holding. Viewed collectively, the requirements of the Code, if followed, should help prevent contamination of foods by handlers and, if foods should happen to be contaminated, prevent outgrowth of *S. aureus* to levels such that enterotoxin production would obtain. Other countries, such as Germany, have set comparable safeguarding food principles and regulation food operators have to follow (see also Chapter 12 for further details). In the EU, standards were agreed to ensure food hygiene; these EU Rules (e.g., Regulation (EC) 852/2004 on the hygiene of foodstuffs; Regulation (EC) 853/2004 laying down specific hygiene rules for food of animal origin; Regulation (EC) 854/2004 laying down specific rules for the organization of official controls on products of animal origin intended for human consumption) regarding food hygiene cover all stages of the production, processing, distribution, and placing on the market of food intended for human consumption.

3.2 DAIRY FOODS

S. aureus may be present in bulk raw milk for a variety of reasons, including direct shedding into the milk from animals with intramammary infections caused by *S. aureus*, shedding from the skin and udder exterior during milk collection, unhygienic practices during milk collection, and the usage of insanitary equipment for milk collection and storage (Asperger and Zangeri, 2011; IDF Bulletin, 2006; Gilmour and Harvey, 1990). *S. aureus* infections have long been considered one of the principal causes of subclinical mastitis in animals used for commercial milk production (Asperger and Zangeri, 2011) and while rates of contamination of bulk milk will vary between species, herds, regions within countries, and countries, it is generally considered that intramammary infections are responsible for the majority of *S. aureus* present in raw milk.

The enterotoxigenicity of strains isolated from milks has been reported with wide variation (Aarestrup et al., 1995; Adesiyun, 1995; Fagundes et al., 2010; Huonga et al., 2010; Jørgensen et al., 2005; Normanno et al., 2005; Neder et al., 2011; Stephan et al., 1999). Enterotoxigenicity of strains recovered from milk has been reported to depend on the type of mastitic condition; for example, DeSantis et al. (2005) reported that *S. aureus* recovered from sheep with subclinical mastitis were enterotoxigenic at a rate of ~34%, whereas with sheep suffering from acute clinical mastitis, isolated strains were enterotoxigenic at a rate of 70%–80%.

Enterotoxigenic *S. aureus* in bulk raw milk, including the genotype associated as being mastitis-causative (genotype B), represent a potential public health hazard, which milk producers should endeavor to control (Johler et al., 2015; Hummerjohann et al., 2014; Kűmmel et al., 2016; Schmid et al., 2009). Control efforts ideally would begin with herd health management to minimize the incidence of *S. aureus*–related intramammary infections within commercial milking herds. Careful attention to milking practices, including the mechanical condition and functionality of the milking system equipment itself, the personal hygiene of milkers, the udder preparation, and postmilking teat dipping practices utilized, will aid in minimizing the incidence of intramammary infections generally, including those caused by *S. aureus*, and will otherwise help to reduce the incidence of *S. aureus* in milk. The prompt cooling of milk postcollection, such that *S. aureus* outgrowth and SE production is minimized, is essential, given that SEs are very heat resistant and only partially inactivated by heat treatment processes ordinarily applied to milk products. Indeed, it is important that once cooled, milk be kept cool until utilized for further processing (e.g., pasteurization) and manufacturing of milk products. For milk, which is ordinarily refrigerated post heat processing, either because it is being stored as intermediate product for further manufacture or for packaging, maintenance of refrigeration conditions is essential as is the prevention of postprocessing contamination because a heat-treated milk represents enhanced growth opportunity for *S. aureus* due to a reduction in competing microflora.

The United States Grade "A" Pasteurized Milk Ordinance (PMO) is an example of how dairy foods are safeguarded, and it is a model ordinance published and utilized by the United States FDA in conjunction with the National Conference on Interstate Milk Shipments to regulate the safety of what are termed Grade "A" milk products (essentially fluid milk products with the addition of certain acidified and fermented milks, i.e., those products that will be defined by US standards of identity (i.e., Title 21 of the United States Code of Federal Regulations, Section 131), as well as cottage cheese and dried versions of the aforesaid products) contain standards for all aspects of commercialized milk production within the United States (the German "Milchgüteverordnung," a national legislation on milk quality lays down similar aspects concerning payment and quality of milk and classifies milk into three different categories "1," "2," and "S," with "S" being the one having the highest quality and being best paid per liter). The US PMO is adopted as the standard for the regulation of Grade "A" milk products by all 50 of the United States and Puerto Rico. Approximately 99% of all commercial bovine milk produced in the United States is produced as Grade "A" milk (USDA NASS, 2016). The PMO contains requirements for all aspects of milk production, collection, storage, transportation, processing and postprocessing storage, and transportation, including standards for animal health, udder preparation and health maintenance, milk collection practices, (including equipment design and cleaning and sanitation standards), postcollection cooling rates, storage temperature requirements, and transportation requirements. With respect to the processing of milk into Grade "A" milk products, the PMO contains requirements for essentially all equipment utilized in the manufacture and packaging of Grade "A" milk and milk products, with particular attention given to the design, operation, and control of pasteurizers. It includes requirements for storage temperature of milk pre- and postprocessing and for milk products packaging equipment and milk products packaging materials. According to the PMO, Grade "A" milk must be cooled to 50°F (10°C) or less within 4 h or less of the commencement of the first milking on any given day and to 45°F (7.2°C) or less within 2 h of completion of daily milking. The blend temperature of the first and second milking may not exceed 50°F (10°C). Once cooled to 45°F (7.2°C) it must be maintained at that temperature until it is processed. With a few very specific exceptions involving acidified or fermented milk products and dried milk products, pasteurized milks and milk products must be immediately cooled to 45°F (7.2°C) prior to filling or packaging. EU Regulation (EC) 854/2004 has set specific rules on hygiene for raw milk and dairy products, too, e.g., on animal health requirements for raw milk production, on hygiene principles to follow at milk production holdings, on criteria for raw milk, on temperature and heat treatment requirements concerning dairy products, and on wrapping, packaging, and labeling of milk and dairy products.

The requirements of the PMO and other national food safeguarding rules will certainly help to minimize the possibility of SEs occurring with products which would be classified as Grade "A" milks or milk products; however, not all milk products are Grade "A" milk products. Most notably, cheeses generally are not considered Grade "A". Cheeses made from raw milk can represent opportunity for *S. aureus* outgrowth and toxin production (Basanisi et al., 2016; Ferreira et al., 2016; Jakobsen et al., 2011; Kérouanton et al., 2007; Lindqvist et al., 2002; Mehli et al., 2017; Ostyn et al., 2010; Rosengren et al., 2010; Walcher et al., 2014) because there is no kill step applied to the main ingredient (and main source of *S. aureus*) and certain steps during the manufacturing process may allow for outgrowth. Additionally, the finished product cheese itself may be conducive to outgrowth, depending on factors such as pH, water activity, ripening temperatures and practices utilized, and whether starter cultures have been utilized during cheese manufacture. For such products, all factors discussed above as mitigating of the presence of *S. aureus* in milk and its outgrowth are of paramount importance, as is control of the cheese manufacturing process itself and management of the acidification, ripening and packaging processes, and postpackaging temperature control.

Raw milk cheese manufacturers who wish to exert a greater degree of control over the safety of their products may wish to investigate the utility of processes such as microfiltration and bactofugation (Gesan-Guiziou, 2010; Amornkul and Henning, 2007). Should it be considered or believed that application of such technologies might have a deleterious impact on the desired organoleptic profiles of finished products; manufacturers might wish to consider whether usage of adjunct starter cultures might be at least somewhat restorative of such (Awad et al., 2010; Bouton et al., 2009).

3.3 SEAFOOD

Pathogenic bacteria including *S. aureus* can be introduced while processing raw fish at various points including the environment, food handlers, utensils and equipment, and cross-contamination between raw and cooked products. The major control step to manage contamination levels is to minimize possible bacterial impurities and maintain temperatures to avoid outgrowth of *S. aureus* as outlined in published guidelines.

As an example, the US FDA's Fish and Fishery Products Hazards and Controls Guidance document, the most current version of which is the Fourth Edition, published in 2011, is presented herewith. This document contains useful guidance on controlling the outgrowth of *S. aureus*. The FDA specifically identifies SEs as a potential process-related hazard for drying of seafood and for batters. The FDA consequently discusses SE formation in hydrated batter mixes, as well as pathogenic bacteria outgrowth and toxin formation because of inadequate drying, including specific discussion of SEs. Hence, it discusses pathogenic bacteria outgrowth and toxin formation because of time and temperature abuse providing particular useful examples of control strategies for unrefrigerated processing control. Additional useful guidance on time and temperature management for controlling outgrowth of *S. aureus* as well as the limiting conditions for *S. aureus* growth and enterotoxin formation are also provided. Hence, FDA's limits for both *S. aureus* and SEs in all fish—as described above and presented in the tables of the Annex—are included in the document.

4. MICROBIAL RISK ASSESSMENT PLANS TO REDUCE *STAPHYLOCOCCUS AUREUS* CONTAMINATION

The microbial risk analysis literature review includes many factors but starts with knowledge of the industry and the processes used to prepare the food. Once this has been recorded the next step would most likely be a literature search. Most risk assessment plans aimed at reducing staphylococcal contamination should contemplate factors necessary for planktonic growth and enterotoxin production in controlled laboratory experiments. These experiments demonstrate the effects of variables such as nutrients, temperature, and pH (Schmitt et al., 1990; Sutherland et al., 1994). Additionally, valuable information is published regarding the growth of *S. aureus* in different food matrices such as milk and cheese (Kűmmel et al., 2016; Lindqvist et al., 2002) or tuna (Kataoko et al., 2016). Thirdly the risk assessment authors must be aware of the persistence of staphylococci on inanimate surfaces (Gutiérrez et al., 2012; Kramer et al., 2006; Kusumaningrum et al., 2003; Todd et al., 2009) and the carriage rate of humans (Argudín et al., 2012; Todd et al., 2008). The risk assessment that analyzes food for potential presence of staphylococci or SEs should also review outbreak investigations especially if reports appear for food products common to the industry of interest. For example, numerous reviews are available describing outbreaks (Hennekinne et al., 2012; Ikeda et al., 2005) that provide valuable information and lessons learned. Once there is a basic understanding of the food preparation the risk assessment authors contemplate practical concepts that focus on hand hygiene, clean equipment and utensils, temperature sensors, and barriers to prevent contamination such as gloves and sneeze guards, but education must accompany these measures. The authors must take notice of reports of food handlers who wear gloves, but fail to wash their hands. Education will help them to become aware of the fact that staphylococci can remain viable on gloves and inanimate objects for extended periods permitting further contamination of the environment and/or food products (Todd et al., 2010). Reports of enterotoxigenic antimicrobial resistant strains have been recovered from gloves, food handlers, and food (Argudín et al., 2012), which clearly demonstrates the necessity of hand hygiene training. Please refer to Chapter 12 of this book for further details on hygiene principles to avoid contamination/cross-contamination in the kitchen environment and during food processing.

Special consideration for a clean environment will be essential to reduce the risk of staphylococcal contamination. This ought to include equipment and utensils plus food preparation areas and nonfood preparation areas. Studies have shown that staphylococci can remain viable on surfaces for 4 days and can readily be transferred to food (Kusumaningrum et al., 2003). Special care must be taken in food-processing plants to prevent areas that may be difficult to clean because the lack of diligence can result in formation of biofilms harboring virulent staphylococci communities that can then serve as a reservoir of contamination (Gutiérrez et al., 2012). It may be wise to survey the food preparation areas and raw ingredients routinely to ensure continued food safety. Once food preparation is initiated the microbial risk assessment will be challenged with real-world scenarios that require periodic evaluation to ensure the plan remains relevant.

5. SUMMARY AND CONCLUSION

Microbial risk analysis has become an important tool for public health officials to use when establishing food safety policy, food safety standards, and regulations designed to protect the consumers, but risk analysis is equally important for the food industry in identifying and assessing hazards to avoid safety issues, especially with the wide variety of new ingredients used, for example, to prepare exotic cuisine options. As will be seen within the standards cited here, newer food preparations may not appear in existing standards and policies established by the national regulatory agencies. Manufacturing firms and food producers should therefore be encouraged to consider an in-depth analysis of the potential hazards, which may exist for their specific foods and their manufacturing environment, and develop preventive controls for the hazards identified, which are reasonably likely to occur to help avoid adverse events.

We have not determined the basis used in establishing the international standards described here, but clearly, many different approaches have been used to assess risk and thereby establish standards. International commerce could benefit if international microbiological standards were established, which are acceptable to the majority of countries trading in foodstuffs.

ACKNOWLEDGMENTS

This project would not have been possible without our fellow experts who happily provided us with the information required for a truly global perspective. Please accept our appreciation for your efforts.

Within FDA:

Dr. Andrew Yeung, Dr. Insook Son, Dr. Socrates Trujillo, Layla Batarseh, Kenneth Nieves

Within USDA:

Kolade Osho, Rachel Vanderberg, Robert Macke

External:

India: Sunil Bakshi (National Dairy Development Board)

New Zealand: Dr. William Jolly, (Ministry for Primary Industries, New Zealand)

Canada: Alison Orr, Wassim Khoury (Canadian Food Inspection Agency)

Japan: Dr. Yuki Morita (Yakult)

South Africa: Guy Kebble (SurePure), Shirley Parring (National Department of Health, South Africa)

ANNEX

Table 1 Parameter Based on Countries That Screen Food Samples for Coagulase-Positive Staphylococci

Country	Food	n	c	m	M
Australia and New Zealand	Butter made from unpasteurized milk and/or unpasteurized milk products	5	1	10	100
Australia and New Zealand	Packaged cooked cured/salted meat	5	1	100	1000
Australia and New Zealand	All comminuted fermented meat that has not been cooked during the production process	5	1	1000	10,000
Australia and New Zealand	Cooked crustacean	5	2	100	1000
Australia and New Zealand	Raw crustacean	5	2	100	1000
Australia and New Zealand	Powdered infant formula products	5	1	0	10/g
European Union	Cheeses made from raw milk	5	2	10,000/g	100,000/g
EU	Cheeses made from milk that has undergone a lower heat treatment than pasteurization and ripened cheeses made from milk or whey that has undergone pasteurization or a stronger heat treatment	5	2	100/g	1000/g
EU	Unripened soft cheeses (fresh cheeses) made from milk or whey that has undergone pasteurization or a stronger heat treatment	5	2	10/g	100/g
EU	Milk powder and whey powder	5	2	10/g	100/g
EU	Shelled and shucked products of cooked crustaceans and molluscan shellfish	5	2	100/g	1000/g
Turkey	Butter and spreadable dairy products and clarified butter	5	2	100/g	1000/g
Turkey	Clotted cream	5	2	100/g	1000/g
Turkey	Milk powder or cream powder, powder mix for ice cream, whey cheese, buttermilk powder and milk-based products, casein, and caseinate	5	2	100/g	1000/g
Turkey	Cheese (all cheese except processed cheese)	5	2	100/g	1000/g
Turkey	Cured and dried bacon, etc.	5	2	100/g	10,000/g
Turkey	Pasta, ravioli and similar products stuffed with meat, vegetables, and other fillings (raw, frozen)	5	2	1000/g	10,000/g
Turkey	Pasta, ravioli and similar products stuffed with meat, vegetables, and other fillings (baked)	5	2	100/g	1000/g
Turkey	Pizza, dough, and dough-based products (frozen, ready-to-cook)	5	2	1000/g	10,000/g

Table 1 Parameter Based on Countries That Screen Food Samples for Coagulase-Positive Staphylococci—cont'd

Country	Food	n	c	m	M
Turkey	Pies and cakes (creamy, chocolate, stuffed, fruit, etc.)	5	2	100/g	1000/g
Turkey	Spices, herbs and/or mixtures thereof (powder, pastes, mixtures, etc.)	5	2	1000/g	10,000/g
Turkey	Mayonnaise and salad dressing containing mayonnaise	5	2	100/g	1000/g
Turkey	Cheeses produced from raw milk	5	2	10,000/g	100,000/g
Turkey	Cheeses produced from heat-treated milk at lower temperatures than pasteurization and seasoned cheeses produced from pasteurized or heat-treated milk at higher temperatures	5	2	100/g	10,000/g
Turkey	Unseasoned cheeses produced from pasteurized or heat-treated milk at higher temperatures (fresh cheeses)	5	2	10/g	100/g
Turkey	Milk powder and whey powder	5	2	10/g	100/g
Turkey	Shelled and shucked products of cooked crustaceans and mollusks	5	2	100/g	1000/g

c, *acceptance number of samples;* M, *maximum tolerance;* m, *minimum tolerance;* n, *total number of samples.*

Table 2 Parameter Based on Countries That Screen Food Samples for Coagulase-Positive *Staphylococcus aureus*

Country	Food	Maximum Permissible Level	n	c	m	M
India	Pasteurized milk/cream/flavored milk		5	0		<10/g
India	Sweetened condensed milk		5	2	10/g	100/g
India	Pasteurized butter		5	2	10/g	50/g
India	Dried powder, milk powder, cream, whey, edible, casein, ice cream mix		5	0		<10/g
India	Ice cream, frozen dessert, milk lolly, ice candy		5	0		<10/g
India	Processed cheese/cheese spread		5	0		<10/g
India	All other cheeses		5	2		
India	Yogurt, dahi, chakka shrikhand		5	2	50/g	100/g
India	Paneer/chhana		5	2	50/g	100/g
India	Khoya		5	2	50/g	100/g
South Africa	Desiccated coconut	NBT				
South Africa	Cooked seawater and freshwater foods such as prawns, shrimps, crayfish, lobsters, crab meat, oysters, mussels, clams, eels, or fish	NBT				

c, *acceptance number of samples;* M, *maximum tolerance;* m, *minimum tolerance;* n, *total number of samples;* NBT, *negative-by-test.*

Table 3 Parameter Based on Countries That Screen Food Samples for *Staphylococcus aureus*

Country	Food	Sub-type	Maximum Permissible Level	n	c	m	M
Canada	Cheese from pasteurized milk			5	2	100	10,000
Canada	Cheese from unpasteurized milk			5	2	1000	10,000
Canada	Instant infant cereal and powdered infant formula			10	1	10	100
Canada	Fresh and dry alimentary paste (noodles, spaghetti and macaroni, etc.)			5	2	500	10,000
Canada	Bakery products			5	2	100	10,000
Canada	Heat-treated fermented sausage			5	1	50	10,000
Canada	Raw fermented sausage			5	1	250	10,000
Canada	Nonfermented ready-to-eat (RTE) sausage			5	2	100	10,000
Canada	Deboned poultry products			5	1	100	10,000
Canada	Dry mixes (gravy, sauce, soup), heat, and serve			5	2	100	10,000
Canada	Soybean products (RTE)			5	2	100	10,000
Canada	Spices (RTE)			5	2	100	10,000
Canada	Raw organ-derived products and herbal products			5	2	100	10,000
Canada	Powdered protein, meal replacements, and dietary supplements			5	2	10	100
China	Meat products	Cooked products		5	1	100	1000
China	Meat products	RTE uncooked products		5	1	100	1000
China	Seafood products	Cooked products		5	1	100	1000
China	Seafood products	RTE uncooked products		5	1	100	1000
China	Seafood products	RTE seaweed product		5	1	100	1000
China	Grain products	Cooked products (including baked category)		5	1	100	1000
China	Grain products	Cooked wheat and rice products with filling		5	1	100	1000
China	Grain products	Instant noodle products		5	1	100	1000

China	RTE soy products	Fermented soy products		5	1	100	1000
China	RTE soy products	Nonfermented soy products		5	1	100	1000
China	RTE fruit and vegetable products including pickled vegetable category			5	1	100	1000
China	Beverages (except packaged drinking water and carbonated drink)			5	1	100	1000
China	Frozen drinks	Ice cream category		5	1	100	1000
China	Frozen drinks	Edible ice, popsicle category		5	1	100	1000
China	RTE seasonings	Soy sauce		5	2	100	10,000
China	RTE seasonings	Paste and paste products		5	2	100	10,000
China	RTE seasonings	Seafood-based seasonings		5	2	100	10,000
China	RTE seasonings	Seasonings (salad dressing, etc)		5	2	100	10,000
China	Fermented alcoholic beverages and its blended alcoholic beverages			5	0	0/25 mL	
China	Cheese			5	2	100	1000
China	Whey powder and whey protein powder			5	2	10	100
China	Evaporated milk, sweetened condensed milk, and formulated condensed milk			5	0	0/25 mL	
China	Honey					0/25g	
China	Quick-frozen wheat and rice products			5	1	100	1000
China	Fermented milk			5	0	0/25g (mL)	
China	Milk powder			5	2	10	100
China	Pasteurized milk			5	0	0/25g (mL)	
China	Cream, butter, and anhydrous milkfat			5	1	10	100

Continued

Table 3 Parameter Based on Countries That Screen Food Samples for *Staphylococcus aureus*—cont'd

Country	Food	Sub-type	Maximum Permissible Level	n	c	m	M
China	Modified milk			5	0	0/25g (mL)	
China	Processed cheese			5	2	100	1000
China	Infant formula			5	2	10	100
China	Special medical purposes formulated foods for infant			5	2	10	100
China	Special medical purposes formulated foods			5	2	10	100
China	Health foods					≤0/25g	
China	Raffinose		less than 3.0 (MPN/g)				
China	Collagen casings			5	1	100	1000
China	Sports nutrition foods			5	2	10	100
Eurasian Customs Union	Food blood sausage products, head cheese, jellied meat products, quick-frozen poultry dishes, pasteurized liquid egg products, frozen ones, dry ones, dry mixtures for omelet		NBT				
Eurasian Customs Union	Quick-frozen meat dishes, pancakes with meat filling, pastes from liver and/or meat, semismoked sausages from poultry, freeze- dried and heat dried chicken minced meat		NBT				
Eurasian Customs Union	Dried products from poultry meat		NBT				
Eurasian Customs Union	Semipreserved, soft-salted foods from fish, hot and cold-smoked fish, culinary roe products cooked frozen products, shell-fish meat, Sturgeon roe, salted salmon roe, roe of other species dried nonfinfish		NBT				
Russia and Eurasian Customs Union	Chilled and frozen fish products		NBT				

Eurasian Customs Union	Raw fish, live fish, shellfish, and other invertebrates		NBT				
Eurasian Customs Union	Bakery products with fillings		NBT				
Eurasian Customs Union	Quick-cooking macaroni products with milk-based additives		NBT				
Eurasian Customs Union	Cakes and pastries with decorations (with custard), sponge rolls with filling		NBT				
Eurasian Customs Union	Cakes and pastries with creamy decorations, cupcakes and rolls in hermetically sealed packages, biscuits with creamy fillings		NBT				
Eurasian Customs Union	Cakes and pastries with decorations including frozen ones (if shelf life is less than 5 days)		NBT				
Eurasian Customs Union	Dried fruit and vegetable desserts		NBT				
Eurasian Customs Union	Salted, smoked, smoke-grilled products from pork fat back, and pork brisket		NBT				
Russia and Eurasian Customs Union	Soy-based drinks, soy protein products (tofu)		NBT				
Eurasian Customs Union	Soy protein isolates, soya flour, texturized soya protein concentrate, dried yeast, cold soups		NBT				
Eurasian Customs Union	Liquid concentrated biologically active additives		NBT				
Eurasian Customs Union	Bulk biologically active additives		NBT				
Eurasian Customs Union	Bulk biologically active additives with amino acids, microelements, mono- and diglycerides		NBT				
Eurasian Customs Union	Products for premature and/or low-birth-weight infants		NBT				
Eurasian Customs Union	Low-lactose and lactose-free products, bulk milk high protein products		NBT				

Continued

Table 3 Parameter Based on Countries That Screen Food Samples for *Staphylococcus aureus*—cont'd

Country	Food	Sub-type	Maximum Permissible Level	n	c	m	M
Eurasian Customs Union	Low-protein products (starches, cereals) and macaroni products		NBT				
Gulf Cooperation Council	Fermented milk products, e.g., yogurt, laban, labena			5	2	10/g (mL)	100/g (mL)
Gulf Cooperation Council	Condensed and sweetened condensed milk			5	1	5/g (mL)	10/g (mL)
Gulf Cooperation Council	Whipped cream			5	1	10/g (mL)	100/g (mL)
Gulf Cooperation Council	Fermented cream			5	1	10/g (mL)	100/g (mL)
Gulf Cooperation Council	Powdered milk, whey			5	1	10/g (mL)	100/g (mL)
Gulf Cooperation Council	Soft cheese (made from pasteurized milk)			5	1	100/g (mL)	1000/g (mL)
Gulf Cooperation Council	Hard and semihard cheese			5	2	100/g (mL)	1000/g (mL)
Gulf Cooperation Council	Processed cheese packed in nonmetal containers			5	1	10/g (mL)	100/g (mL)
Gulf Cooperation Council	Caseinate			5	1	10/g (mL)	100/g (mL)
Gulf Cooperation Council	Edible ices, ice cream, ice milk, water ice			5	1	10/g (mL)	100/g (mL)
Gulf Cooperation Council	Milkshakes			5	2	10/g (mL)	100/g (mL)
Gulf Cooperation Council	Biscuits			5	1	10/g (mL)	100/g (mL)
Gulf Cooperation Council	Shelf-stable dried biscuits coated or filled with chocolate or others			5	1	10/g (mL)	100/g (mL)
Gulf Cooperation Council	Dried and instant products requiring reconstitution			5	0	0	
Gulf Cooperation Council	Cereal-based foods for infants			5	1	10/g (mL)	100/g (mL)
Gulf Cooperation Council	Powdered infant formula, including those with lactic acid producing cultures			5	1	10/g (mL)	100/g (mL)

Gulf Cooperation Council	Dietetic foods to be eaten by high risk category			10	1	10/g (mL)	100/g (mL)
Gulf Cooperation Council	Body-building foods			5	0	0	
Gulf Cooperation Council	Raw minced meat and poultry; chilled/frozen			5	2	100/g (mL)	1000/g (mL)
Gulf Cooperation Council	Raw minced/pieces of meat (chilled/frozen) with soy or marinated (kubba; meatballs, fresh sausage, meat burgers)			5	2	500/g (mL)	1000/g (mL)
Gulf Cooperation Council	Cured and/or smoked meat; mortadella luncheon meat, basturma			5	2	500/g (mL)	5000/g (mL)
Gulf Cooperation Council	Cured and/or smoked poultry meat; mortadella frankfurters, turkey, smoked turkey breast			10	2	1000/g (mL)	10,000/g (mL)
Gulf Cooperation Council	Cooked sausages			5	1	100/g (mL)	1000/g (mL)
Gulf Cooperation Council	Cooked poultry meat, e.g., prepared frozen meals, chicken burgers, chicken/turkey rolls			5	1	1000/g (mL)	10,000/g (mL)
Gulf Cooperation Council	Dehydrated meat or meat components; protein concentrates from meats			5	3	100/g (mL)	1000/g (mL)
Gulf Cooperation Council	Vacuum-packed semipreserved but perishable meat and poultry products			5	2	100/g (mL)	1000/g (mL)
Gulf Cooperation Council	Raw (chilled/frozen) crustaceans			5	2	100/g (mL)	1000/g (mL)
Gulf Cooperation Council	Frozen/chilled breaded fish, crustaceans, and mollusks			5	1	1000/g (mL)	10,000/g (mL)
Gulf Cooperation Council	Smoked fish			5	2	1000/g (mL)	10,000/g (mL)
Gulf Cooperation Council	Dried seafood			5	1	100/g (mL)	1000/g (mL)
Gulf Cooperation Council	Salted and/or fermented fish			5	1	10/g (mL)	100/g (mL)
Gulf Cooperation Council	Cooked (chilled/frozen) crustaceans, molluscans			5	1	100/g (mL)	1000/g (mL)

Continued

Table 3 Parameter Based on Countries That Screen Food Samples for *Staphylococcus aureus*—cont'd

Country	Food	Sub-type	Maximum Permissible Level	n	c	m	M
Gulf Cooperation Council	Pudding with egg (powders)			5	1	10/g (mL)	1000/g (mL)
Gulf Cooperation Council	Egg mix dehydrated			5	0	10/g (mL)	
Gulf Cooperation Council	Dried cake mixes with high egg content			5	1	100/g (mL)	1000/g (mL)
Gulf Cooperation Council	Butter (salted and unsalted)			5	0	0	
Gulf Cooperation Council	Ghee (butter oil), fats from milk			5	1	0	10/g (mL)
Gulf Cooperation Council	Margarine			5	0	0	
Gulf Cooperation Council	Mayonnaise, mustard, salad sauce, and other sauces			5	1	10/g (mL)	100/g (mL)
Gulf Cooperation Council	Dried herbs and spices, ready-to-eat (RTE) herbs and spices			5	1	100/g (mL)	1000/g (mL)
Gulf Cooperation Council	Starch and starch-containing products (custard)			5	2	10/g (mL)	100/g (mL)
Gulf Cooperation Council	Special breads sweetened with egg, milk			5	1	10/g (mL)	100/g (mL)
Gulf Cooperation Council	Cakes and bakery products			5	1	10/g (mL)	100/g (mL)
Gulf Cooperation Council	Cakes, desserts, and bakery products (frozen or dehydrated)			5	2	10/g (mL)	100/g (mL)
Gulf Cooperation Council	Malt, malt derivatives			5	1	100/g (mL)	1000/g (mL)
Gulf Cooperation Council	Fresh fruit and vegetables (precut and crudités) to be consumed raw			5	2	100/g (mL)	1000/g (mL)
Gulf Cooperation Council	Dehydrated desserts (bonbons, caramels, and similar products)			5	2	10/g (mL)	1000/g (mL)
Gulf Cooperation Council	Arabic sweets			5	0	0	

Gulf Cooperation Council	Gelatine			5	1	100/g (mL)	1000/g (mL)
Gulf Cooperation Council	Drink powder (dry)			5	0	0	
Gulf Cooperation Council	Sterilized soya drink			5	0	0	
Gulf Cooperation Council	Sandwiches and filled rolls without salad			5	1	20/g (mL)	100/g (mL)
Gulf Cooperation Council	Coleslaw (cabbage)			5	1	100/g (mL)	10,000/g (mL)
Gulf Cooperation Council	Sandwiches and filled rolls with cheese (RTE meals, pasta, pizza)			5	1	20/g (mL)	100/g (mL)
Gulf Cooperation Council	Rice			5	1	20/g (mL)	100/g (mL)
Gulf Cooperation Council	Tofu (not ultraheat treated)			5	2	100/g (mL)	1000/g (mL)
Gulf Cooperation Council	Sesame seed products (Tahini, Halwa)			5	1	10/g (mL)	100/g (mL)
India	Frozen shrimps or prawns	Raw	100/g				
India	Frozen shrimps or prawns	Cooked	NBT/25g				
India	Frozen lobsters	Raw	100/g				
India	Frozen lobsters	Cooked	NBT/25g				
India	Frozen squid		100/g				
India	Frozen finfish		100/g				
India	Frozen fish fillets, minced fish flesh, or mixtures		100/g				
India	Dried shark fins		100/g				
India	Salted fish/dried salted fish		100/g				
India	Canned finfish		NBT/25g				
India	Canned shrimp		NBT/25g				
India	Canned sardines or sardine-type products		NBT/25g				
India	Canned salmon		NBT/25g				
India	Canned crab meat		NBT/25g				
India	Canned tuna and Bonito		NBT/25g				

Continued

Table 3 Parameter Based on Countries That Screen Food Samples for *Staphylococcus aureus*—cont'd

Country	Food	Sub-type	Maximum Permissible Level	n	c	m	M
India	All fruits and vegetable products and ready to serve beverages including fruit beverages and synthetic products, table olives, raisins, pistachio nuts, dates, dry fruits and nuts, vinegars		NBT/25g (mL)				
Japan	Not heat-treated meat products		1000 CFU/g				
Japan	Heat-treated meat products		1000 CFU/g				
Japan	Heat-treated meat products sterilized after packaging		1000 CFU/g				
Japan	Lunch boxes, prepared side dishes (such as omelets, fried foods)		NBT				
Japan	Fresh pasta, boiled pasta, heat-treated toppings, and other heat-treated ingredients (such as tempura and soup)		NBT				
Japan	Fresh cakes and pastries		NBT				
Mexico	Infant formula		NBT				
Mexico	Dry-salted and smoked fish products		500 CFU/g				
Mexico	Fresh cheese		1000 CFU/g				
Mexico	Aged cheeses		100 CFU/g				
Mexico	Processed cheese		<100 CFU/g				
Mexico	Beef		1000 CFU/g				
Mexico	Shellfish, fresh, refrigerated, and frozen		1000 CFU/g				
Mexico	Fish, fresh, refrigerated, and frozen		1000 CFU/g				
Mexico	Butter		<100 CFU/g				
Mexico	Cream (not including ultra-heat–treated or sterilized)		<100 CFU/g				
Mexico	Sweetened condensed milk		<100 CFU/g				
Mexico	Fermented and acidified milk products		<100 CFU/g				
Mexico	Milk-based sweets		<100 CFU/g				

Mexico	Milk-based infant formula and mixed milk products, pasteurized products		<10 CFU/g				
Mexico	Eggs and egg products		<100 CFU/g				
Mexico	Pan dulce		<100 CFU/g				
Mexico	Pasteles		100 CFU/g				
Mexico	Pays		100 CFU/g				
Peru	RTE nonacidified milk–based desserts (flans, puddings)			5	1	10/g	100/g
Peru	Unaged cheeses (queso fresco, soft cheese, ricotta, cream, petit Swiss)			5	2	10/g	100/g
Peru	Aged cheeses (Camembert, Roquefort, Brie, Gorgonzola, Emmental, Gruyere, Cheddar)			5	1	10/g	100/g
Peru	Processed cheeses			5	1	10/g	100/g
Peru	Milk-based ice creams			5	1	10/g	100/g
Peru	Desserts with milk ice cream base			5	1	10/g	100/g
Peru	Butters and margarines			5	2	10/g	100/g
Peru	Dried soups, broths, creams, salsas, and instant mashed potatoes			5	1	10/g	100/g
Peru	Instant dry mixes (gelatins, jellies, creams)			5	1	10/g	100/g
Peru	Dry mixtures requiring cooking (puddings, flans, etc.)			5	1	10/g	100/g
Peru	Pastas and fresh doughs and/or precooked without filling, refrigerated, or frozen			5	1	100/g	1000/g
Peru	Pastas and doughs, fresh and/or precooked with filling, refrigerated, or frozen			5	1	100/g	1000/g
Peru	Dried noodles or pasta with or without filling			5	1	100/g	1000/g
Peru	Soft or hard confectionery nougat, cereal bars			5	1	10/g	100/g

Continued

Table 3 Parameter Based on Countries That Screen Food Samples for *Staphylococcus aureus*—cont'd

Country	Food	Sub-type	Maximum Permissible Level	n	c	m	M
Peru	Bakery products with or without filling and/or frosting that do not require refrigeration (bread, enriched cookies or breads, cupcakes, muffins, panettone, pancakes)			5	1	10/g	100/g
Peru	Sweet or savory bakery products that require refrigeration (cakes, tarts, empanadas, etc.)			5	1	10/g	100/g
Peru	Infant powders (infant formulas and breast milk substitutes)			5	1	<3/g	10/g
Peru	Instantized cooked products for infants between 6 and 36 months (purees and similar)			5	1	10/g	100/g
Peru	Instantized cooked products such as fortified milk products, milk substitutes, fortified blends, etc.			5	1	10/g	100/g
Peru	Dietary products needing reconstitution before consumption			5	1	<3/g	10/g
Peru	Dietary products needing cooking before consumption			5	1	<3/g	10/g
Peru	RTE dietary products not included in other categories			5	1	<3/g	10/g
Peru	Precooked frozen fowl meat needing reheating before consumption			5	1	1000/g	10,000/g
Peru	Ground or chopped raw meat			5	2	100/g	1000/g
Peru	Refrigerated or frozen meat (hamburgers, breaded meat, croquettes, and other dressing covered meats)			5	1	100/g	1000/g
Peru	Dried, dry-salted meats			5	1	100/g	1000/g
Peru	Cold cuts (chorizo, franks, etc) and cured meats (cooked and raw hams, prosciutto, etc.)			5	1	100/g	1000/g
Peru	Aged raw cold cuts (salami, sausage, etc.)			5	1	10/g	100/g

Peru	Heat-treated cold cuts (cured: English ham, bacon, ribs, pork chops, etc., blanched: hot dogs, Spanish sausage, etc., cooked: pork cheese, blood sausage, pork rinds, pate, etc.)			5	1	10/g	100/g
Peru	Raw fish and shellfish products (fresh, refrigerated, frozen, salted, cold-smoked)			5	2	100/g	1000/g
Peru	Precooked and cooked fish and shellfish products (frozen or refrigerated) RTE (final product)			5	1	100/g	1000/g
Peru	Raw mollusks and crustaceans (fresh, refrigerated, and frozen)			5	2	100/g	1000/g
Peru	Precooked and cooked mollusks and crustaceans (refrigerated or frozen)			5	2	300/g	1000/g
Peru	Heat-smoked fish and shellfish products			5	1	10/g	100/g
Peru	Frozen breaded raw fish and shellfish products			5	2	100/g	1000/g
Peru	Precooked and cooked frozen breaded fish and shellfish products			5	1	100/g	1000/g
Peru	Mayonnaise and other egg-based sauces			5	1	10/g	100/g
Peru	Prepared foods without heat treatment (fresh salads, potato sauce, dressings, juice, home-style yoghurt, etc.), prepared foods made with ingredients with or without heat treatment (mixed salads, stuffed avocado, sandwiches, ceviche, desserts, soft drinks, etc.)			5	2	10/g	100/g
Peru	Heat-treated prepared foods (cooked salads, stews, rice dishes, cooked desserts, rice pudding, porridge, etc.)			5	1	10/g	100/g

Continued

Table 3 Parameter Based on Countries That Screen Food Samples for *Staphylococcus aureus*—cont'd

Country	Food	Sub-type	Maximum Permissible Level	n	c	m	M
Peru	Semipreserves with pH>4.6			5	1	10/g	100/g
South Africa	Partly cooked or uncooked seawater and freshwater foods such as prawns, shrimps, crayfish, lobsters, crab meat, eels, or fish		10 CFU/g				
South Africa	Cooked poultry		NBT				
South Africa	Dried spices and aromatic plants		NBT				
South Africa	Pasteurized or irradiated egg products		NBT				
South Africa	Milk and dairy products		NBT				
South Korea	Bread or rice cake (applicable only to cream buns)		NBT				
South Korea	Seasoned dried shellfish and fish fillet		<100/g				
South Korea	RTE foods		<100/g				
South Korea	Unheated and unprocessed seafood products		<100/g				
South Korea	Cooked ham and sausage			5	1	10/g	100/g
South Korea	Cooked ham and sausage	(Pasteurized)	NBT	5	0	0/25g	
South Korea	Raw meat (excluding raw material to be used for processing) and livestock products that are eaten without pasteurization, sterilization, or further processing, including cooking		NBT	5	0	0/25g	
The United States	Most FDA-regulated foods, with the exception of low-acid canned food		10,000 CFU/g				

c, *acceptance number of samples;* CFU, *colony-forming units;* M, *maximum tolerance;* m, *minimum tolerance;* MPN, *most probable number;* n, *total number of samples;* NBT, *negative-by-test;* RTE, *Ready-to-eat.*

Table 4 Parameter Based on Countries That Screen Food Samples for Staphylococcal Enterotoxins

Country	Food	Maximum Permissible Level	n	c
Russia and Eurasian Customs Union	Cheeses and cheese products, cheese pastes, including for baby food; dry baby food products on the basis of milk (except for dry cereals), including specialized food products for dietary therapeutic nutrition	NBT		
European Union	Cheeses,[a] milk powder, whey powder	NBT/25g	5	0
Turkey	Processed cheeses and processed cheese products	NBT/25g	5	0
Turkey	All types of ready-to-eat (cooked) meat and vegetable meals, etc.	NBT/25g	5	0
Turkey	All types of RTE salads, delicatessen products and cold side dishes, etc.	NBT/25g	5	0
Turkey	All types of RTE (cooked) baked products (pasta, all varieties of pastry, pizza, ravioli, etc.)	NBT/25g	5	0
Turkey	All types of RTE (cooked) sweet pudding, cream, Noah's pudding, etc.	NBT/25g	5	0

[a]*Including three different categories (cheese made from raw milk; cheese made from milk that has undergone a lower heat treatment than pasteurization and ripened cheese made from milk or whey that has undergone pasteurization or a stronger heat treatment; unripened soft cheese made from milk or whey that has undergone pasteurization or a stronger heat treatment) as described before—see Table 1.*
c, *acceptance number of samples;* n, *total number of samples;* NBT, *negative-by-test.*

REFERENCES

21 U.S.C. Chapter 9, Section342. https://www.law.cornell.edu/uscode/text/21/342.

Aarestrup, F.M., Andersen, J.K., Jensen, N.E., 1995. Lack of *staphylococcal* enterotoxin production among strains of *Staphylococcus aureus* from bovine mastitis in Denmark. Acta Vet. Scand. 36, 273–275.

Adesiyun, A.A., 1995. Characteristics of *Staphylococcus aureus* isolates isolated from bovine mastitic milk: bacteriophage and antimicrobial agent susceptibility and enterotoxigenicity. Zentralbl. Veterinaermed. B 42, 129–139.

American Meat Institute Foundation, 1997. Good Manufacturing Practices for Fermented Dry and Semi-Dry Sausage Products. https://meathaccp.wisc.edu/Model_Haccp_Plans/assets/GMP%20Dry%20Sausage.pdf.

Amornkul, Y., Henning, D.R., 2007. Utilization of microfiltration or lactoperoxidase system or both for manufacturing of cheddar cheese from raw milk. J. Dairy Sci. 90, 4988–5000.

Argudín, M.A., Mendoza, M.C., González-Hevia, M.A., Bances, M., Guerra, B., Rodicio, M.R., 2012. Genotypes, exotoxin gene content, and antimicrobial resistance of *Staphylococcus aureus* strains recovered from foods and food handlers. Appl. Environ. Microbiol. 78 (8), 2930–2935. http://doi.org/10.1128/AEM.07487-11.

Asperger, H., Zangeri, P., 2011. *Staphylococcus aureus* – Dairy, Encyclopedia of Dairy Sciences, second ed. pp. 111–116.

Australia New Zealand Food Standards Code Schedule 27 Microbiological limits in food. http://www.foodstandards.gov.au/code/Documents/Sched%2027%20Micro%20limits%20v157.pdf.

Awad, S., Ahmed, N., El Soda, M., 2010. Influence of microfiltration and adjunct culture on quality of domiati cheese. J. Dairy Sci. 93, 1807–1814.

Basanisi, M.G., Nobili, G., La Bella, G., Russo, R., Spano, G., Normanno, G., et al., 2016. Molecular characterization of *Staphylococcus aureus* isolated from sheep and goat cheeses in southern Italy. Small Rumin. Res. 135, 17–19.

Borneman, D.L., Ingham, S.C., Ane, C., 2009. Predicting growth/no-growth of *Staphylococcus aureus* on vacuum-packaged ready-to-eat meats. J. Food Prot. 72, 539–548.

Bouton, Y., Buchin, S., Duboz, G., Pochet, S., Beuvier, E., 2009. Effect of mesophilic lactobacilli and enterococci adjunct cultures on the final characteristics of a microfiltered milk Swiss-type cheese. Food Microbiol. 26, 183–191.

Buchanan, R.L., Smith, J.L., Long, W., 2000. Microbial risk assessment: dose-response relations and risk characterization. Int. J. Food Microbiol. 58, 159–172.

Bulletin of the International Dairy Federation, 2006. *Staphylococcus Aureus* Intramammary Infections, 408.

Burnham, G.M., Hanson, D.J., Koshick, C.M., Ingham, S.C., 2008. Death of *Salmonella* serovars, *Escherichia coli* O157:H7, *Staphylococcus aureus*, and *Listeria* monocytogenes during the drying of meat: a case study using biltong and droewors. J. Food Saf. 28, 198–209.

China Standards GB 29921-2013: Meat Products (cooked, RTE uncooked products), Seafood Products (cooked, RTE uncooked, RTE seaweed products), Grain Product (cooked-including baked category, cooked wheat and rice products with filling, instant noodle products), RTE Soy products (fermented soy products, non-fermented soy products), RTE Fruit and Vegetable Products (including pickled vegetable category), Beverages (except packaged drinking water and carbonated drink), Frozen drinks (ice cream category, edible ice, popsicle category), RTE seasonings (soy sauce, paste and paste products, seafood based seasonings, seasonings), GB 2758-2012: Fermented alcoholic beverages and its blended alcoholic beverages, GB 5420-2010: Cheese, GB 11674-2010: Whey powder and whey protein powder, GB 13102-2010: Evaporated milk, sweetened condensed milk and formulated condensed milk, GB 14693-2011: Honey, GB 19295-2011: Quick frozen wheat and rice products, GB 19302-2010: Fermented milk, GB 19644-2010: Milk Powder, GB 19645-2010: Pasteurized milk, GB 19646-2010: Cream, butter and anhydrous milkfat, GB 25191-2010: Modified milk, GB 25192-2010: Processed cheese, GB 10765-2010: Infant formula, GB 25596-2010: Special medical purposes formulated foods for infant, GB 29922-2013: Special medical purposes formulated foods, GB 16740-2014: Health foods, GB 31618-2014: Raffinose, GB 14967-2015: Collagen casings, GB 24154-2015: Sports nutrition food.

CIA World Fact Book. https://www.cia.gov/library/publications/the-world-factbook/rankorder/2119rank.html.

Commission Regulation (EC) No 2073/2005 on Microbiological Criteria for Foodstuffs. November 15, 2005. http://eur-lex.europa.eu/legal-content/EN/TXT/HTML/?uri=CELEX:32005R2073&from=EN.

DeSantis, E., Mureddu, A., Mazzette, R., Scarano, C., Bes, M., 2005. Detection of enterotoxins and virulence genes in *Staphylococcus aureus* strains isolated from sheep with subclinical mastitis. In: Hogeveen, H. (Ed.), Mastitis in Dairy Production. Wageningen Academic Press Publishers, The Netherlands, pp. 504–510.

European Commission Biological Safety of the Food Chain, EU Regulation (EC) 854/2004. https://ec.europa.eu/food/safety/biosafety/food_hygiene/legislation_en.

Fagundes, H., Barchesi, L., Filho, A.N., Ferreira, L.M., Oliveira, C.A.F., 2010. Occurrence of *Staphylococcus aureus* in raw milk produced in dairy farms in São Paulo state, Brazil. Braz. J. Microbiol. 41 (2), 376–380. http://doi.org/10.1590/S1517-83822010000200018.

FDA Food Code. http://www.fda.gov/downloads/Food/GuidanceRegulation/RetailFoodProtection/FoodCode/UCM374510.pdf.

Ferreira, M.A., Bernardo, L.G., Neves, L.S., Campos, M.R., Lamaro-Cardoso, J., André, M.C., 2016. Virulence profile and genetic variability of *Staphylococcus aureus* isolated from artisanal cheese. J. Dairy Sci. 99, 8589–8597.

Food and Drug Administration Compliance Policy Guide for Dairy Products – Microbial Contaminants and Alkaline Phosphatase Activity, Section 527.300. http://www.fda.gov/downloads/ICECI/ComplianceManuals/CompliancePolicyGuidanceManual/UCM238480.pdf.

Food and Drug Administration Compliance Program Guidance Manual, Program 7303.803 (Domestic Food Safety). http://www.fda.gov/downloads/Food/ComplianceEnforcement/ucm072848.pdf.

Food and Drug Administration Compliance Program Guidance Manual, Program 7303.844 (Import Seafood Products). http://www.fda.gov/downloads/Food/ComplianceEnforcement/UCM219993.pdf.

Food and Drug Administration Compliance Program Guidance Manual, Program 7321.002 (Medical Foods). http://www.fda.gov/downloads/Food/ComplianceEnforcement/ucm073339.pdf.

Food and Drug Administration Compliance Program Guidance Manual, Program 7321.006 (Infant Formula). http://www.fda.gov/downloads/Food/ComplianceEnforcement/ucm073349.pdf.

Food and Drug Administration Compliance Program Manual, Program 7303.819 (Import Food Safety). http://www.fda.gov/downloads/Food/ComplianceEnforcement/ucm073108.pdf.

Food and Drug Administration Compliance Program Manual, Program, Program 7303.037 (Domestic and Imported Cheese and Cheese Products). http://www.fda.gov/downloads/Food/ComplianceEnforcement/FoodCompliancePrograms/UCM456592.pdf.

Food and Drug Administration, 2011. Fish and Fishery Products Hazards and Controls Guidance, fourth ed. http://www.fda.gov/downloads/Food/GuidanceRegulation/UCM251970.pdf.

FSIS Compliance Guideline for Meat and Poultry Jerky Produced by Small and Very Small Establishments. 2014. https://meathaccp.wisc.edu/doc_support/asset/Compliance-Guideline-Jerky-2014.pdf.

Gesan-Guiziou, G., 2010. Removal of bacteria spores and somatic cells from milk by centrifugation and microfiltration techniques. In: Griffiths (Ed.), Improving the Safety and Quality of Milk. Milk Production and Processing, vol. 1. CRC.

Gilmour, A., Harvey, J., 1990. Staphylococci in milk and milk products. Soc. Appl. Bacteriol. Symp. Ser. 19, 147S–166S.

Grade A Pasteurized Milk Ordinance, 2015. Revision.

Guidelines for Environmental health Officers on the Interpretation of Microbiological Analysis Data of Food, Republic of South Africa, Department of Health, Directorate: Food Control. https://foodsolutions2-public.sharepoint.com/SiteAssets/legislation-standards/Environmental%20health%20officers%20on%20the%20interpretation%20of%20microbiological%20analysis%20data%20of%20food.pdf.

Gutiérrez, D., Delgado, S., Vázquez-Sánchez, D., Martínez, B., Cabo, M.L., Rodríguez, A., García, P., 2012. Incidence of *Staphylococcus aureus* and analysis of associated bacterial communities on food industry surfaces. Appl. Environ. Microbiol. 78 (24), 8547–8554. http://doi.org/10.1128/AEM.02045-12.

Haeghebaert, S., Le Querrec, F., Gallay, A., Bouvet, P., Gomez, M., Vaillant, V., 2002. Les toxi-infections alimentaires collectives en France, en 1999 et 2000. Bull. Epidémiol. Hebd. 23, 105–109.

Hennekinne, J.A., De Buyser, M.L., Dragacci, S., 2012. *Staphylococcus aureus* and its food poisoning toxins: characterization and outbreak investigation. FEMS Microbiol. Rev. 36, 815–836.

Hoffmann, S., 2010. Ensuring food safety around the globe: the many roles of risk analysis from risk ranking to microbial risk assessment. Risk Anal. 30, 711–714. http://dx.doi.org/10.1111/j.1539-6924.2010.01437.x.

Hummerjohann, J., Naskova, J., Baumgartner, A., Graber, H.U., 2014. Enterotoxin-producing *Staphylococcus aureus* genotype B as a major contaminant in Swiss raw milk cheese. J. Dairy Sci. 97, 1305–1312.

Huonga, B.T.M., Mahmuda, Z.H., Neogic, S.B., Kassua, A., Nhiena, N.V., Mohammada, A., Yamatoa, M., Otaa, F., Lamb, N.T., Daob, H.T.A., Khanb, N.C., 2010. Toxigenicity and genetic diversity of *Staphylococcus aureus* isolated from Vietnamese ready-to-eat foods. Food Control 21, 166–171.

Ikeda, T., Tamate, N., Yamaguchi, K., Makino, S., 2005. Mass outbreak of food poisoning disease caused by small amounts of staphylococcal enterotoxins A and H. Appl. Environ. Microbiol. 71, 2793–2795.

India Food Safety and Standards Regulation. 2010. http://indiaenvironmentportal.org.in/files/finalregualtion.pdf.

International Commission on Microbiological Specifications for Foods (ICMSF), 1996. Microorganisms in Foods 5 – Microbiological Characteristics of Food Pathogens. Blackie Academic & Professional, NY, p. 148.

International Commission on Microbiological Specifications for Foods (ICMSF), 2002. Microorganisms in Foods 7. Microbiological Testing in Food Safety Management. Kluwer Academic/Plenum Publishers New York, New York.

Jakobsen, R.A., Heggebø, R., Sunde, E.B., Skjervheim, M., 2011. *Staphylococcus aureus* and *Listeria monocytogenes* in Norwegian raw milk cheese production. Food Microbiol. 28, 492–496.

Japan, Food Hygiene Regulation, 1959. Food Standards, Article 11.

Japan, Hygiene Norms, MHLW.

Johler, S., Weder, D., Bridy, C., Huguenin, M.C., Robert, L., Hummerjohann, J., Stephan, R., 2015. Outbreak of staphylococcal food poisoning among children and staff at a Swiss boarding school due to soft cheese made from raw milk. J. Dairy Sci. 98, 2944–2948.

Jørgensen, H.J., Mørk, T., Høgåsen, H.R., Rørvik, L.M., 2005. Enterotoxigenic *Staphylococcus aureus* in bulk milk in Norway. J. Appl. Microbiol. 99, 158–166.

Kataoka, A.I., Enache, E., Napier, C., Hayman, M., Weddig, L., 2016. Effect of storage temperature on the outgrowth and toxin production of *Staphylococcus aureus* in freeze-thawed precooked tuna meat. J. Food Prot. 70, 620–627.

Kérouanton, A., Hennekinne, J.A., Letertre, C., Petit, L., Chesneau, O., Brisabois, A., De Buyser, M.L., 2007. Characterization of *Staphylococcus aureus* strains associated with food poisoning outbreaks in France. Int. J. Food Microbiol. 115, 369–375.

Korea Food Code (see standards at menu at http://www.mfds.go.kr/eng/index.do?nMenuCode=63).

Kramer, A., Schwebke, I., Kampf, G., 2006. How long do nosocomial pathogens persist on inanimate surfaces? A systematic review. BMC Infect. Dis. 6, 130–138.

Kűmmel, J., Atessl, B., Gonano, M., Walcher, G., Bereuter, O., Fricker, M., Grunert, T., Wagner, M., Ehling-Schulz, M., 2016. *Staphylococcus aureus* entrance into the dairy chain: tracking *S. aureus* from dairy cow to cheese. Front. Microbiol. 7, 1–11.

Kusumaningrum, H.D., Riboldi, G., Hazeleger, W.C., Beumer, R.R., 2003. Survival of foodborne pathogens on stainless steel surfaces and cross-contamination to foods. Int. J. Food Microbiol. 85, 227–236.

Lammerding, A.M., 1997. An overview of microbial food safety risk assessment. J. Food Prot. 60 (11), 1420–1425.

Lindqvist, R., Sylvén, S., Vågsholm, I., 2002. Quantitative microbial risk assessment exemplified by *Staphylococcus aureus* in unripened cheese made from raw milk. Int. J. Food Microbiol. 78, 155–170.

Martin, S., Iandolao, J.J., 2014. *Staphylococcus aureus*. In: Encyclopedia of Food Microbiology, vol. 3, second ed. pp. 501–507.

May, F.J., Polkinghorne, B.G., 2016. Epidemiology of bacterial toxin-mediated foodborne gastroententeritis outbreaks in Australia, 2001 to 2013. Commun. Dis. Intell. 40 (4), E460–E469.

Mehli, L., Hoel, S., Thomassen, G.M.B., Jakobsen, A.N., Karlsen, H., 2017. The prevelance, genetic diversity and antibiotic resistance of *Staphylococcus aureus* in milk, whey, and cheese from artisan farm dairies. Int. Dairy J. 65, 20–27.

Mexico standards: Infant formula (http://www.salud.gob.mx/unidades/cdi/nom/131ssa15.html), Dried-salted and smoked fish products (http://www.salud.gob.mx/unidades/cdi/nom/129ssa15.html), Cheese (http://www.salud.gob.mx/unidades/cdi/nom/121ssa14.html), Ground beef (http://www.salud.gob.mx/unidades/cdi/nom/034ssa13.html), Shellfish, fresh, refrigerated, and frozen (http://www.salud.gob.mx/unidades/cdi/nom/029ssa13.html), Fish, fresh, refrigerated, and frozen (http://www.salud.gob.mx/unidades/cdi/nom/027ssa13.html), Dairy products (http://www.salud.gob.mx/unidades/cdi/nom/185ssa12.html), Milk, milk-based infant formula, and mixed milk products (http://www.salud.gob.mx/unidades/cdi/nom/184ssa12.html), Egg and egg products (http://www.salud.gob.mx/unidades/cdi/nom/159ssa16.html), Cereals and flour (http://www.salud.gob.mx/unidades/cdi/nom/147ssa16.html).

Microbiological Criteria for Foodstuffs, GCC Standardization Organization, GSO 1016/2015.

National Advisory Committee on Microbiological Criteria for Foods (NACMCF), 2015. Response to Questions Posed by the Department of Defense Regarding Microbiological Criteria as Indicators of Process Control or Insanitary Conditions. https://www.fsis.usda.gov/wps/wcm/connect/2ea3f473-cd12-4333-a28e-b2385454c967/NACMCF-Report-Process-Control-061015.pdf?MOD=AJPERES.

Neder, V.E., Canavesio, V.R., Calvinho, L.F., 2011. Presence of enterotoxigenic *Staphylococcus aureus* in bulk tank milk from Argentine dairy farms. Rev. Argent. Microbiol. 43, 104–106.

Normanno, G., Firinu, A., Virgilio, S., Mula, G., Dambrosio, A., Poggiu, A., Decastelli, L., Mioni, R., Scuota, S., Bolzoni, G., Di Giannatale, E., Salinetti, A.P., La Salandra, G., Bartoli, M., Zuccon, F., Pirino, T., Sias, S., Parisi, A., Quaglia, N.C., Celano, G.V., 2005. Coagulase-positive Staphylococci and *Staphylococcus aureus* in food products marketed in Italy. Int. J. Food Microbiol. 98, 73–79.

Ostyn, A., De Buyser, M.L., Guillier, F., Groult, J., Félix, B., Salah, S., Delmas, G., Hennekinne, J.A., 2010. First evidence of a food poisoning outbreak due to staphylococcal enterotoxin type E, France. Euro Surveill. 15pii: 19528.

Peru NTS No. 071, MINSA/DIGESA-V.01, Norma Sanitaria Que Establece Los Criterios Microbiologicos de Calidad Sanitaria e Inocuidad para Los Alimentos Y Bebidas de Consumo Humano.

Population Reference Bureau, 2016 World Population Data Sheet. http://www.prb.org/pdf16/prb-wpds2016-web-2016.pdf.

Raccach, M., 1981. Control of *Staphylococcus aureus* in dry sausage by a newly developed meat starter culture and phenolic-type antioxidants. J. Food Prot. 44, 665–669.

Regulation on Turkish Food Codex Microbiological Criteria, Official Gazette of Publication 29.12.2011-28157. http://www.tarim.gov.tr/Belgeler/ENG/Legislation/regulation_microbiological_criteria.pdf.

Reynolds, A.E., Harrison, M.A., Rose-Morrow, R., Lyon, C.E., 2001. Validation of dry cured ham process for control of pathogens. J. Food Microbiol. Saf. 66, 1373–1379.

Rosengren, A., Fabricius, A., Guss, B., Sylvén, S., Lindqvist, R., 2010. Occurrence of foodborne pathogens and characterization of *Staphylococcus aureus* in cheese produced on farm-dairies. Int. J. Food Microbiol. 144, 263–269.

Scallan, E., Hoekstra, R.M., Angulo, F.J., Tauxe, R.V., Widdowson, M.A., Roy, S.L., Jones, J.L., Griffin, P.M., 2011. Foodborne illness acquired in the United States—major pathogens. Emerg. Infect. Dis. 17 (1), 7–15. http://doi.org/10.3201/eid.1701.P11101.

Schelin, J., Wallin-Carlquist, N., Thorup Cohn, M., Lindqvist, R., Barker, G.C., Rådström, P., 2011. The formation of *Staphylococcus aureus* enterotoxin in food environments and advances in risk assessment. Virulence. 2 (6), 580–592. http://doi.org/10.4161/viru.2.6.18122.

Schmid, D., Fretz, R., Winter, P., Mann, M., Höger, G., Stöger, A., Ruppitsch, W., Ladstätter, J., Mayer, N., de Martin, A., Allerberger, F., 2009. Outbreak of staphylococcal food intoxication after consumption of pasteurized milk products, June, 2007, Austria. Wien. Klin. Wochenschr. 121, 126–131.

Schmitt, M., Schuler-Schmid, U., Schmidt-Lorenz, W., 1990. Temperature limits of growth, TNase and enterotoxin production of *Staphylococcus aureus* strains isolated from foods. Int. J. Food Microbiol. 11, 1–20.

South Africa GNR 1551, 1997.

South Africa GNR 692, 1997.

Standards and Guidelines for Microbiological Safety of Foods, an Interpretive Summary, 2008. Food Directorate, Health Products and Food Branch, Government of Canada.

Stephan, R., Dura, U., Untermann, F., 1999. Resistance situation and enterotoxin production capacity of *Staphylococcus aureus* strains from bovine mastitis samples. Schweiz. Arch. Tierheilkd. 141, 287–290.

Sutherland, J.P., Bayliss, A.J., Roberts, T.A., 1994. Predictive modelling of growth of *Staphylococcus aureus*: the effects of temperature, pH and sodium chloride. Int. J. Food 21, 217–236.

The Technical Regulation of the Russia-Kazakhstan-Belarus Customs Union on Food Safety (TR TS 021/2011).

The Technical Regulation of the Russia-Kazakhstan-Belarus Customs Union on the Safety of Meat and Meat Products (TR TS 034/2013).

The Technical Regulation of the Russia-Kazakhstan-Belarus Customs Union on Safety of Milk and Dairy Products (TR TS 033/2013).

Tilkens, B.L., King, A.M., Glass, K.A., Sindelar, J.J., 2015. Validating the inhibition of *Staphylococcus aureus* in shelf-stable, ready-to-eat snack sausages with varying combinations of pH and water activity. J. Food Prot. 78 (6), 1215–1220.

Todd, E.C.D., 2014. Bacteria: *Staphylococcus aureus*. In: Encyclopedia of Food Safety, vol. 1, pp. 530–534.

Todd, E.C.D., Grieg, J.D., Bartleson, C.A., Micaels, B.S., 2007. Outbreaks where food workers have been implicated in the spread of foodborne disease. Part 3. Factors contributing to outbreaks and description of outbreak categories. J. Food Prot. 70 (9), 2199–2217.

Todd, E.C.D., Grieg, J.D., Bartleson, C.A., Micaels, B.S., 2009. Outbreaks where food workers have been implicated in the spread of foodborne disease. Part 6. Transmission and survval of pathogens in the food processing and preparation enviornment. J. Food Prot. 72 (1), 202–219.

Todd, E.C.D., Grieg, J.D., Bartleson, C.A., Micaels, B.S., 2008. Outbreaks where food workers have been implicated in the spread of foodborne disease. Part 4. Infective doses and pathogen carriage. J. Food Prot. 71 (11), 2339–2373.

Todd, E.C.D., Michaels, B.S., Holah, J., Smith, D., Greig, J.D., Bartleson, C.A., 2010. Outbreaks where food workers have been implicated in the spread of foodborne disease. Part 10. Alcohol-based antiseptics for hand disinfection and a comparison of their effectiveness with soaps. J. Food Prot. 73, 2128–2140.

Toyofuku, H., 2006. Harmonization of international risk assessment protocol. Mar. Pollut. Bull. 53, 579–590.

United States Code of Federal Regulations, Title 21 Food and Drugs Chapter 1 – Food and Drug Administration Department of Health and Humans Services, Subchapter B-Food for Human Consumption Part 131 Milk and Cream. http://www.accessdata.fda.gov/scripts/cdrh/cfdocs/cfcfr/CFRSearch.cfm?CFRPart=131.

United States Egg Products Inspection Act. https://www.fsis.usda.gov/wps/portal/fsis/topics/rulemaking/egg-products-inspection-act/EPIA.

United States Federal Meat Inspection Act. https://www.fsis.usda.gov/wps/portal/fsis/topics/rulemaking/federal-meat-inspection-act.

United States Poultry Products Inspection Act. https://www.fsis.usda.gov/wps/portal/fsis/topics/rulemaking/poultry-products-inspection-acts.

USDA FSIS Baseline Data. https://www.fsis.usda.gov/wps/portal/fsis/topics/data-collection-and-reports/microbiology/baseline/baseline.

USDA NASS, April 2016. Milk Production, Disposition and Income, 2015 Summary. http://usda.mannlib.cornell.edu/usda/nass/MilkProdDi//2010s/2016/MilkProdDi-04-28-2016.pdf.

Walcher, G., Gonano, M., Kümmel, J., Barker, G.C., Lebl, K., Bereuter, O., Ehling-Schulz, M., Wagner, M., Stessl, B., 2014. *Staphylococcus aureus* reservoirs during traditional Austrian raw milk cheese production. J. Dairy Res. 81, 462–470.

FURTHER READING

Federal Ministry of Justice, Germany Milchgüteverordnung. https://www.gesetze-im-internet.de/bundesrecht/milchg_v/gesamt.pdf.

Index

'*Note*: Page numbers followed by "f" indicate figures and "t" indicate tables.'

E

F

R

S

Printed in the United States
By Bookmasters